5/8/89

# The Heredity-Environment
# Controversy,    1900-1941

THE JOHNS HOPKINS UNIVERSITY PRESS / 1988
Baltimore and London

# THE TRIUMPH
# OF EVOLUTION

*Hamilton Cravens*

*For Joan*

*Originally published in 1978 by the University of*
*Pennsylvania Press as* The Triumph of Evolution:
American Scientists and the Heredity-Environment
Controversy, 1900–1941. *Johns Hopkins edition*
*published by arrangement with The University of*
*Pennsylvania Press.*

Johns Hopkins Paperbacks edition, 1988

The Johns Hopkins University Press
701 West 40th Street
Baltimore, Maryland 21211
The Johns Hopkins Press Ltd., London

Library of Congress Cataloging-in-Publication Data

Cravens, Hamilton.
    The triumph of evolution.

    Reprint. Originally published: Philadelphia :
University of Pennsylvania Press, 1978.

    Bibliography: p.
    Includes index.
    1.  Nature and nurture—History.  I.  Title.
QH438.5.C7   1988        575.01        88-45394
ISBN 0-8018-3742-1

# CONTENTS

# PREFACE

WHEN *The Triumph of Evolution* was published a decade ago, I was keenly aware of the scientific and popular controversies on the moral and social policy implications of the heredity-environment conundrum. Apparently these debates had been triggered by the publication of Arthur R. Jensen's famous article in the *Harvard Educational Review* in 1969. In that piece, Jensen questioned whether blacks would ever be able to equal the performance of whites in the schools, primarily on the basis of what he dubbed a large gap in the average IQs of whites over blacks.[1] The debates involved more than the question of "average" racial IQs, to be sure, but somehow that problem was always central, if not necessarily to the scientific then certainly to the moral and policy debates. This was especially true for my generation, for many of us had been excited by the possibilities that the civil rights movement of the 1950s and 1960s held for a decent social order.

If in one sense *The Triumph of Evolution*'s publication was timely, nevertheless I thought of my effort as historical scholarship, not current history or social advocacy. I tried very hard to keep the contemporary debates out of my research and writing. I thought I could make my contribution to a scholarly and historical understanding of the heredity-environment controversy of the early twentieth century. Such demanding and difficult historical reconstruction obligates one to understand the past, understand in the sense to comprehend rather than to forgive. I was not interested in arguing with the past, in choosing sides, as it were, but with the far more humbling task of attempting to grasp the pastness of the past, the otherness of that age from ours. Happily, *The Triumph of Evolution* went on to win its spurs as a contribution to scholarship and has been of continuing utility to teachers, scholars, researchers, and the general public.

When the opportunity for a paperback edition arose, it seemed that sufficient time had passed that a brief preface would be feasible, in which I would offer a sketch of the history of the current debates. This would also provide some perspective on both controversies. Presumably in time a diligent scholar will plow through the avalanche of paper which the current debates have generated and write a large and definitive scholarly book. In that volume's absence, some tentative remarks can be made.

Perhaps the most important comment is that the two controversies were products of their own distinct ages, each with its own particular notions and definitions of natural and social reality.[2] Manifestations of these differences certainly have included events in economic, social, and political history. It was no accident that the debates over heredity and environment took place in the early twentieth century at a time of severe class, racial, and ethnocultural tension, polarization, and virtual group warfare. Heated debates took place over whether the white immigrant groups, especially those from southern and eastern Europe, could ever be assimilated into the larger American culture. By contrast, there was minimal discussion over the conditions and prospects of American blacks within the white community. Most whites, whether scientists or not, simply assumed that blacks, like other nonwhite groups, were naturally inferior to whites in every respect. And, furthermore, there was little debate, outside feminist circles at least, of the power of gender and the status of women as separate and distinct from men.[3]

In our own time these matters have received very different emphases. The recent debates have not focused on European nativity groups, but instead on comparisons of whites and nonwhites, especially whites and blacks, and on sex or gender, the latter meaning the extent to which biology or culture does or does not make men and women "equal." As much as power, equity, and politics have been involved in these controversies, however, the structures and processes of science have mattered as well. The interactionist paradigm that crystalized as a result of the heredity-environment controversy spelled a reunification of biological and social theory on terms very different than those that had reigned since the late nineteenth century. The earlier model of biological and social theory assumed more than the well-known subordination of the social or the cultural to the biological. It also projected their distinctiveness, their separateness—their segregation from each other. In such a formulation, in other words, the whole was no more than the sum of its parts. In the new interactionist model, however, the whole was greater than the sum of its parts. The whole consisted of many—indeed, virtually infinite—elements or parts that were distinct (that is, cultural or biological), but also that were interconnected and which functioned together. In the new picture, it was the system's intricate dynamics that mattered, whereas in the old, it was the system's structure. Thus the mirroring of the transformation from one age to the next.

What emerged from the heredity-environment controversy of the early twentieth century, then, was a complex, intricate, and somewhat contradictory model (as models commonly are) of the interaction of nature and culture. In this blueprint there were elements of probability and chance, especially in the direction that the system might be perceived to be moving. But there were also strong components of determinism, particularly as in the causes of events within that system. All progress was relative—a rising tide carried all boats—but, to continue this homely figure of speech, each boat maintained its position in relation to every other in the larger fleet. Thus the behaviorists, for example, following John B. Watson's lead, and that of traditional developmental thinking, more generally, assumed that growth was predetermined. The environmental stimuli simply "awakened" the mute anatomical structures that were a part of the phylogenetic endowment of the species to which the individual belonged. And those interested in intelligence assumed that it was fixed from the time of birth on in the life cycle, and that different groups in the social and occupational pecking order usually had different average IQs that were characteristic of those groups. Put another way, what biology did not cause, culture did. From this perspective, this was a deterministic science of man.

Nor was this all. The interactionist paradigm projected a specific notion of the meaning of group identity for the individual.

Individuals did not exist apart from the group or groups to which they belonged or were assigned, whether group designations signified class, race, ethnicity, gender, or the like. Indeed, an individual's traits could only deviate from the mean for the group or social category to which "science" and "society" had "assigned" that person. In practical terms, social mobility was, as implied above, relative and group-specific. Consider, for instance, the classic longitudinal study of that era, that of Lewis M. Terman and his associates at Stanford University. They studied California children whom the Stanford-Binet test identified as geniuses. Terman's technique was to take data from the individuals in the sample at given times and to present the material cross-sectionally (that is, representations of group "averages," or norms, at particular points in time). Terman discussed individual differences and peculiarities, often at length, but only within the context of deviation from, or approximation toward, the "mean" of the group—in this instance, the group being the geniuses. In such a study an individual cannot exist apart from a predetermined "group." So powerful was this notion of group identity for the individual that even those researchers whose results challenged the notion of a fixed IQ, such as George Stoddard and his colleagues at the Iowa Child Welfare Research Station, did not realize that they were challenging that notion in their studies of orphan children. They did not recognize that they were "tracking" individuals over time, rather than divining group averages at a moment in time.[4]

After World War II there was growing recognition of the validity of the "individualistic" point of view in the scientific community. In the late 1940s, for example, researchers at the Institute of Child Welfare at the University of California began to report findings from their longitudinal studies which in some sense undermined the interactionist model of determinism. Thus Nancy Bayley found considerable instability and inconstancy in the IQs of the children in her longitudinal study at various times from birth to eighteen years—considerably more variation than the deterministic model allowed.[5] Nor was she alone. Other scientists, at Berkeley and elsewhere, reported similar findings.[6] A different group of scientists questioned the notion of predetermined development. Thus some researchers on animals insisted that neurological structures did not develop without the environment. From a slightly different perspective, other investigators argued that animals had brains in which they processed information; they did not merely "respond" to "stimuli," as mainstream behaviorism would have it.[7]

By the early 1960s, then, a new school of thought on human and animal development had emerged in the scientific community. Its members attacked various elements of the interactionist paradigm. In 1961, J. McVicker Hunt, a developmental psychologist at the University of Illinois, published the new school's manifesto, *Intelligence and Experience*. Hunt reviewed much of the scientific literature over the last half-century on the IQ, on development, and, ultimately, on intelligence. He provided numerous scientific arguments to support the social policy conclusions that individuals did not vary by group at birth, and perhaps poverty, discrimination, and other forms of social prejudice and oppression were unjustified.[8] In particular his work had much influence among teachers, researchers, and students in the related fields of child development and early childhood education. Many in the more traditional fields tended to ignore or take issue with it.

During the 1950s and 1960s the growing civil rights movement seized the nation's imagination—and anxieties. The movement's ultimate message was equality of all individuals. Its morality play featured the oppression of the individual by the system, an interesting perspective given the intellectual changes then occurring in the behavioral sciences, as with the work of Hunt and others.[9] The modern feminist movement crystalized when writer Betty Friedan published *The Feminine Mystique*, a book that spoke for many American women, especially middle-class white women, about their status as women—as members of a particular group who could not "vary" beyond its confines or become independent individuals. Friedan portrayed a drama which resembled that of Dr. Martin Luther King, Jr., and other civil rights activists—of the oppression of the individual by the system or the larger whole, of the necessity of the liberation of the self, of the individual.[10]

By the late 1960s there was a growing chorus of liberation for the individual, a profound sense of alienation among certain groups in the population with the status quo, and an increasingly fervent polarization of American society. It was in the context of such 1960s political and social scientific liberationism that the Head Start programs of compensatory education started. It was these programs and their intellectual and scientific rationales that

Arthur Jensen criticized in his famous article in the *Harvard Educational Review.*

Jensen essentially used arguments that had been deployed in the 1920s, with some modern technics here and there. Thus Jensen insisted that IQ was fixed from birth, that there were average IQs for specific groups (that is, races and classes), and that the primary reason for the difference between average IQs for such human "groups" was biological inheritance. He noted, correctly, that in most comparative studies of black and white IQs done since the early twentieth century there was a gap of fifteen IQ points, or one standard deviation, between the average white IQs and black IQs. He insisted that 80 percent of the gap's causes were due to heredity, and only 20 percent to environment—his well known estimate of the heritability of the IQ.[11]

The question of the persistent gap in IQ points between blacks and whites in American society has been taken up by some social scientists from a fresh perspective—the individualistic taxonomy of the natural and social reality of our own age. There have been some novelties and innovations in the scientific arguments.

Professor John Ogbu, an anthropologist at the University of California, Berkeley, has been one of the chief architects of the new approach. He has argued that many blacks in America are in a social context similar to that of other "castelike" minorities in other societies throughout the world. The gap between white and black IQ scores in America and that of majority and minority castes in other cultures, regardless of "race," has been virtually identical; that is, about fifteen IQ points, or one standard deviation. This suggests, he has insisted, that the structural features of majority-minority caste relations in these cultures have been responsible for this IQ gap understood as a difference between average IQs for different "races."[12]

Other scholars and scientists have attacked the notion, another remnant from the interactionist paradigm, that poor children are inherently culturally deprived, either as a result of "race" or as a consequence of "culture." They have also attacked the IQ as a meaningless score or, at least, have insisted it can only measure limited things—especially given the lack of a coherent definition of intelligence among testers.[13] Still other scientists have studied trends in average IQ scores for various European nations, for example, and have found there the reappearance of the gap of approximately fifteen IQ points from previous to fu-

ture generations. Put positively, average IQ scores have been rising about fifteen points for each generation of white Europeans.[14]
In summary, what unites all of these scholars is a belief in the
inherent equality of individuals, at least so far as potentiality is
concerned, to make their way in society as good citizens. The gap
between privileged and nonprivileged in IQ scores and school
performance is chiefly the result of such social or cultural or environmental influences as racial or religious prejudice, low expectations, bad schools and worse teaching, plus well known
"ceilings" on post-scholastic opportunities for jobs and decent
living in the larger society. These scholars do not necessarily
deny that individuals inherit the potentiality for certain traits.
Rather, they assert that this is not as meaningful as the fact of
implacable and near-universal social oppression, which grinds
individual members of such minority "castes" into desperate
circumstances.[15]

The IQ is not the only measure some psychologists have used
for the purposes of making intellectual comparisons among the
various "races." For example, Professor George A. Chambers of
the University of Iowa has used the ACT test scores of white
and nonwhite individuals. In matching individuals from various
"races" in all other attributes besides the ACT test scores, such
as social class, education of parents, and so forth, psychologists
find that the test scores of the individuals, regardless of race, also
are very close. In other words, when one matches pairs of individuals in this manner, racial differences dissipate.[16]

The heredity-environment debate has, of course, also involved
another significant group in the culture—females. Betty Friedan
argued in *The Feminine Mystique* for the equality of all individuals regardless of "sex" (later "gender"). Culture made us what
we were, in the main, with biology responsible for certain obvious male and female differences. In the years between the
world wars there had been, as Rosalind Rosenberg has noted in a
useful book, something of a tradition of feminist social science
research, whose champions (such as Margaret Mead) insisted
that women did not belong to a separate group from men and that
science in some sense upheld feminism—the notion that men
and women were best understood as individuals with their own
characteristics *as individuals,* with the obvious exception of male-
ness and femaleness, as in the sexual organs and so forth.[17] As
feminists interpreted the new social science, the system was op-

pressive of individuality. That was indeed the message of Friedan, who was herself quite learned in the social sciences and who had picked up some of the anti-deterministic arguments of social science in the interwar years, including those of Kurt Lewin of the Iowa Child Welfare Research Station.[18]

This line of argumentation was revived in the social sciences in the 1970s. Of particular interest was *The Psychology of Sex Differences* (1974), the massive work of two psychologists at Stanford, Eleanor Maccoby and Caroline Nagy Jacklin. In this volume they evaluated all behavioral science studies of male and female psychology and behavior that had been done in the America since the mid-1960s. They argued that anatomy was not destiny. Social institutions and practices were not shaped by biology; a variety of social arrangements could easily function within the existing biological framework, and it "is up to human beings to select those that foster the life styles they most value."[19] It was perhaps to be expected that challenges would arise to this position. From within the temples of behavioral science, there were female scientists who insisted that not all female behavior was so similar to male behavior as was insisted, and perhaps there were ineluctable biological differences between men and women—in a different voice, as psychologist Carol Gilligan put it. A more radical challenge, and one that mocked much of the apparatus of male culture and, presumably, male science, was the work of theological and feminist writer Mary Daly of Boston College, whose works since the early 1970s have drawn not simply on a radical feminist perspective but also on our age's pervasive sensibility for the physical, the body, the animal in our species.[20] Apparently the heredity-environment issue is a frequent topic of controversy in the general field of male and female behavior, revolving around both whether the individual or the group perspective is superior, and that of nature versus culture.

The other challenge to equalitarianism or individualistic feminism in the social or human sciences has come from sociobiology and its cousins, the various popular "aggression" books of the 1960s, such as Robert Ardrey's *The Territorial Imperative* (1967) and Desmond Morris's *The Naked Ape* (1967).[21] As a field, sociobiology is distinct from these more accessible works. To the extent that sociobiologists—such as Edward O. Wilson of Harvard, the field's leader and founder—insist on the inheritance of patterns of behavior, such as aggression, and identify them as male, and other traits (for example, submission) as female, sociobiology is obviously a field that is opposed to contemporary feminist theory and ideology.[22] Yet that would oversimplify what sociobiologists say. Sociobiology as a field seems perhaps as prosperous as the social science of sex differences—or contemporary feminist science—yet it seems not to have the sensational media attention it did a decade and more ago. It is perhaps also noteworthy that, as might have been expected, the majority of authors and writers in sociobiology are male, and those in the psychology of sex differences are female.

In conclusion, the heredity-environment dichotomy is in some sense timeless, a puzzle that people in each age regard as central to other issues and problems.

### Notes

1. Arthur R. Jensen, "How Much Can We Boost IQ and Scholastic Achievement?" *Harvard Educational Review* 39 (1969): 1–123; reprinted in Arthur R. Jensen, ed., *Genetics and Education* (New York, 1972), pp. 69–204.
2. Hamilton Cravens, "Recent Controversy in Human Development: A Historical View," *Human Development* 30 (1987): 325–35; Cravens, "History of the Social Sciences," in Sally Gregory Kohlstedt and Margaret W. Rossiter, eds., *Historical Writing on American Science: Perspectives and Prospects* (Baltimore, 1986), pp. 183–207.
3. John Higham, *Strangers in the Land: Patterns of American Nativism, 1860–1925* (New Brunswick, N.J., 1955); August Meier and Elliot Rudwick, *From Plantation to Ghetto*, 3d ed. (New York, 1976); Carl N. Degler, *At Odds: Women and the Family in America from the Revolution to the Present* (New York, 1980). The scholarly literature on these phenomena is massive: the books cited above are useful introductions.
4. Hamilton Cravens, "The Wandering IQ: American Culture and Mental Testing," *Human Development* 28 (1985): 113–30; Cravens, "Child-Saving in the Age of Professionalism, 1915–1930," in Joseph M. Hawes and N. Ray Hiner, eds., *American Childhood* (Westport, Conn., 1985), pp. 415–88.
5. Nancy Bayley, "Consistency and Variability in the Growth from Birth to Eighteen Years," *Journal of Genetic Psychology* 75 (1949): 165–96.
6. Marjorie P. Honzig, Jean W. Macfarlane, and L. Allen, "The Stability of Mental Test Performance between Two and Eighteen Years," *Journal of Experimental Education* 4 (1948): 309–24; Harold E. Jones, "The Environment and Mental Development," in Leonard Carmichael, ed., *Handbook of Child Psychology* (New York, 1954).
7. See, for example, H. G. Birch, "Sources of Order in Maternal Behavior of Animals," *American Journal of Orthopsychiatry* 26 (1956): 279–84; Wayne Dennis and Pergrouhi Najarian, "Infant Development under Environmental Handicap," *Psychological Monographs*

71, no. 7 (1957); Dennis, "Causes of Retardation among Institutional Children," *Journal of Genetic Psychology* 96 (1960): 47–59.

8. J. McVicker Hunt, *Intelligence and Experience* (New York, 1961).
9. David J. Carrow, *Bearing the Cross: Martin Luther King, Jr., and the Southern Christian Leadership Conference* (New York, 1986).
10. Betty Friedan, *The Feminine Mystique* (1963). 20th anniversary ed. (New York, 1983).
11. Jensen, "How Much Can We Boost IQ and Scholastic Achievement?"
12. Signithia Fordham and John U. Ogbu, "Black Students and the Burden of 'Acting White,' " *Urban Review* 18 (1986): 176–206; see also John U. Ogbu, *Minority Education and Caste: The American System in Cross-Cultural Perspective* (New York, 1978).
13. Herbert Ginsburg, *The Myth of the Deprived Child: Poor Children's Intellect and Education* (Englewood Cliffs, N.J., 1972).
14. James R. Flynn, "Massive IQ Gains in Fourteen Nations: What IQ Tests Really Measure," *Psychological Bulletin* 101 (1987): 171–91.
15. Ulric Neisser, ed., *The School Achievement of Minority Children: New Perspectives* (Hillsdale, N.J., 1986), is a handy guide to this school and its scientific literature.
16. George A. Chambers, "All of America's Children: Variants in ACT Scores—What Principals Need to Know," unpublished paper, dated 7 March 1988; see also the doctoral dissertation of one of Professor Chambers's students, Alan Keith Whitworth, "A Comparison of ACT Assessment Scores for Similarly Situated Black and White Students," Ph.D. dissertation, University of Iowa, 1987.
17. Rosalind Rosenberg, *Beyond Separate Spheres: Intellectual Roots of Modern Feminism* (New Haven, 1982); Nancy F. Cott, *The Grounding of Modern Feminism* (New Haven, 1987).
18. Friedan, *Feminine Mystique,* p. 14.
19. Eleanor Emmons Maccoby and Caroline Nagy Jacklin, *The Psychology of Sex Difference* (Stanford, 1974), p. 374.
20. Mary Daly, *Beyond God the Father: Toward a Philosophy of Women's Liberation* (Boston, 1973); Daly, *The Church and the Second Sex* (New York, 1975); Daly, *Gyn/ecology: The Metaethics of Radical Feminism* (Boston, 1978).
21. Robert Ardrey, *The Territorial Imperative* (London, 1967); Desmond Morris, *The Naked Ape* (London, 1967); Lionel Tiger, *Men in Groups,* 2d ed. (New York, 1984).
22. Edward O. Wilson, *Sociobiology: The New Synthesis* (Cambridge, 1975); Wilson, *On Human Nature* (Cambridge, 1978). Critical of sociobiology are Marshall Sahlins, *The Use and Abuse of Biology: An Anthropological Critique of Sociobiology* (Ann Arbor, 1976), and R. C. Lewontin, Steven Rose, and Leon J. Kamin, *Not In Our Genes: Biology, Ideology, and Human Nature* (New York, 1984).

# PREFACE TO THE FIRST EDITION

Most informed Americans would agree that few scientific questions possess as many implications for our beliefs and practices as a democracy as the heredity-environment issue. Apparently the problem of whether human nature and conduct are shaped more by innate endowment or social environment is one of those classic dichotomies of our public policy discussions that in turn helps shape the positions many of us take on a host of less philosophical and more immediate matters. Perhaps inevitably the ideals and the customs denoted by such words as equality, freedom, opportunity, and democracy mean different things for the champions of the hereditarian than for the advocates of the environmental explanation of human behavior. Usually most Americans have not explicated their positions on the nature-nurture issue in public policy discourse. But there have been a few episodes in our post-Appomattox cultural history in which our commitments to heredity or to environment have come out into the open for all to recognize.

This book is about one of those episodes: the heredity-environment controversy among natural and social scientists in the years between the two World Wars. By focusing upon that particular episode I have assumed that it possesses sufficient historical identity and integrity to be the subject of an independent volume.

I do not mean to imply that the nature-nurture controversy had no relationship to the intriguing (and complex) transatlantic scientific discussion of the role of heredity and environment in the making of species and individuals that took place up to the 1890s. Nor do I wish to leave the impression that the resolution of the heredity-environment controversy in the late 1930s and early 1940s had no impact upon scientific and lay discussions of nature and nurture thereafter. But I think it will be perfectly clear in the following pages that the nature-nurture controversy may be considered on its own, as a phenomenon in American science and culture. If we define the nature-nurture controversy merely as an episode in the history of scientific ideas, it certainly was related to the scientific discussion of the role of heredity and environment in evolution that has continued from Darwin's day to our own. After all, heredity and environment have always been the two general conceptual categories of causal explanation in evolutionary theory. And if this were my interest, the focus of this book would be quite different than it is.

On closer inspection the nature-nurture controversy stands apart from the scientific discussions of heredity and environment in evolution that came before it and have continued since its resolution. The controversy's origins may be traced back to the 1890s, with the sudden rise of experimental biology and psychology. Starting in that decade, the advances in the new experimental work made it possible for the first time for natural scientists to disentangle heredity from environment in their descriptions and explanations of evolutionary phenomena. Within a few years, experimental natural scientists were able to create a very different model of heredity and environment than scientists had traditionally employed since the mid-nineteenth century—a conception in which nature and nurture were clearly separated and defined as distinct, independent variables. It was in American culture, on this side of the Atlantic, that the new generation of natural scientists imposed a radically hereditarian interpretation upon the new experimental discoveries that conceded little causal role to environment. And it was in America that biologists and psychologists also pointed the most conspicuously to the exciting possibilities inherent in evolutionary science for the creation of an experimental science of human nature and conduct whose practitioners could predict, if not control, man's behavior in society. Quite obviously these extreme hereditarian views, and the larger notions of a science of social control, owed much to historical circumstances and developments peculiar to America. And when a new, post-1900 generation of American social scientists attacked the natural scientists' hereditarian interpretation of man between 1915 and 1920, and thus sparked the controversy, their actions, their motives, their ideas, and their very behavior as professional scientists reflected much that was attributable to conditions and changes in American science, culture, and society. I do not wish to imply that there were no relationships between science in the Old World and the New, but merely that the controversy was largely an American phenomenon.

In the dozen or so years following the stock market crash of 1929 American natural and social scientists resolved the heredity-environment controversy. They did so in certain ways that opened the door to a new era in the scientific discussion of the roles of heredity and environment in evolution. What the critics of the extreme hereditarian interpretation of human nature and conduct objected to was not the legitimate place of innate endowment in the evolution of species and individuals but the natural scientists' one-sided explanatory model of nature over nurture. In the 1930s, thanks to new developments in the natural and social sciences, a new model of evolution—and hence a new picture of heredity and environment—gained wide acceptance in the scientific community. According to the proponents of the new theory, nature and nurture were distinct but *interdependent* variables. One could not speak of nature without nurture. This applied with equal force to ontogeny (the development of the individual) and to phylogeny (the evolution of species). The new theory made it possible, furthermore, to explain ontogeny without resorting to conceptions of group or racial endowment and experience, just as it made it feasible to account for phylogeny without invoking the typological models of the nineteenth century theorists of unilinear racial development. In other words, new perspectives on ontogeny and phylogeny emerged that had the effect of defining them as distinct levels of phenomena. And the picture of the interaction of heredity and environment as interdependent variables led to the articulation of a new paradigm of evolutionary science. In this view, man as a species is the product of two evolutionary processes—biological evolution, which has endowed man with those characteristics as a species that distinguish him from the animals and grant him the facility to be a cultural species; and cultural evolution, or the development of human cultures throughout the world in which men acquire particular habits and roles as the consequence of being members of a human society. In both his biological and his cultural evolution, nature and nurture were interdependent variables. But obviously it was man's particular cultural traits— the specific domain of the social sciences—that had the most to do with his behavior as a member of a culture.

We still live in a time in which these ideas have an important influence upon our culture. In the immediate sense, the practical result of the nature-nurture controversy on the scientific discussion of heredity and environment was to shift attention away from the either-or-statements of earlier times to more quantitative questions of how much nature and nurture mattered in a particular biological or cultural phenomenon. But in a more general sense, it is manifestly apparent that the long-term practical result of the nature-nurture controversy was to grant a fresh lease on life to evolutionary science in American culture. For what the new theory of interdependency permitted, when scientists fully elaborated it, was the creation of a rationale for the further development of that exciting science of human nature and conduct—now a truly interdisciplinary endeavour in which natural scientists and social scientists could amicably resolve their jurisdictional disputes and forge ahead with the development of a science of man. In a culture and a society that persistently faced revolutionary challenges in foreign affairs and social and economic instability at home, this was indeed a noble and electrifying prospect. In recent years, some scientists have begun to question certain critical parts of the interdependency model of nature and nurture, and have insisted that perhaps the weight of scientific evidence, at least with regard to racial intelligence quotients, is better explained by a more precise restatement of the older nature or nurture model of the early twentieth century. The chronicling of these contemporary events has no place in this book. I would note, that so nearly as I am aware, these recent developments do not mean that the goal of a science of man has been surrendered; if anything, they would seem strong evidence of a renewed call for its further application, and, at most, a technical revision in the explanatory conceptions natural scientists use. Those of my fellow citizens who share my hopes for a more gentle and tolerant society might ask themselves whether they truly wish public policy matters of such moment to depend upon both highly complex technical questions in nature and upon the dreams and illusions that seem to emanate so inevitably from such questions.

It is a pleasant ritual to thank those institutions and individuals who have helped and encouraged me in so many ways in the preparation of this book. Stow Persons provided much help in the early stages. For more than a decade Henry D. Shapiro and John C. Burnham have proved superb discussants of intellectual problems and extraordinary friends. They helped me perceive the flaws in many of my ideas, as did George Daniels, Robert R. Dykstra, James Gilbert, Austin Kerr, Walter Rundell, Jr., and Robert A. Skotheim. After I had prepared a different manuscript, Garland E. Allen, Stephen G. Brush, Merle E. Curti, Louis G. Geiger, Willard F. Hollander, Michael M. Sokal, and Wallace A. Russell subjected it to painstaking criticism and provided encouragement at critical times. The departments of history of the University of Iowa, the Ohio State University, the University of Maryland, and Iowa State University offered indispensable intellectual and institutional support. The University of Iowa awarded me a University fellowship; the National Endowment for the Humanities and the Sciences and Humanities Research Institute of Iowa State University made possible two summers in which to prepare the manuscript; over the years the University Research Grants Program of Iowa State University has generously supported the costs of research in archives across the country. It would be impossible to thank all the librarians and archivists who helped me for the genuine assistance and unfailing courtesy I always received. Suffice to say that the librarians of the above named universities made it possible and pleasant for me to use the voluminous printed sources, and the archivists employed at the manuscripts depositories listed in the bibliographical essay made the life of the peripatetic scholar truly enjoyable. Mrs. Susie Ulrickson, my department's expert manuscript typist, prepared several versions of the book with outstanding precision, efficiency, and cheerfulness. I am grateful to her and to my department for this marvellous service.

My largest debt goes to my wife and children, who always understood the importance of scholarship to me and gave me continual support.

# ACKNOWLEDGMENTS

I would like to thank Mrs. Jessie Bernard and the State Historical Society of Wisconsin for permission to quote from letters of Luther Lee Bernard in the Ross Papers; Miss Franziska Boas and the American Philosophical Society for permission to quote from Franz Boas' correspondence in the Boas Papers, and Miss Boas and Columbia University for permission to quote from Franz Boas' correspondence in the Columbia University Central Files; Columbia University to quote from Nicholas Murray Butler's correspondence; the Library of Congress to quote from James McKeen Cattell's letters; Miss Isabel Conklin and Princeton University to quote from Edwin Grant Conklin's letters in the Conklin Papers and Miss Conklin and The Director, The Mount Wilson and Mount Palomar Observatories, to quote from a Conklin letter in the George Ellery Hale Papers; Harvard University Archives and Mr. Charles W. Eliot II to quote from the Charles W. Eliot correspondence; Duke University Library for permission to quote from the Charles A. Ellwood correspondence; the Honorable Gerhard Gesell and the Library of Congress for permission to quote from the correspondence of Arnold Gesell; Mr. Alexander R. James and the Houghton Library, Harvard University, for permission to quote from the letters of William James; Mrs. Theodora Kroeber Quinn and the Bancroft Library, University of California, Berkeley, to quote from the letters of

Alfred L. Kroeber; the Bancroft Library to quote from the letters of Robert H. Lowie; the Library of Congress to quote from the letters of Jacques Loeb; Mr. Frank A. Ross and the State Historical Society of Wisconsin to quote from the letters of Edward A. Ross; Dr. Frederick A. Terman to quote from the letters of Lewis M. Terman; Mrs. Theodore Baird and Cornell University Library for permission to quote from letters of Edward Bradford Titchener; Mrs. Roberta Yerkes Blanshard and the Historical Division, Yale University Medical Library, for permission to quote from the letters of Robert M. Yerkes; Mrs. Laughlin Campbell and the Smithsonian Institution Archives for permission to quote from a letter of William H. Holmes in the Charles D. Walcott Papers.

Some material in this book has been previously published with different emphases. I would like to thank the editor of *American Studies* and the Mid-Continent American Studies Association for permission to use again parts of "The Abandonment of Evolutionary Social Theory in America: The Impact of Academic Professionalization Upon American Sociological Theory," 12 (Fall 1971): 5–20; the editor of *American Quarterly*, the Trustees of the University of Pennsylvania, and my coauthor John C. Burnham to use again passages from "Psychology and Evolutionary Naturalism in American Thought 1890–1940," 23 (December 1971): 635–57; the editor of *The Science Teacher* and the National Science Teachers Association to use again parts of "The Role of Universities in the Rise of Experimental Biology," 44 (January 1977): 33–37.

# THE TRIUMPH OF EVOLUTION

# INTRODUCTION

IN 1874 the executive committee of the New York Prison Association delegated Richard L. Dugdale, one of its members, to make a tour of inspection of thirteen county jails in upstate New York. No sooner had this Manhattan merchant and social reformer embarked upon his assigned task than he was struck by a peculiar fact: many of the inmates were related by birth or marriage. So impressed was Dugdale by the consanguinity of the majority of the prisoners that he decided to make a scientific study of one such family group, to which he gave the pseudonym the Jukes family. The next year Dugdale published the results of his statistical investigation of the Jukes family in a report for the Prison Association, and in 1877, two years later, his work appeared under the title *The Jukes: A Study in Crime, Pauperism, Disease and Heredity* with the imprint of G. P. Putnam's Sons of New York.

Dugdale's study of the Jukes family soon made a sensation, especially among those middle class Americans interested in reform, or simply those appalled by the ever increasing problems of a disorderly industrial society. Indeed, what Dugdale found by digging into local records and using statistical analysis was enough to horrify all but the most blasé and cynical of readers. In his extensive research into local records and interviews of local citizens, Dugdale was able to identify at least 709 persons

3

in the Jukes clan stetching back to the era of the American Revolution, 540 of Juke blood, and 169 other persons associated through cohabitation or marriage. It was indeed an awesome record of dissipation and social pathology. One hundred and forty had been convicted of criminal offenses. One hundred and eighty had been in the almshouse or received other forms of charity for the staggering total of 800 years. Sixty were lifelong thieves. Fifty were ordinary prostitutes. Forty Jukes women had given venereal diseases to at least 440 other persons. There had been 30 prosecutions for bastardy. And 7 had been murdered. Ever the scientific student of society and the social statistician, Dugdale estimated that the Jukes family had cost New York taxpayers at least $1,308,000.

In clear, graphic prose, accompanied by impressive tables and charts of seemingly unimpeachable integrity and accuracy, Dugdale outlined the real costs, human and financial, of what he called the *debris* of civilization. Dugdale's concern was as deep as it was genuine; it emanated from his experiences and his commitments to social reform. Born in Paris in 1841, the son of an English manufacturer and journalist, Dugdale came to New York City at the age of ten with his parents. He attended public school until he was fourteen; then he worked for a sculptor and as a stenographer for several years. He attended evening classes at Cooper Union, where he developed a strong interest in debating contemporary public issues, and it was this that convinced him to devote his life to social reform. Yet in mid-century America there were no foundations or institutes of social reform, so that Dugdale became a businessman to learn the livelihood that would in turn enable him to study and reform society. Even before the close of the Civil War, Dugdale had achieved an estimable reputation in New York as a reformer, and in 1868 the New York Prison Association made him a member of its executive committee in recognition of his commitment to the Association's good works. It was from these experiences and value commitments that Dugdale undertook his assignment in 1874 for the Prison Association; indeed, until his death from congenital heart disease in 1883 Dugdale was ever the activist social reformer and student of social problems.

What was remarkable, historically, about Dugdale's book was the approach he took to his subject—his attitude toward the possibilities of harnessing science in the aid of reform, the methods he used, the analysis he made, and general explanatory framework he employed. He wrote not as a religious reformer of the ante bellum era who sought to lead his downtrodden and sinful fellow countrymen into a state of Christian bliss, but as a secular student of the science of society, fully inspired by the new natural and medical and demographic science of the age, who used the most modern and up-to-date conceptions and methods of science, and who was supremely confident that such objective instruments of modern, rationalistic, scientific culture would lead first to the discovery of tentative truths, next to their verification, and, finally to the improvement of society by their dissemination to the electorate that would in turn influence the reconstruction of law and public policy in line with the dictates of reason and science. By the use of local records and interviews, Dugdale was able to amass a considerable amount of information about the Jukes clan, and, indeed, he argued that it was necessary to make a minute study of as many individual lives of offenders as possible before understanding the general social causes of crime and dissipation. Then, through the use of genealogical charts, statistical tables, and individual case studies, Dugdale presented both the facts and what he believed were the causes of the facts he narrated. Using data from several generations of Jukes, he argued, would make it possible to separate the influence of heredity from that of environment in the formation of persons of high and low moral standards of conduct. Heredity, he insisted, fixed the organic characteristics of the individual, whereas environment modified that heredity. Dugdale's method of study, he declared, enables "us to estimate the cumulative effects of any condition which has operated through successive generations; heredity giving us those elements of character which are derived from the parent as a birthright, environment all the events and conditions occurring after birth which have contributed to shape the individual career or deflect its primitive tendency."[1]

Dugdale was convinced, furthermore, that the problem was not so intractable as the doctrine of original sin suggested or as easy to remedy as the dogma of Christian perfectionism implied. The two basic causes of human behavior were heredity and environment, and, because these were scientifically knowable and understandable, they seemed more amenable to human manipulation than the cosmic forces called up by religious tradition. Dugdale was frankly optimistic. When the causes of immoral conduct could be laid to idiocy, insanity, or some other kind of mental debility, the causes were hereditary; but here improvement could be made through the environment. When on the other hand the reasons for crime and dissipation depended upon the individual's knowledge of moral obligation, Dugdale continued, the environment was more influential than the heredity, because the moral sense was a postnatal development. Dugdale noted further that heredity tended to create an environment perpetuating that heredity; the remedy was to reform the environment. Finally, environment tended to invent habits that become inherited, and the very cerebral tissue would become changed if environment pressure were sufficiently constant. Dugdale concluded that "the whole question of the control of crime and pauperism become possible, within wide limits, if the necessary training can be made to reach over two or three generations."[2] As an event in the history of nineteenth-century ideas, Dugdale's study is reasonably easy to identify and understand. Dugdale was clearly inspired by contemporary evolutionary science. His omission of any reference to Christian truth or tradition; his emphasis upon the scientific method; his use of the causal categories of nature and nurture (a direct borrowing from evolutionary theory); and his references to the principle of the inheritance of acquired characters, both an artifact of traditional folk wisdom and of contemporary natural science theory, all indicate to which niche he belongs.

And if our concern were merely a simple linear intellectual history (or what is often referred to as intellectual history) we would need to go little further in our discussion and analysis of his book. His book, and the reputation it quickly acquired in late nineteenth-century America, must also be understood as a phenomenon reflecting larger currents in American civilization.[3] More was involved than a shift from traditional Christian attitudes to modern, secular approaches to social problems; more was involved, in other words, than the content of Dugdale's ideas, as important as they were. What Dugdale wrote, as artifact, was important too, for it was a manifestation of a larger ideological movement in the America of his day to use the instruments and explanatory conceptions of contemporary science as guides to the control of a disorderly, messy, and unstable social order in transit from a traditional to a modern era. The obvious point about Dugdale, then, was that he was an integral part of the Social Darwinist movement of the later nineteenth century.

More important than the answers Dugdale and other Social Darwinist intellectuals provided were the questions they asked, the ways in which they perceived the world, and the possibilities they saw in a science of society. Clearly the answers they provided represented a way of thinking that the intellectual historian might well define as evolutionary naturalism. Like many constructs of the intellectual historian, this is a difficult and ambiguous term.[4] What is meant by it is, first, the assumption that man is an animal who has evolved in the same fashion as the animals. An obvious corollary is that man was as subject to the dictates of natural law and phenomena as the animals. A third point is that human nature and conduct were determined by natural forces beyond the control of the individual. And indeed most, if not all, of the Social Darwinist intellectuals in late nineteenth century America accepted these assumptions. Yet because we as historians have focused our attention upon the answers the Social Darwinist intellectuals provided their readers, rather than the questions they asked themselves, we have missed an important part of the Social Darwinists' legacy to their own time and to those eras that succeeded theirs. For the questions Social Darwinist intellectuals, including Dugdale, asked themselves revolved around the possibilities of contemporary science for the prediction and the control of human behavior, as Dugdale clearly indicated. If our only concern with Social Darwinists in late nineteenth-century America were the elaborate social theories based on analogies between nature and society that they contrived, then, quite apparently, they constituted a trivial or marginal phenomenon in our cultural history, or so the work of several scholars strongly suggests.[5] But if we are also interested in the Social Darwinists, Dugdale included, as representing a growing segment of Americans who wished to create a science of man that would lead to the prediction and control of human behavior in a society undergoing modernization, then it can be argued that Social Darwinism, and Dugdale, were indeed manifestations of a major shift in American civilization.

Such a shift did not occur in the abstract. Disembodied ideas did not merely call forth more theories. The structure of American society was changing rapidly in the later nineteenth century, and these changing social processes helped produce the sense of crisis and urgency that men such as Dugdale experienced when they looked at the world around them. It is a tiresome cliché to say that in the later nineteenth century, America became transformed from a traditional, rural, individualistic society, into a modern, industrial, urban, and increasingly organized society. But tedious though the generalization may be, vague though it certainly is, it is nevertheless true. The problems middle class Americans of Northwestern European ancestry faced at the end of the nineteenth and the beginning of the twentieth centuries were not all conducive to the questions answerable by a science of social control, at least not directly. But there were several social policy issues that both attracted the attention of these middle class Americans and which seemed within the purview of the new secular evolutionary social science of the day. In turn, these issues arose because of changes in the structure and the functions of American society. In the closing decades of the nineteenth century three issues in particular—immigration, segregation, and social competency—arose and seemingly challenged the older, white, Anglo-Saxon Protestant America and apparently necessitated the creation of new instruments of social control in a rapidly changing society. Especially after 1880 the number of immigrants entering the country who were from Southern and Eastern Europe, and hence not as assimilable into American culture as those from Northern and Western Europe, constituted the largest fraction of immigrants in the continual migration to America. Perhaps understandably, perhaps not, many such middle class Americans regarded these immigrants as unwholesome and perhaps dangerous additions to the body politic, and unstable elements in the social fabric. Ethnic conflict and tension between the "old" and the "new" Americans became a staple feature of public life in many American communities, urban and rural, before the close of the nineteenth century. These ethnocultural conflicts became even more serious after 1900, thanks to both the labor strikes of the 1890s and later to religious and other cultural tensions among particular ethnic groups. Toward the close of the progressive era, American society was becoming polarized along ethnocultural lines.[6]

And in the closing decades of the nineteenth century the condition of black Americans, never an enviable one, took a turn for the worse. Historians of the black experience have eloquently narrated the shifts in southern politics that produced formal, rigid, segregation in the South after 1890, beginning with disfranchisement, and ending with the segregation of water fountains, restrooms, and other public facilities.[7] The creation of a highly structured caste society in the South, and the economic, social, and technological forces that pushed blacks off agricultural land in the South, helped stimulate a mass migration of black people to cities, especially in the North. In many northern cities, black ghettoes crystallized rather quickly; the lines of separation were cleanly drawn by whites; and soon informal demarcations of caste had given way to a national society in which segregation was an obvious fact of life. Race consciousness heightened; prophets of separatism arose within the black community as the consequence of the emergence of advocates of segregation among whites and of their successes. By the end of World War I, as race riots broke out in a number of American cities, relations between whites and blacks were tense and hostile.[8]

The issue of social competency, or, conversely put, social dependency, was of course the one that Richard Dugdale addressed himself to in his famous study of the horrifying Jukes family, and it meshed with the issues of immigration and segregation. Before the emergence of an urban society, and before the development of the factory system, most Americans did not worry so much about poverty, dependency, and crime in their midst, preferring to ascribe their causes to the failures of particular individuals. But in the new urban-industrial culture the problem was more conspicuous. (Whether the proportional incidence of such social phenomena had increased was another question.) Hence the new sense of urgency. And in a sense, from the perspective of many middle-class Americans who were participants in and members of WASP culture, the three problems came together, or seemed different aspects of a larger problem: the quality of the peoples living within American civilization.

And here was precisely the contribution that a science of man, based on the precepts of evolutionary science, could make to social stability and control. It could determine what were the natural causes of human conduct among the different peoples in the American population, whether they were white, nonwhite, native-born, immigrants from abroad, paupers, criminals, merchant chieftains, or saints. Indeed, that was clearly Dugdale's message: now men of reason and scientific training could ask wholly new questions about nature and man. And for that still emerging science of man, Dugdale even provided helpful clues: that science must focus upon the heredity and the environment of the individuals in society.

Toward the close of the nineteenth century it seemingly became possible for men and women of science in America to begin to ask the kinds of questions that men such as Dugdale obviously thought were so important. The new possibilities opened up precisely because of the coalescence of an institutional revolution in American science and higher education and an intellectual revolution in the natural sciences. In the closing decades of the ninteenth century the emergence of the new graduate universities, and their satellite institutions, the land-grant colleges, made possible the creation of a new social role for the American scientist, the social role of the academically trained researcher who identified professionally with a particular science, who was encouraged by the institution that employed him full time to pursue independent investigation, and who could have the opportunity for original research and for the development of a professional identity. The changes inherent in American higher education enabled other American scientific institutions—societies, government agencies, and networks of philanthropy for scientific research, principally—to change their own standards and expectations of the social role of the American scientist as well. Even before 1900 this process of academic professionalization was becoming obvious in at least the more prestigious independent and public graduate universities, together with the sprinkling of land-grant colleges.[9] And the rise of expertise as a social role in science had implications, even if they were indirect, for the internal structures and processes of science itself. Toward the end of the nineteenth century the natural sciences suddenly shifted from being historical and observational disciplines to sciences that focused on experimental measurement of ongoing natural processes. The reasons for this shift were complex, but certainly much of the appeal of experimental natural science was the relative certainty of the answers that experimental inquiry would apparently provide. One could observe nature at work. That was, or seemed to be, infinitely superior to arid and philosophical speculations. And the new graduate universities made possible, not inevitable, the opportunities for American scientists to become practicing experimental scientists.[10]

It was in this context, at the dawning of the twentieth century, that the new experimental biology and psychology suddenly began to produce spectacular and brilliant explanations of some of the most complex and intractable mysteries of evolution, heredity, and variation, together with the articulation of seemingly well established explanations of human nature, intelligence, and conduct. Since the vast majority of American scientists were from the middle class, WASP culture, and since whatever tendencies they had had toward hereditarian interpretations of man were reinforced by their scientific training, it was not surprising that in the midst of an era most historians have labelled as one of reform, a powerful and authoritative group of American scientists loudly proclaimed that they had unlocked the secrets to a science of social control based upon the latest discoveries of evolutionary science. What this book is about is how and why those natural scientists took up advocacy for such a science of man, as well as for the particular hereditarian explanations of human conduct they read from their data, and the circumstances under which rivalling groups of social scientists challenged their beliefs in the power of original nature, but not in the electrifying possibilities of an evolutionary science of man. Usually we refer to those circumstances as the heredity-environment controversy.

# 1 *The Discovery of Nature 1890-1920*

# 1 THE NEW BIOLOGY

*1*

IN 1904 Charles B. Davenport announced the implications of the new experimental biology in the pages of *Science,* the prestigious weekly publication of the American Association for the Advancement of Science:

Today biology has to recognize that its individuals are . . .diverse combinations of units—relatively very numerous—which. . .we call . . . 'characteristics'. Characteristics are thus to individuals what atoms are to molecules. As the qualities and behavior of molecules are determined by their constituent atoms, so the essence of the individuals is determined by its constituent characteristics. And as we may construct new substances at will by making new characteristics of atoms, so we may produce new species at will by making new combinations of characteristics.[1]

This was, of course, an astounding message; but the progressive era was a time of considerable ferment and excitement in many fields of American culture—including the biological sciences.

Davenport could speak authoritatively of the new trends in natural science. He was a member of the new class of university scientific researchers that had taken shape in the closing decades of the nineteenth century. And he was one of a handful of the new breed of experimental biologists in America who were exploring the ramifications of Gregor Mendel's rules of inheritance and testing their applicability to all forms of life. Davenport's experiences helped determine that he would be a leading exponent of the new science of genetics. Born in Stamford, Connecticut, in 1866, this son of a long line of Yankees foresook the rural way of life and the traditional piety of his fathers for a career in the new industrial society and a life in science. At first, he wanted to be an engineer, or perhaps an agronomist. After graduating from Brooklyn Polytechnic Institute in 1886, he found no suitable employment. He then entered Harvard to study biology; there he discovered that the study of living things could be as precise as the science of inanimate nature. Even before he finished his Ph.D. in zoology in 1892, he had won an instructorship. He remained instructor at Harvard until 1899 and was doubtless the most inspiring teacher of the new experimental biology in the zoology department. His star rose rapidly in the 1890s and thereafter. He published papers then regarded as important contributions in experimental biology; he became a correspondent of such internationally known scientists as Sir Francis Galton of England, who founded the science of biometrics, the statistical study of inheritance, and the Dutch botanist Hugo de Vries, who was one of the rediscoverers of Mendel's laws of heredity. Davenport also became an important figure in several scientific societies. In 1904 he resigned his associate professorship at Chicago to become director of the Station for Experimental Evolution. Located at Cold Spring Harbor, on Long Island, this Station, supported by the Carnegie Institution of Washington, was for many years a leading genetics research institute.[2]

The message Davenport brought to the readers of *Science* had several implications. On one level Davenport celebrated the stunning triumphs experimental biologists had achieved in recent years in shedding new light on some of the more intractable riddles of evolution—how and why inheritance, variation, and evolution worked. The rediscovery of Mendel's principles of heredity in 1900 inspired experimental biologists in Europe and America, for it seemed to promise that they could quickly make the biological sciences as precise and rigorous as the physical sciences. Davenport's message was also important as a manifestation of a new emphasis upon heredity biologists increasingly made in the early 1900s. Although the new work did not necessarily justify an overemphasis on heredity apart from the environment, and not all biologists subscribed to such views, the new discoveries focused attention on inheritance, which gave the impression that heredity was far more important than environment. It was easy for many biologists—and those laymen who heard of their work—to assume that recent developments somehow justified the attribution of more explanatory power to nature and less to nurture. Such attention to inheritance was understandable in any event for biologists trained to assume that physical heredity was the material link between generations. And in retrospect it is clear that the context of the times and their particular experiences as middle class Americans of Northwestern European ancestry reinforced their technical reasons for believing that heredity was more important than environment. Finally, Davenport's message addressed itself to an important issue of public policy in an era of social and political unrest. With all the authority of the up-to-date scientific expert Davenport revived the ancient idea of breeding a better race. As historian Charles Rosenberg has pointed out, this idea had plenty of support among influential nineteenth-century Americans who involved themselves in public health, social reform, and the application of biological science to social problems.[3]

Davenport's message, then, was an important clue in understanding the origins of the heredity-environment controversy, for the ideas he expressed were both symbol and substance of a new vogue of hereditarian thinking among natural scientists in the progressive era. These new ideas—or representations of scientific ideas—gave the miniscule American eugenics movement a body of current scientific principles, and a fresh aura of scientific legitimacy, and a new lease on life as well. In the next thirty years the American eugenics movement would disseminate hereditarian ideas in the mass media and the educational system, fan the fires of the nature-nurture controversy, and even leave its imprint on federal and state laws. Furthermore, the bulk of America's working natural scientists came to assume that heredity was very important, or, at least, their writings often left that impression. Much of this impression was created by the simple fact that inheritance and variation, as research topics, were very fashionable in the early 1900s and created quite a stir within and without professional biology. But many working biologists interpreted the new discoveries as justifying a strong, if not an extreme, hereditarian point of view. Later generations of natural scientists would note, more in sorrow than in anger, that the new discoveries of the early 1900s did not support such interpretations of the data. This was partly a matter of later generations learning more concrete information about how nature works than did earlier generations. Yet in the main the new work did not prove or disprove anything about the relative potency of heredity and environment. It constituted a more precise description of how heredity and variation worked than had been heretofore known. It may be impossible to reconstruct a definitive answer as to why so many well-trained and intelligent scientists construed the new work in the way they did. But most probably we may assume that their scientific training and their cultural experiences and values reinforced a belief in heredity; that the newness of experimental biology, and the spectacular successes of experimental research, encouraged many to assume science had proven more than it really had; and that the sense of participating in thrilling discoveries with important ramifications for nature and society may have blinded them to what later generations would regard as exaggeration and error.

The rise of hereditarian thinking in the biological sciences, then, had important ramifications for the origins of the nature-nurture controversy. The new line of thought gave the impression that heredity was all powerful and environment was of little significance in the creation of species and individuals. That impression influenced educated Americans beyond the biological fraternity—especially the psychologists, as we shall see in the next chapter. But first we must understand how and why experimental biology came to America from Europe in the later nineteenth century, for it was from American scientific institutions that the new hereditarian ideas were so effectively and authoritatively disseminated.

## II

On the evening of 18 October 1895, Edwin Grant Conklin spoke before the Woman's Club of Evanston, Illinois, on "Recent Work in Biology." This assistant professor of zoology from Northwestern University celebrated the dawn of a new age in the history of biology. Thanks to the verification of the evolutionary hypothesis, he exclaimed, biology now was an independent science in its own right, equivalent to, and possibly superior to, the physical sciences in its explanatory power, its generality of application, and its promise for human betterment. At long last biologists could unlock life's deepest secrets. They could study natural processes in the laboratory. Best of all, he declared, there was a new spirit among biologists that boded well for the future of biology. Biologists wanted facts and empirical explanations— not outmoded philosophies of evolution and life. "In a word, the spirit of modern Biology is reactionary," he declared, "a revolt from theories to facts, from idealism to realism. . . . There is everywhere a spirit of questioning and doubting—no assumed fact goes unchallenged, no theory escapes a fiery trial. The whole atmosphere breathes of skepticism—a condition inimical to faith but indispensable to the advancement of science."[4] Conklin had reason to be so enthusiastic. The late 1890s were an exciting time for young experimental biologists such as he in America. Experimental research held out to them the promise of unlocking nature's secrets on a far grander scale than the old-time natural history; and rapid development of experimentalism in America also suggested that perhaps the Americans could seize leadership away from the Europeans in this science. Conklin's generation was a fortunate one. They came to the study of biology just as experimental methods began to triumph over the more traditional methods of natural history—with apparently superior results. And these young men and women entered careers in science just as the expansion of American higher education was increasing the number of positions in which they could pursue their research. In a very real sense, the new graduate universities and the new experimental biology grew up together in late nineteenth-century America. The graduate universities provided much of the institutional support for experimental biology, support the old-time denominational colleges could not supply, such as laboratories, libraries, opportunities for specialized teaching, and, perhaps most importantly, the institutionalized expectation that all faculty would involve themselves in original research. The graduate universities furthermore did not operate under the same religious restraints that the old-time denominational colleges did, and they had growing enrollments and expanding financial resources. The private or independent graduate universities developed experimental biology far more rapidly than did the public or land grant graduate universities, because the private institutions did not have to curry favor with political special interest groups and they had stronger financial support.[5] Experimental biology grew later at the public institutions. This academic professionalization of biology caused a wholesale institutional revolution, both within and without the new universities—as the graduate universities came to staff more and more faculty with advanced degrees and an academic identification with biology, these scientists in turn founded extramural professional institutions, such as societies, journals, and research resources, to serve their needs as professional academic scientists. To understand the flowering of experimental biology, however, we must first grasp its progress from the private to the public graduate universities.

Harvard rapidly became a leading center of the new biology. Harvard's long tradition in natural history and the biological sciences helped pave the way for its conversion to experimentalism in the later nineteenth century. Even before the Civil War, such distinguished natural scientists as the anatomist Jeffries Wyman, the botanist Asa Gray, and the zoologist-geologist Louis Agassiz taught at Harvard; indeed, the organization of the Lawrence Scientific School in 1847, the creation of autonomous departments for the various natural history disciplines, and the founding of the Museum of Comparative Zoology all testified to Harvard's commitment to advanced work in the life sciences. Louis Agassiz, for example, trained several generations of zoologists and geologists to be specialists capable of independent research; some became eminent in their fields.[6] In 1869 Harvard entered a new phase of its history, for in that year the young chemist Charles W. Eliot became president. Eliot instituted many reforms at Harvard that transformed America's oldest university into a recognizably modern university; for our purposes, the most important thing he did was further scientific specialization. In the early 1870s the Harvard zoology department began offering the Ph.D. Just before his death, Agassiz founded a small marine laboratory, which he intended as an institution of natural history instruction, and which died with him in 1873. But a marine laboratory possessed many advantages for exper-

imental work as well as for natural history, and Harvard's zoologists continued to work at the world's two major marine
laboratories—the Zoological Station in Naples, Italy, and the
United States Fish Commission's facility at Woods Hole, Massachusetts. In the 1880s Harvard zoologists took another step
toward specialization when they started a *Contributions* series in
which to publish the research of faculty and graduate students.

By the early 1890s the new experimental biology was beginning to replace natural history at Harvard. In 1892, for example,
the year Davenport won his doctorate, the department of zoology
dropped its natural history courses for offerings in such particular
fields of experimental biology as microscopical anatomy, experimental morphology, and statistical methods in biology (the latter
two taught by Davenport). Indeed the younger men in the
department—above all, Davenport—pushed for these changes.
They taught the experimental courses. They imparted to future
generations of students the breathtaking promises of experimental
biology. Thus Davenport trained his graduate students to discover
the laws of inheritance, variation, and environmental influence
by using breeding experiments and statistical methods, arguing
that in these ways the inheritance and development of traits could
be scientifically and systematically studied. Davenport corresponded with the English mathematician Karl Pearson, a leading
founder of the science of biometrics—and a champion of the
eugenics movement in the United Kingdom. By the later 1890s
the Harvard zoologists were deeply committed to the new experimental work in their research as well as their teaching. They
performed the first breeding experiments with the common fruit
fly, later the classic subject of countless genetics experiments.[7]
And the new biology was pursued with equal, if not superior,
fervor and results at the Bussey Institution, which had been
Harvard's undergraduate college of practical agriculture since
1871. When the Bussey was reorganized in 1908 as a graduate
school of applied biology, it had already become an important
center of work in the new science of genetics, thanks to the
efforts of William E. Castle, a mammalian geneticist who had
studied with Davenport in the 1890s, and Edward M. East, a
chemist turned geneticist from the University of Illinois whose
work on hybrid corn rapidly earned him an important niche in
the history of agronomy.[8]

From its opening in 1876 the Johns Hopkins University became distinguished in the new biology. The Hopkins could appoint able faculty, such as Henry Newell Martin in physiology, a former associate of Thomas Henry Huxley, and William Keith Brooks, one of Agassiz's Ph.D.s in natural history. And because the Hopkins represented the new American graduate university, it attracted perhaps more than its share of brilliant young men who would later leave their mark on American culture and politics: Frederick Jackson Turner in history, Woodrow Wilson in political science, Edward A. Ross in economics and later sociology, and John Dewey in philosophy and psychology. The biological disciplines benefitted from the presence of graduate students who would later become famous; in addition to E. G. Conklin, Edmund B. Wilson, perhaps the preeminent cytologist of his era, Ross G. Harrison, a distinguished anatomist, and Thomas H. Morgan, eventually a Nobel laureate in genetics, all took their doctorates at Johns Hopkins. Ironically, they worked not with Martin (who was an experimentalist), because Martin's chief duty was teaching medical students, but with Brooks, who was mainly interested in developing his own philosophical statement of biology. Yet even if Brooks seemed an improbable mentor for young doctoral candidates eager to become specialists in experimental work, in the final analysis he was not so unlikely, for he left his graduate students alone, he encouraged them to take advantage of the University's splendid resources for advanced work, and his deep philosophical interests stimulated them to think of organisms in terms of causes and functions, not merely static forms and structures.[9]

In the 1890s Columbia University nurtured a strong zoology department with a profoundly experimental focus. How a formerly undistinguished commuter college for the scions of Manhattan's wealthy became a great university was achieved in large measure because of the able leadership Presidents Seth Low and Nicholas Murray Butler provided from the 1890s on. Low was a prominent local businessman and reformer who upgraded many of the academic departments and who worked out a highly imaginative staff-sharing arrangement with the American Museum of Natural History under which both institutions would develop certain fields in biological science and ethnology with the same specialists; the first such appointment was Henry Fairfield Osborn in vertebrate paleontology in 1891. When Low stepped down from the presidency to enter municipal politics as a reformer in 1899, Nicholas Murray Butler, former head of the philosophy department, succeeded him, and continued Low's policies. In 1892, while Low was still president, the department of zoology was formally organized. Although Osborn was a distinguished member of that department for many years in his own field, the department's real strength was in experimental work. From his appointment in 1891, Edmund B. Wilson became a leading cytologist; he did much to reorient American biologists toward the cellular approach to heredity and development in general and the cytological perspective in particular. Thus he helped prepare himself and his professional colleagues on Morningside Heights and elsewhere in American science for the necessary background to grasp the significance of Mendel's laws of inheritance. One of Wilson's doctoral students, Gary Nathan Calkins, remained in the department for the rest of his career; Calkin's work on the cytology of the protozoa soon won him fame and distinction among experimentalists in America. Columbia's zoologists were active researchers, partly because they were self-motivated to do research, and partly because the University provided a congenial environment for scientific professionalism. Columbia's zoologists had important connections with the international biological science community; they corresponded with leading biologists the world over, and, through the department's program for visiting lecturers, they arranged for such famous biologists as Edward B. Poulton and C. Lloyd Morgan of England to deliver public addresses on their work at Columbia. In the 1890s the department graduated seven Ph.D.s; in the next ten years it trained another eighteen. In many respects a new phase in the department's history began in 1904, when Thomas H. Morgan of Bryn Mawr College, a rising star in experimental morphology and embryology, became professor of experimental zoology at Columbia. Around 1907 Morgan and his students began to extend and elaborate Mendel's laws of heredity on the basis of their brilliant research with the common fruit fly. In time Morgan and his associates founded a leading school of Mendelian genetics.[10]

The University of Chicago soon rivalled Harvard, Johns Hopkins, and Columbia in experimental biology. Founded in

1891, Chicago was destined for instant greatness; it was a new institution, with few ties to the past; it had impressive financial backing; and it was run by a dynamic president, William Rainey Harper. Even before Chicago opened its doors to students, Harper was recruiting faculty, raiding European and American universities for the best possible professors in all fields. He offered phenomenal terms: light teaching loads, salaries double the national average at each rank, opportunities for graduate teaching, ample research facilities and funding, and even a university press. So attractive were Harper's blandishments that he even lured no less than seven college presidents back into faculty ranks. Harper promoted the biological sciences handsomely at Chicago. He hired such outstanding faculty as the zoologist Charles O. Whitman and the neurologist Henry H. Donaldson. He permitted the founding of separate departments of anatomy, botany, neurology, physiology, and zoology, thus recognizing and sanctifying the several experimental disciplines. Each department had its own building, including spacious laboratory facilities, its own graduate program, its own funding for research expenses and technical assistants, and, above all, the avid support of Harper and the trustees.[11]

Other private universities followed suit to the extent that they were able. At Cornell University, an independent university with some public funding, the natural history tradition continued strong into the twentieth century, but the prestige of experimentation was so powerful, so compelling and attractive, that natural history there was developed along experimental lines whenever possible.[12] At Clark University, in Worcester, Massachusetts, President G. Stanley Hall, a psychologist, attempted to implement his dream of a graduate school in the basic sciences from the time the university opened in 1889. He appointed a brilliant faculty, including Franz Boas in anthropology, Arthur Michael in chemistry, Albert A. Michelson in physics, Charles O. Whitman in zoology, and Henry H. Donaldson in neurology. But Hall turned out to be a devious and imperious president, partly because of his personality, partly because of the financial restraints the University's main benefactor (and source of revenues) Jonas Clark imposed upon him. In the spring of 1892, many of his most distinguished faculty, including Whitman and Donaldson, resigned in disgust to take positions elsewhere.

This was a disaster for Clark University, which thereafter retained its reputation in psychology. Yet Hall was a psychologist much intrigued by the new biology; he managed to keep biology alive, if only on a shoestring, and the remaining biologists were committed to the new biology and managed to keep up with the new work.[13]

By the early 1900s the diffusion of experimental biology in American science entered a second stage as state universities and land grant colleges began to incorporate it into their offerings. This was at least as important as the first phase, when the more prestigious private or independent graduate universities began to support the new work, because although for many years to come most leading researchers in the new biology took their doctorates at these private universities, the state universities and land grant colleges provided full time appointments for the graduates of the private university doctoral programs and enrolled far more undergraduate students. Thus, the public institutions helped diffuse experimental biology to a broader audience than the private universities could, and some of the state universities soon rivalled the private universities in prestige.

The processes of institutional transmission and diffusion were complex. They were dependent upon many circumstances and influences: the availability of funds, the interest of institutional administrators, the curricular needs of individual colleges, the receptivity of particular colleges and universities to new ideas, the political sensitivities of the various states, the importance of the evolution controversy, and the like. In many instances, a bright young Ph.D. from an American graduate program established a colony for the new biology when he took his first teaching or extension appointment in a college or land grant institution. Often these colonies operated under penurious circumstances, for most land grant colleges had tight budgets. Undoubtedly few young American Ph.D.s who took jobs in the early 1900s at land grant colleges were as fortunate as Alfred F. Blakeslee. Blakeslee took his Ph.D. from Harvard in 1904 as a specialist in botanical genetics. He wanted a job in which he could devote all, or almost all, of his time to research. Mendelism and the mutation theory were of great interest to him; he did not want to spend his time teaching elementary botany to hordes of freshmen of varying skills. Botanical genetics had important "practical" applications for agricultural hybridization; and therein lay Blakeslee's good fortune. In 1907 he won appointment at Connecticut Agricultural College. He had much time free for his own work, and, of course, he brought the latest work with him to Storrs. The practical applications of his work gave him the authority and the credibility to persuade College officials to expand the institution into a thriving center of genetics research where bright young geneticists spent their apprenticeships before moving to more broadly-based universities.[14]

Upon occasion, an ambitious administrator (usually a president) imported a whole discipline, lock, stock, and barrel. In 1902, for example, Benjamin Ide Wheeler, the aggressive president of the University of California, lured the brilliant physiologist Jacques Loeb from the University of Chicago to Berkeley with an irresistible offer. Loeb was to have a handsome salary raise, of course, but he was also to have the funds necessary for his whole department: a building, a junior professor or two, several graduate fellows, laboratory technicians, and even a janitor! Loeb could have a free hand in selecting all staff and in planning the building. "We intend to make the fullest provision for research," Wheeler told Loeb, "and indeed to found the department upon the idea of research."[15] Few presidents in the early 1900s had the financial resources to imitate Wheeler's example; and, for that matter, Wheeler could not recruit every department in the University in so lavish a manner. Importation was an infrequent, if not rare, phenomenon. Far more commonly new men gradually converted established programs, often with the assistance of administrators anxious to raise their institution's prestige. In the later nineteenth century the University of Wisconsin was one of the stronger land grant universities. By the late 1880s the University had faculty competent to teach anatomy, histology, and embryology as undergraduate service offerings to students in such career programs as agriculture and medicine. In the 1890s University officials stressed the earned doctorate for faculty and inaugurated graduate programs in a number of departments including the biological sciences. In 1893 William Stanley Marshall came to Madison fresh from the University of Leipzig; he published papers on histological and embryological aspects of insects and trained several doctoral students. The appointments of vertebrate embryologist Bennet Mills Allen and zoological geneticist Samuel Jackson Holmes, two young Ph.D.s from Chicago, in the early 1900s furthered both specialization in biology and the number of experimentalists at Madison. The new biology grew there during this time, partly because the biologists taught service courses for students in occupational and professional programs, but also partly because they were now capable of offering advanced graduate work in new and exotic fields such as cytology and genetics. Before the 1920s the biological sciences had grown sufficiently to be housed in a new biological sciences building, complete with spacious laboratories and the most modern equipment, with more faculty and graduate students.[16]

The institutional changes within American higher education that helped create and solidify the social role of the scientist-researcher for biology and many other sciences soon led to changes in those scientific institutions that were extramural to universities and colleges—scientific societies, journals, and research institutes. The new experimental biologists soon came to dominate these extramural institutions, thus completing the process of disseminating their ideas and approaches in American science. In the three decades following 1880 scientists in many disciplines started national scientific societies. In most instances these national scientific organizations were limited to a single discipline, for the founders of each wanted to promote communication and self-consciousness among all practitioners of their discipline. Symptomatic of the new trend for scientific societies to be based on explicit disciplinary affiliations was the reform of the American Association for the Advancement of Science in 1882. Until that year the Association had met once a year (sometimes more before the Civil War) for the delivery of papers in but two "sections," or general meetings, one for the natural sciences, the other for the physical sciences. In 1882 the Association, at the behest of its leaders (who were, largely speaking, specialized researchers in their respective sciences) voted henceforth to meet in nine "lettered" sections, segregated according to discipline so that specialists could easily communicate with one another and exchange information and views. Ever since, the Association has made further refinements of this general scheme, but has not altered it fundamentally.[17]

The Association's reforms did not satisfy all scientists; probably that was not possible, for the Association still had the flavor of a general-science convention. In 1883 a group of professors of natural history organized the American Society of Naturalists, largely because the reformed American Association did not provide what they deemed sufficient opportunity for professional discourse, and in any event, the Association's meetings were held but once a year and were still open to all regardless of expertise. The Society's founders consciously rejected as a model for their new organization the example of the municipal natural history society or academy of the nineteenth century, which included as members any and all persons interested in natural history.[18] The Society's pioneers wanted a national society limited to specialists in natural history. They also desired a membership of persons who followed a full-time career directly related to natural history. Prominent among the Society's founders were the former students of Louis Agassiz and men who had taken advanced work abroad. Membership in the Society was by invitation and election only. Charles S. Minot, a young professor of embryology at Harvard, issued the original invitations, corresponding with approximately one hundred and fifty "professional naturalists of the country" in 1882.[19] Virtually all of these gentlemen became elected members. In the 1880s and 1890s, the Society's leaders carefully restricted membership to "professional naturalists," which meant, for them, persons who pursued full time natural history careers in colleges or government service, individuals who for all practical purposes in those years were scientific specialists who identified professionally with a particular discipline. The rise of the new biology as a system of ideas and as a network of personal and institutional relationships left its mark on the American Society, for ironically in the 1890s and thereafter the natural historians once prominent in the Society's affairs increasingly found themselves displaced by the new experimentalists, who took over the Society and imposed their standards and priorities in research upon it. As early as 1890, one-third of the Society's members held an earned doctorate; by 1911 over three-fourths of the membership possessed that degree. And most of the new members were experimental biologists, not natural historians. As more and more experimentalists joined, the natural historians appeared less and less frequently as readers of papers or as officers of the Society. After 1910 the Society's membership standards included, in addition to all the other criteria, evidence of original research in published form.[20]

By the early 1900s, organizations even more specialized than the American Society were founded, in many instances by the new experimentalists. Thus a group of professors of zoology active in the Society organized the American Society of Zoologists in 1903. Entomologists in government service and higher education launched the Entomological Society of America in 1906. Geneticists formed the American Genetic Association in 1913 as a splinter group from the American Breeders' Association, whose membership included too many stock breeders for their purposes. In most of these new societies membership was by election only. An existing group of specialists sat in judgment on the scientific qualifications of prospective applicants. And, in any case, the specialized interests of these organizations, and the formidable academic credentials of their members, probably acted as an effective bar against amateurs' intrusions. That this was so was suggested when lay natural historians founded the American Nature Study Association in the early 1900s, thus recognizing the split between generalists and specialists in biological study. Most probably the Society for Experimental Biology and Medicine, organized in 1904, had the most severe standards of all the new biology societies. Membership was by invitation and election; the Society for Experimental Biology and Medicine went further than many scientific organizations of its day by insisting as a requirement of maintaining membership that each member had to present a paper based on original research at least once every two years.[21] This was perhaps an unusual example, but it underlined how thoroughly the new specialized professoriate had taken hold in American biology, and how much attraction experimentalism held for biologists, by the early twentieth century—a reflection of the impact of the new universities. The more prestigious universities, and the scientific societies with the most visibility, were becoming enclaves of the new biology—and the new biologists.

Much the same transition took place in biology journals. The need for journals more specialized than Benjamin Silliman's *American Journal of Science* or E. L. Youman's *Popular Science Monthly,* definitely existed in the Gilded Age, before the emergence of experimental biology in America. There were enough specialists in natural history then to support *The American Naturalist*, founded by several natural historians, including the Philadelphia paleontologist Edward Drinker Cope. These men, with Cope as their editorial and financial leader, launched the journal with their own funds as an entrepreneurial venture to promote specialized work in natural history.[22] *The American Naturalist* became the major national periodical for natural historians in America. Even such scientific magazines as *Science* and *Popular Science Monthly* did not carry the same quantity and quality of contributions. Somehow, Cope and his associates always managed to keep the county sheriff from seizing the journal's assets for nonpayment of the printer's bill at the last minute. By the 1890s, however, the space available in *The American Naturalist* could not accommodate the growing number of potential contributions. Consequently a group of experimental neurologists and zoologists founded *The Journal of Comparative Neurology and Zoology* in 1891 and opened its pages to the new work. Cope died in 1897, and the *Naturalist* passed into the hands (and the ownership) of James McKeen Cattell, an able young experimental psychologist at Columbia University who was wholly sympathetic to the new biological work and who in any case was something of a builder of scientific institutions and organizations in America. Cattell willingly took articles in experimental biology. But that did not accommodate all worthy authors. Experimentalists associated with the Marine Biological Laboratory in Woods Hole, Massachusetts, launched the *Biological Bulletin* in 1899. Still, many experimental biologists believed there were not enough journals in America. As Herbert Spencer Jennings, a prominent experimental protozoologist, put it in 1902, "It would be helpful if a first class journal could be established in America, in which the results of American investigations could be published in good form, without unreasonable delay, and without expense to the author."[23] The next year a group of experimental biologists, including Jennings, mounted a campaign to start the *Journal of Experimental Zoology*. They sent a prospectus to each member of the new national societies, and special appeals to leading professors to secure institutional support. The *Journal* was to be limited to contributions with an experimental focus. And the editors' values were those of the national academic expert, the professor-researcher with universalistic rather than localistic or particularistic standards of competence. The *Journal,* they declared in their prospectus, "will not be identified with or managed by local interests, but will be representative, and, as far as possible, national in character. . . , it is proposed to form, in addition to the editorial committee, a board of collaborators, representing a large number of institutions."[24] The *Journal* became a major organ of the new experimental biology. Together with such journals as *Science, Popular Science Monthly,* the *American Naturalist,* the *Biological Bulletin,* and others, it helped spread the new biology to American scientists and provided publication outlets for American experimentalists. As the subdiscipline of genetics became a coherent specialization in the wake of the rediscovery of Mendel's laws of heredity after 1900, geneticists founded even more journals, most notably the *Journal of Heredity,* aimed at breeders as well as geneticists, and *Genetics,* whose intended audience was obviously advanced specialists in that field.

The final step was taken in the institutional revolution that established both scientific professionalization and experimental biology in American science when biologists were able to direct the development of research institutions and thus gain some control over their research resources and priorities. The first such permanent institute was the Marine Biological Laboratory, located at Woods Hole, Massachusetts. In the 1870s, Spencer F. Baird, Assistant Secretary of the Smithsonian Institution and United States Fish Commissioner, patiently laid the groundwork for the ideal of a research institute in the biological sciences. The Fish Commission had a facility at Woods Hole; in the 1870s and 1880s, Baird, in his capacity as Fish Commissioner, permitted natural historians to work at the facility during the summers, collecting and classifying specimens and in many ways training themselves and their students. Baird realized that his dream of a marine laboratory required more than a community of specialists; it necessitated in particular political and financial support from the business community. So as Fish Commissioner he gave countless public talks before businessmen in New England in particular and the nation at large, spreading the gospel that proper fish culture was vital to the economy of New England and the nation, and fish culture could be advanced by proper marine research. By the mid-1880s he had a flourishing marine laboratory at Woods

Hole in all but official title and secure administrative and fiscal structure. By then he could also point to the growing number of biologists (natural historians and experimentalists alike) who commonly worked at Woods Hole all summer. And he had gathered much support from important politicians and business-men for marine research.[25] Baird was by now an old man, but he had prepared the ground for a group of academic biologists in the Boston area to persuade local philanthropists and entrepre-neurs to raise funds for an endowment for a new laboratory at Woods Hole. The professors organized a national campaign to secure funds and endorsements. In their prospectus, the profes-sors took care to point out the new institute would not be a parochial one; it would not be controlled by "any college, uni-versity, or other institution," but it "should be truly national in character and. . . should invite the cooperation of all. . . inter-ested in the advancement of the science of Biology."[26] The Laboratory held its first summer session in 1887; the distin-guished zoologist Charles O. Whitman was appointed director. Under his leadership the Laboratory became an interdisciplinary summer institute where professors and students from all over the nation could meet, do research together, attend special lectures, take courses on new subjects, become friends, discuss the latest discoveries, and, above all, develop a strong sense of professional identity. Within a few years the experimentalists were prominent in the Laboratory's affairs as researchers, as contributors to its *Biological Bulletin,* as teachers of short courses, and as trustees.[27]

Even more spectacular was the Carnegie Institution of Wash-ington's commitment to experimental biology, which Charles B. Davenport helped encourage. Industrialist Andrew Carnegie established the Institution in 1902 with a $10 million endow-ment, stipulating that the Institution support basic research, especially in the sciences. Davenport, then teaching at Chicago, conducted a lengthy campaign to persuade the trustees of the Institution to found a biological laboratory at Cold Spring Harbor, Long Island. (Davenport was director of Chicago's summer school at Cold Spring Harbor.) So effective an advocate was Davenport that the trustees established the Station for Experimental Evolution as a department of the Institution and, in 1904, appointed Davenport director. He proved an able pro-moter of his new enterprise. Within two years he had convinced the Institution's trustees to provide recurring funding for six full time resident genetics investigators, occasional visiting lecturers, symposia, a handsome physical facility, laboratory technicians, custodians, and special research grants for specialists to work there in the summers, not to mention support for a monograph series and his own salary as director. Quite rapidly the Station became one of the world's leading centers of genetics research, even though one of Davenport's close associates at the time remained convinced that Davenport did not really have a systematic research program worked out.[28] But Davenport was an imaginative promoter; he never missed an opportunity to tell the Institution's trustees, in glowing terms, of the importance of the Station's work and its ramifications for science and society. In his annual report of 1907, for example, he wrote that the "direction that our work has already taken us has been determined by the conviction that the most important definite question to answer is, how may the course of the stream of germ plasm that has come down to us from remote ages be controlled in its onward course?"[29]

By the first decade of the twentieth century, then, the experimental biologists had established themselves in American educational and scientific institutions, and had overshadowed the old-time natural historians. By and large the experimentalists now held the best university appointments, dominated the leading scientific societies, published heavily in the major journals in the field, and controlled the research institutes. The prestige, the limelight, and the glory went largely to the experimentalists who overshadowed the natural historians. Now the spotlight was on the experimentalists; their ideas were perhaps given even more attention than they might otherwise have been, thanks to the rather complete control they had of the most prestigious biological science institutions. In a sense the reasons for the triumph of the experientialists over the natural historians had much to do with the seemingly superior explanatory power of the new work, with, in short, the ideas of experimental biology; but institutional developments mattered too, as the means for the diffusion of a certain group of ideas about the processes of inheritance, variation, and evolution in living things. Those ideas seemed to suggest to all but the most careful observer that heredity was far more important than environment.

*III*

By the early 1900s, many experimental biologists had come to believe that biology stood on the threshold of a new era in its history. They were persuaded that the application of experimental methods and statistical techniques would enable them to extract definitive answers to some of the most difficult questions of biology, thus elevating biology to the precision and the rigor of the physical sciences. This was an exciting prospect, to say the least. The distinguished physiologist Jacques Loeb aptly expressed the optimism of this first generation of American experimentalists when he insisted in 1904 that "the work of Mendel and de Vries and their successors marks the beginning of a real theory of heredity and variation."[30] Loeb's generation was so enthusiastic because, in the three decades since the scientific acceptance of Darwin's theory of evolution, biologists had not arrived at a common understanding of how heredity and variation worked. It was difficult, if not impossible, to explain the causes of evolution without at the same time having a precise grasp of how traits were inherited and how they varied from generation to generation. Now it appeared that experimental biology was on the verge of unlocking these mysteries of nature. At the same time, the new theories of particulate inheritance, of variation, and of evolution through mutation placed much emphasis upon the importance of physical inheritance, and gave the impression, especially to persons outside the profession of biology, that heredity was of far greater causal importance than environment. Many within the profession wrote and spoke in glowing terms of the importance of the new work, even before they understood it clearly or before all the evidence was in. And many biologists disagreed among themselves over their understandings of heredity, variation, and evolution. Yet they gave the impression in their writings before the general scientific community that heredity was far more important than environment, and that physical and mental traits were inherited and could be studied scientifically. This dissonance between a general hereditarian consensus and often wide disagreement over particular technical issues cannot be underscored too greatly. It gave educated Americans outside the biology profession the idea that these scientific questions had been resolved, when in fact most conscientious experimental biologists knew they had not. This curious situation was the result, largely speaking, of a particular turn of events in the history of biological science in late nineteenth-century America.

All evolutionary biologists understood perfectly well that it was essential to explain heredity before they could grasp the causes of evolution. Physical inheritance was the linchpin of evolutionary theory, the device the evolutionary theorist used to reconcile continuity and change in organic nature. In the later nineteenth century many biologists attempted to work out a satisfactory theory of heredity. Yet heredity remained an intractable mystery, capable of inspiring various theories but not a commonly agreed upon explanation, so long as biologists continued to use the observational techniques of the traditional natural historian. Among American biologists in the period between the acceptance of the evolutionary hypothesis and the rise of experimental biology probably the leading theory of heredity was the Neo-Lamarckian theory of the inheritance of acquired characters. The so-called American school of Neo-Lamarckians dressed up the ancient idea of the inheritance of acquired characters in the modern garb of the evolutionary ideas of the French zoologist Jean Baptiste Lamarck. The Neo-Lamarckians embraced the theory of evolution. But they vigorously criticized Charles Darwin's explanation of the dynamics of evolution through natural selection; the Neo-Lamarckians believed that evolution was a progressive and unilinear process in which organisms adapted to their environments—not the cold, messy, and perhaps haphazard process Darwin had outlined in his writings.[31] They were repelled by the implications of Darwin's world view, which apparently denied the existence of a higher design or purpose in nature. And the Neo-Lamarckians attacked Darwin's theory of natural selection on the technical ground that it did not explain the origins of traits and the causes of variations in traits—which was perfectly true. The Neo-Lamarckians used the principle of the inheritance of acquired characters to support their particular teleological views of evolution and to account for the origins of traits and the causes of variations. According to the theory of the inheritance of acquired characters, new traits arose when an organism, faced with new environment pressures, would adapt itself to those pressures to survive. The resulting internal changes in the physical constitution of the organism would then become a part of the organism's congenital endowment, and the new trait would be passed on to future generations. This would work for species as well as for individuals provided that the environmental pressures were sufficiently forceful and general in their application. In this way new traits arose and variations appeared. And the acquired characters theory also permitted the Neo-Lamarckians their cherished progressive view of evolution through particular stages toward that perfect result of evolution, man.[32]

The Neo-Lamarckians were able to disseminate their particular interpretations so effectively in the American scientific community in large measure because their leaders, such as Edward Drinker Cope, Boston zoologist Alpheus Hyatt, and William H. Dall, invertebrate paleontologist long associated with the U.S. Coastal Survey and the U.S. Geological Survey, occupied important positions and were respected scientists in their day. Several owned and edited *The American Naturalist*. A number held major offices in such prestigious scientific organizations as the American Philosophical Society, the American Society of Naturalists, the Academy of Natural Sciences of Philadelphia, and the Boston Society of Natural History. As a school they were prolific scholars who published hundreds of specialized articles and books.

By virtue of their strategic position in the scientific community in the Gilded Age, the Neo-Lamarckians bequeathed to their successors, the experimentalists, an abstract and generalized model of nature that was hereditarian. It attributed all causal power for the evolution of species and of individuals to biological inheritance. According to the American school, all environment could do was exert pressure for adaptation. The evolutionary process itself was biological. The vehicle of evolutionary change was physical inheritance. By definition, environment was of transient importance, merely the catalyst of evolutionary adaptation; the change caused by environment had no role in either ontogeny or phylogeny unless it became a part of original endowment and was transmitted to succeeding generations.

Further evidence that the Neo-Lamarckians emphasized the importance of heredity over environment may be found in their interpretation of the evolution of mind. They argued that mind (or consciousness) was the innate adaptive instrument of all sentient species. In their view, man stood at the apex of the evolutionary pyramid; therefore, man's mind was the most innate in nature. The Neo-Lamarckians thus laid the groundwork for a thoroughgoing hereditarian interpretation of the evolution of mind as applied to all species, and particularly to man.[33]

The Neo-Lamarckians' specific ideas did not survive the nineteenth century. The rise of experimental biology made inevitable the testing of many biological theories. And a prime candidate for such scrutiny was the Neo-Lamarckians' acquired characters idea. The scientist who challenged the acquired characters theory was August Weismann, a specialist in cytology at the University of Freiburg. Weismann attacked the theory in part because he strongly supported Darwin's theory of natural selection. But at least as important a consideration for Weismann was his own experimentation, which persuaded him of the fallaciousness inherent in the acquired characters concept. Time and again he found in his experiments evidence that environmentally caused changes in the somatic constitution of an animal were not transmitted to its offspring. During the ensuing controversy over the acquired characters idea, which involved American and European scientists, Weismann refined his ideas on heredity and evolution. Weismann's theory of heredity was subtle and complex. According to his "germ plasm" theory of heredity, determiners existed for each somatic trait in the reproductive germ plasm in all sexually-reproducing species. It was these determiners that gave rise to the developed traits in the organism; the only kind of environmental pressure that could affect the traits of future generations had to influence the determiners in the germ plasm. Weismann went further. He argued that there was a radical difference between the somatic or body cells, which died with the individual organism, and the germinal cells (or determiners) that gave rise to the somatic cells and which were therefore continuous from generation to generation.[34]

It is important to understand that Weismann did not say that all traits of an organism were inherited biologically. He insisted that all biologically transmitted traits were inherited as determiners in the reproductive germ plasm at the time of conception. This was a far cry from an extreme hereditarian determinism. Indeed, as an avid champion of Darwin's theory of natural selection, Weismann could not exclude environment as a causal factor in evolution. Unlike the Neo-Lamarckians, who insisted that environmental adaptations had to be transmitted biologically to be of evolutionary significance (and thus blurred the distinction between nature and nurture), Weismann distinguished between heredity and environment as different kinds of causal factors. Indeed, Weismann insisted that many traits were not biologically inherited. Thus he flatly rejected the argument many Neo-Lamarckians (and many who subscribed to theories of racial superiority) advanced, that man's mind, and therefore his culture, were biologically inherited. Weismann anticipated modern cultural anthropology's argument that man's mind and culture were independent of biology and depended upon tradition and other social processes.[35] Yet it was easy to misinterpret Weismann. His theory was a description of the workings of heredity, not an argument for the potency of heredity over environment; yet his emphasis on the role of the germ plasm seemed to create the impression he was insisting on an extreme hereditarian position. Certainly few scientists, in the heat of the Neo-Lamarckian controversy, caught his careful distinctions between culture and nature. And Weismann insisted that traits were relatively stable over time, which implied to many of his scientific contemporaries in America that it was possible to study evolution by tracing characteristics through controlled breeding experiments. It was natural for most American biologists to approach the issues of the Neo-Lamarckian controversy from a largely biological perspective, for such a point of view was powerfully reinforced by both the intellectual traditions of their science and their cultural backgrounds as middle class, Anglo-Saxon Protestants. The controversy's participants dwelt on whether experimental evidence existed for the acquired characters theory, thus taking for granted the larger assumption that evolution was fundamentally caused by heredity rather than by the interaction of heredity and environment. By the late 1890s the controversy was ending, largely because the Neo-Lamarckians could not satisfy the rising generation of experimentalists that there was a single scrap of independently verifiable evidence for their cherished acquired characters theory.[36] Ironically, the experimentalists distrusted Weismann's natural selectionism for the same reason they rejected acquired characters—a lacuna of experimental proof and an excess of vague philosophizing.

But the new experimentalists did accept Weismann's basic idea that determiners in the germ plasm were the fundamental cause of inherited traits, or some version of that idea, chiefly because the emerging cytological view of heredity seemed to verify it, at least in broad outline. Thus in 1895, Edmund B. Wilson, an extraordinarily able young cytologist at Columbia University, declared in an important monograph that the nucleus of the cell was the basis for physical heredity, and that heredity itself might be "affected by the physical transmission of a particular chemical compound from parent to offspring."[37] What Wilson wrote carried great weight within the biology profession. A number of biologists continued to say that Weismann's germ plasm theory indicated that all traits were biologically inherited for more than a decade after the Neo-Lamarckian controversy.[38]

In the early 1900s, furthermore, more discoveries followed closely on the heels of the Neo-Lamarckian controversy, encouraging the impression that Weismann's germ plasm theory sanctioned the supremacy of heredity over environment. In 1900 three European experimentalists, Hugo de Vries, Carl Correns, and Erich von Tschermak-Seysenegg independently "rediscovered" Mendel's laws of heredity. And over the next several years, de Vries erected the mutation theory of evolution as an alternative to both Darwinian natural selection and the Neo-Lamarckian acquired characters theory of evolution. These new discoveries were, of course, the product of a new approach biologists now took toward the explanation of life—the experimental and the statistical view of ongoing natural processes, rather than the traditional methods of historical reconstruction. Thus the "rediscovery" of Mendelism meant that biologists were finally able to think in statistical terms. And the appeal of both Mendelism and the mutation theory was based on the possibility that they could be independently tested by many investigators—what biologists imagined physical scientists always did. Mendel provided a statistical description of the segregation and reassortment of determiners for traits over time in species. What he demonstrated were seemingly predictable and constant laws of inheritance of traits from one generation to the next.[39] Mendel's segregating and recombining determiners were the same as Weismann's determiners in the germ plasm, or so many American biologists assumed. Thus Charles Davenport was not out of line with his professional colleagues when he declared in 1901 that in "the study of hybrids, we must. . .pay primary attention to the behavior of the peculiar characters by which the crossed individuals and their ancestors are distinguishable. For each of these somatic characters corresponds to some peculiarity of the germ plasm."[40] Not all professional biologists would have made so definite a statement as this, and some objected. But many professional biologists regarded this as a likely hypothesis.

In the first decade of this century Mendelism won over the vast majority of professional biologists in America. A number of bright young experimentalists tested the Mendelian rules through breeding experiments on a wide variety of plant and animal species. Generally speaking they found the Mendelian rules and ratios substantiated in their work. Occasionally a worker would find an exception to the Mendelian ratios, but not to the phenomena of segregation and reassortment.[41] A number of leading experimentalists remained quite cautious, far more so than Davenport, until the end of the decade. Cytologist Edmund B. Wilson, for example, remarked in 1908 that Mendelism and the germ plasm theory were still a "working hypothesis." He encouraged other scientists to continue to test its validity, and he gave it a cautious endorsement.[42] And there was some loud opposition to Mendelism as well. Thomas H. Morgan, an experimental embryologist, and a close friend and colleague of Wilson's, was reluctant to endorse Mendelism for some years. In 1907, for example, he questioned the Mendelian assumption of independent and immutable unit characters in the germ plasm. In 1909 he seemed to back away from Mendelism even more. Then, in 1910, three years after he had launched the genetics research program with *Drosophila melanogaster* that would eventually win him a Nobel Prize, he became converted to Mendelism as the result of the work he and his associates did in their laboratory.[43] While it would be inaccurate to say that most American experimentalists concluded as the result of the general acceptance of Mendelism by 1910 or so that heredity was all powerful and environment of no consequence, it was nevertheless true that heredity occupied a much more prominent place than environment in their writings. When most experimentalists thought of environment, it was in the sense that the now-discredited Neo-Lamarckians discussed it; and the declining fortunes of the acquired characters theory, in addition to the rising reputation of the new germ plasm and Mendelian interpretation of heredity, brought inheritance to the forefront of professional biology's literature in the early twentieth century. Many biologists wrote as if for each trait there existed a determiner in the germ plasm; environment was of negligible importance because acquired characters could not be inherited; the determiners were "unit characters" and were inherited and varied according to the famous Mendelian ratios.[44] This point of view received favorable endorsement from a number of British biologists—William Bateson, J. Arthur Thompson, and R. C. Punnett, for example—which had the effect of reinforcing what the American experimentalists said.[45]

Hugo de Vries' mutation theory of evolution placed far more emphasis upon the role of heredity than did the work of Weismann and Mendel. Indeed, if any of the new work may be justifiably interpreted as sanctioning a hereditarian interpretation of the development of individuals and species, it was de Vries' mutation theory. In the early 1900s, de Vries presented his mutation theory of evolution as an alternative to Darwin's theory of natural selection. He had worked out the mutation theory as the consequence of his breeding experiments with several species of the evening primrose, *Oenothera lamarckiana,* which seemed to show sudden, dramatic, and discontinuous mutations constituting new strains. In other words, new species arose through large mutations in the genetic endowment of old species. Actually de Vries found several kinds of mutations; some were radical enough to give birth to a new species immediately, whereas in other instances it might take a number of these saltations to produce a new species.[46] In today's terminology, the mutations to which de Vries referred were macromutations, or far larger mutations than the micromutations commonly recognized by the modern biologist.

The mutation theory received much approval from professional biologists in Europe and America, especially from the younger men and women. To a considerable extent this was because of the apparent experimental verification of the mutation theory. As an anonymous writer put it in the British journal *Nature* in 1901, thanks to de Vries, "new life has been infused into the problem of evolution and. . .tangible facts are now available and experiments which will replace a good deal of rather empty theorizing and hollow controversy between rival speculations."[47] In America a number of biologists quickly assumed that Mendelism and the mutation theory were part of a more general mosaic of interpretation. Thus William E. Castle, a young zoologist attracted to genetics who was teaching in Harvard's Bussey Institution, summarized the results of his breeding experiments in 1903 by declaring that, "Mendel's principles strengthen the view that species arise by discontinuous variation. They explain why new types are especially variable, how one variation causes others, and why certain variations are so persistent in their occurrence."[48] This is not to say that the mutation theory swept all opposition in European and American scientific circles. Far from it. Men of science on both sides of the Atlantic continued to champion natural selection—August Weismann in Germany, A. R. Wallace and George John Romanes in England, and the Americans David Starr Jordan, O. C. Marsh, and V. L. Kellogg, for example.[49] In England the influential school of biometricians (who had opposed Mendelism, and whose ideas had developed from the work of Sir Francis Galton and Karl Pearson) continued to stress the Darwinian idea of evolution by small, continuous variations rather than by the large, discontinuous leaps de Vries and his followers emphasized. Some distinguished natural historians in America opposed the mutation theory; thus C. Hart Merriam, the long time head of the United States Biological Survey, concluded from a careful study of over a thousand species and subspecies of North American mammals that mutations were so rare as to be trivial factors in evolution.[50] Merriam meant, of course, de Vries' macromutations, not the modern biologist's micromutations.

Yet probably most American biologists opposed Darwinian natural selection, even before de Vries publicized the mutation theory in European and American scientific forums. The Neo-Lamarckians, of course, criticized natural selection. And many of the experimentalists who succeeded them regarded Weismann's restatement of natural selectionism (known in biological circles as "Neo-Darwinism") as too metaphysical. In 1907 the Stanford University entomologist Vernon L. Kellogg published a book in which he assessed the reputation of the natural selection theory; quite probably he was correct in estimating that most biologists did not accept it.[51]

The mutation theory won rapid acceptance among American experimentalists in the early 1900s. The reasons were complex and mutually reinforcing. As historian of biology Garland Allen has perceptively pointed out, the mutation theory resolved, or appeared to resolve, a number of logical and technical problems in the Darwinian natural selection theory: the character of variation, the operation of selection, the problem of "swamping," the issue of time and the earth's age (raised by the British physicist William Thomson, later Lord Kelvin), the phenomenon of isolation, and the highly confused problem of speciation.[52] For many American experimentalists the mutation theory provided a logically attractive alternative to Darwinian natural selection. And in subtle ways it appealed to their beliefs that biology should not be based on mysticism (as they thought natural selection was; natural selection implied to them purposeful adaptation), that biological theories should be verified by direct experimentation; and the mutation theory assumed that physical inheritance was the instrument of evolution and individual development, somewhat as the Neo-Lamarckian theory of evolution had.

Important also in the acceptance of the mutation theory among American experimentalists were the exertions of de Vries and his American champions to win converts. A man of considerable personal charm, de Vries corresponded tirelessly with a number of leading American experimentalists. "I have been so fortunate as to secure what Professor de Vries considers as the first actually authenticated mutations observed in America," wrote D. T. MacDougal, the assistant director of the New York Botanical Garden, to his friend Charles Davenport in 1904.[53] In that year de Vries came to America on a lecture tour. On 2 September 1904, he gave the convocation address at the University of Chicago—on the mutation theory. It was the metaphysical quality of late nineteenth century evolutionary theory that had finally persuaded him to experiment. "I concluded that the only way to get out of the prevailing confusion was to return to the method of direct experimental inquiry. . . . I have brought it into my own garden, and here, under my very eyes, the production of new species has been going on. . . . At once it rendered superfluous all considerations and all more or less fantastical explanations, replacing them by simple facts."[54] Some of de Vries' success in America may be attributed to his friendships with a number of American biologists; but far more important in explaining de Vries' appeal in the early 1900s was the conjunction of the rise of experimental biology in American science and the experimental character of de Vries' mutation theory. And de Vries' role as one of the "rediscoverers" of Mendelism gave him added prestige among many American experimentalists.

There were soon signs of the mutation theory's quick diffusion in the American biology profession. The American Society of Naturalists sponsored an important panel discussion of the mutation theory at its 1905 meeting in Philadelphia. The panelists, all prominent experimentalists, supported the mutation theory largely because the recent work—the germ plasm theory, Mendelism, and the mutation theory—seemed to them to constitute a coherent interpretation of life. Participant Edwin Grant Conklin, who now taught at the University of Pennsylvania, declared that "the evolution of animals and plants must be accompanied by an evolution of their germ cells. . . the principal problem of evolution is not how modifications are produced in adults, but how they arise in the germ." He continued: the germ cells "contain visible morphological elements, which have a particular role in hereditary transmission. . . which suggests that modifications of these elements. . . are the real causes of evolution."[55] And at a number of universities across the nation the mutation theory became the focus of research for faculty in the biological sciences. Jacques Loeb at the University of California and Thomas H. Morgan at Columbia, for example, performed important experiments to test the mutation theory. In Morgan's case it was this research that eventually converted him to the Mendelian theory of heredity.[56] Clifton Hodge, a professor of physiology and neurology at Clark University, made the mutation theory the topic of his year-long graduate seminar in biology during the 1905–6 academic year.[57] In 1906 Charles Davenport was instrumental in making the arrangements for de Vries' extensive lecture tour of many American universities. The Dutch plant physiologist spoke, sometimes formally before general audiences, at other times among faculty and advanced students in more informal settings.[58] De Vries' personal exposure to American scientists undoubtedly helped convince doubters. Skeptics and opponents remained. Many of them, such as Charles O. Whitman, professor at Chicago and director of the Marine Biological Laboratory at Woods Hole, were distinguished and respected in the biology profession.[59] Yet de Vries obviously had many more supporters than critics.[60] "I have just returned from our scientific meetings at New Haven and Chicago," Charles Davenport wrote de Vries in January 1908, "and you would have been delighted to see what interest in both places has been awakened in the new lines of study which you have started."[61] Davenport was correct.

The implications of these developments were very important for the heredity-environment issue. The crystallization of a new point of view, a fresh interpretation, was taking place. According to this new line of thinking, there was strong evidence that heredity mattered far more than environment in the evolution of species and the development of individuals. The evidence came directly from the new techniques of experimental biology. While it was true that the emerging cytological view of physical inheritance of Weismann, Mendel, and others, merely provided a precise and compelling description of heredity, and did not, by themselves, justify the belief that nature was more powerful than nurture, the new view, by focusing on inheritance, seemed to imply that heredity was very important. This emphasis on heredity as distinct from and superior to environment owed much to the Neo-Lamarckians. But it owed even more, most probably, to the mutation theory, with its emphasis upon heredity as the cause of new species and variations, as the vehicle of evolution itself. This was because many biologists tended to consider other technical issues from the perspective of the most generally accepted model, or paradigm, of nature. Given the fact that the mutation theory appeared to many American experimentalists the consequence of brilliant experimental research, its rapid acceptance by them was perhaps a foregone conclusion. And doubtless the rise of the mutation theory, when understood within the context of other intellectual and institutional changes sweeping American biology around 1900, skewed the scientific discussion of the nature-nurture problem toward the conclusion that heredity was all powerful and environment was of little consequence.

*IV*

The new biology had an undeniable impact upon the American eugenics movement. In the early twentieth century American eugenists quickly seized upon the ideas of Weismann, Mendel, de Vries, and other experimental biologists. They made those ideas an integral part of the movement's political and scientific credo, and they were more extreme than professional biologists. Indeed, between the early 1900s and the late 1920s the most visible eugenics leaders recited virtually the same interpretations of heredity, variation, and evolution. They took scant notice of the momentous intellectual revolutions in the science of genetics in those years. Eugenists commonly argued that heredity was all powerful, and environment was of little or no effect in man as in the lower animals. "All life is conditioned by the same fundamental laws of nature," wrote eugenist H. E. Jordan, a professor of anatomy at the University of Virginia. "It would seem . . . that the same methods that man now employs in producing a high quality breed of dogs . . . he must apply to himself. . . . Whence and by what authority is it given that man can violate with impunity the laws of heredity?"[62] This aggressive assertion of naturalistic ideology—that man is a part of nature and is subject to the same natural laws as all animals— was characteristic of virtually all, if not all, publications written by American eugenists. American eugenists were also profoundly convinced that heredity was far more important than environment. David Starr Jordan, the eminent president of Stanford University, and leader of the American peace movement, declared in 1914 that to "have really good children, the parents must be of good stock themselves. Bad fruit is born mainly by bad trees, and the inheritance of badness springs from inherent tendencies."[63] American eugenists commonly insisted that all human traits, mental and physical, were Mendelian unit-characters in the germ plasm, were transmitted as independent, nonreducible characters, and constituted the basis for all traits, desirable and undesirable, in man. As one of them put it in 1912:

Studies in heredity indicate that every man is an aggregation of large numbers of certain physical and mental characters, and that

these characters are not reducible to simpler forms. They are therefore called unit characters; and they are transmitted through the
germ plasm as separable units. Furthermore, the inheritance of these
unit characters seems to follow Mendel's law and the presence or
absence of desirable and undesirable characteristics marks the differences in the character of the men and women about us.[64]

Before World War I the American eugenics movement recruited almost all of its members from the professional and
business classes of the native born Americans of Anglo-Saxon
Protestant ancestry. Correspondingly eugenists reflected different perspectives and points of view from within that subculture.
The main intellectual theme uniting this rather heterogeneous
movement was the prospect of using modern science to improve
society. As often happens in the history of American reform
movements, eugenists could not even agree among themselves
upon the proper eugenic programs. Some believed in "positive
eugenics"—that is, selectively breeding a better race, even if
that meant increasing the powers of the state. Others believed
in what was known within the movement as negative eugenics—
merely preventing the "unfit" from continuing to reproduce
their own kind, and keeping the powers of government as minimal as possible. Obviously each implied rather different political
and legal mechanisms. Men and women from the right, the
center, and the left in American politics belonged to the eugenics
movement before the World War. In turn this meant that different eugenists were interested in different issues, and disagreed
among one another on these issues. Thus Margaret Sanger joined
the eugenics movement because of her commitment to birth
control. Other eugenists such as Charles B. Davenport opposed
birth control—and many of Mrs. Sanger's leftish views.[65] David
Starr Jordan came to the eugenics movement mainly as the
consequence of his deep involvement in the prewar American
peace movement. Certainly his professional background as a
zoologist in the 1870s and 1880s made it easy for him to sympa-
thize with the ideas of eugenics. But by the early 1900s the
Stanford University president was even more interested in
peace thanks to the Spanish-American War. He argued that war
had a dysgenic impact on nations that practiced it. The men
with the best germ plasm were those who fought bravely and

died in the greatest numbers in war. Those with lesser gifts avoided combat and propagated their own.[66] Yet many other eugenists had little sympathy for the peace movement and did not share Jordan's horror of war.

A large fraction of American eugenists were far more interested in the related social issues of dependency and deviance. They focused their attention on the poor, the criminal, the feeble-minded, or the delinquent members of society whose anti-social behavior, they believed, constituted a threat to the stability of the urban industrial social order—and perhaps an affront to traditional American values of hard work, sobriety, and discipline. Many of these eugenists were not businessmen or members of the traditional professions. Rather they belonged to the emerging "semi-professions" of the new industrial society —social work, applied psychology, and the like. Perhaps their most conspicuous spokesman was the psychologist Henry H. Goddard. As research psychologist on the staff of the New Jersey Training School for the Feeble-Minded, in Vineland, New Jersey, Goddard became interested in the Binet-Simon scaled mental test as an instrument to measure the abilities of the school's inmates. Indeed, Goddard published the earliest adaptation of the Binet-Simon test in America. He conducted an ambitious testing program with his version of this examination on persons of normal and subnormal intelligence. In a number of books, speeches, and articles he maintained that differences in levels of intelligence were inborn—an argument for which he appealed to the new biology and the test results. In his lurid and well known study, *The Kallikak Family* (1912), Goddard traced the social pathology of a particular family of social deviants that had spawned whole generations of troublemakers. The cause of all these troubles, he assured his readers, was the innate low intelligence of the members of the Kallikak family. He assumed furthermore that mental deficiency was inherited according to Mendelian ratios. The difficulty, he concluded, was that these unfortunate individuals were so crippled by their inferior heredity that they could not tell the difference between right and wrong.[67] Goddard believed that moral ethics were inherited according to the Mendelian expectations of dominance and recessiveness, dependent solely upon the quality of the germ plasm involved.

After 1910 the eugenics movement became increasingly concerned with immigration restriction; other issues, such as peace, birth control, and dependency, took a back seat. This reflected the growing importance of restrictionism in American politics, the obvious intellectual and emotional ties between eugenics and restrictionism, and the sudden emergence of restrictionists within the ranks of the eugenics movement. Immigration restriction—and the distaste of the native born Americans of white, Anglo-Saxon Protestant heritage for the "new" immigrants from Southern and Eastern Europe, not to mention nonwhites—drew together men and women of various political hues. Thus Madison Grant, the conservative New York lawyer, and Edward A. Ross, the liberal Populist-Progressive sociologist, joined political forces to clamour for immigration restriction and eugenics.[68] By the early 1920s the eugenics movement seemed obsessed by the restriction issue and was dominated by men such as Grant and Boston lawyer Lothrop Stoddard, who argued that the Nordic races were superior to all others and should be the basis of the American nation.

The American eugenics movement attracted the support of a modest number of recognized biologists. Perhaps Charles Davenport was the professional biologist who became the most deeply involved in the movement. His participation was important in several ways. As director of the Station for Experimental Evolution, Davenport had enormous prestige within the scientific community in the early twentieth century, which by his advocacy of eugenics he transferred to the eugenics movement. He also promoted eugenics institutions. By 1907 he was becoming intensely interested in human eugenics and genetics; with his wife he published several articles on Mendelian inheritance in man. He also tirelessly corresponded with other eugenists: Henry H. Goddard, of the Vineland Training School, psychologist Edward Lee Thorndike at Columbia, ultimately famous for his work in mental measurement, animal psychologist Robert M. Yerkes of Harvard, later the organizing genius of the Army mental testing program during the World War, biologist Frederick Adams Woods of Massachusetts Institute of Technology, and Robert DeC. Ward, a leader of the Immigration Restriction League. Davenport was ever the visionary promoter of institutions, and by 1910 he was far more interested in eugenics than genetics. Now he set out to raise money for a eugenics institute. A happy combination of luck and determination enabled him to persuade Mrs. E. H. Harriman to underwrite the costs of his planned Eugenics Record Office. Between 1910 and 1919 (when the Carnegie Institution combined the Eugenics Record Office and the Station for Experimental Evolution and took over financial responsibility for both), Mrs. Harriman donated more than half a million dollars for Davenport's project.[69] The Office and the Station were both located in Cold Spring Harbor. Now Davenport directed the major research institutes for genetics and eugenics. From the Eugenics Record Office came a growing number of eugenics pamphlets and studies, some written by Davenport, others by his proteges. In his widely read *Heredity in Relation to Eugenics* (1911) and his manual for eugenics field workers, *The Trait Book* (1912), Davenport argued that most, if not all, human traits were inherited in the germ plasm as Mendelian unit characters; this was true for mental and physical traits. Davenport—and most eugenists—believed that human progress was biological; what mattered in the long run was the innate endowment. With some alarm Davenport argued that certain pathological and degenerative characteristics could be explained according to his interpretations of Mendelian heredity. Thus feeble-mindedness was caused by the absence of a "unit character" for normal intelligence. Such conditions as violent temper and the wandering impulse were transmitted strictly according to Mendelian expectations.[70]

A handful of professional biologists joined Davenport in advocating the goals of eugenics in the years between the rediscovery of Mendelism and the World War. They did not go to the lengths he did; they did not lose their professional identification with genetics research, nor did they follow Davenport into the tricky field of human genetics. Yet they joined eugenics societies and they published approving statements about the goals of eugenics, and many of them probably accepted Davenport's conservative political ideas. But they went no further. Thus Davenport's old friend William E. Castle published a number of books and articles in which he applauded the eugenics movement, insisting that because the germ plasm carried all the determiners for traits, environment could do no more than permit their normal development from the genetic determiners.[71] Castle's friend and colleague, Edward M. East, supported eugenics, too. In 1917, for example, he cited Henry H. Goddard as authority for the statement that feeble-mindedness was inherited and represented a genuine threat to the stability of society.[72] In a careful monograph, *Inbreeding and Outbreeding* (1919), written with another geneticist, Donald F. Jones, East devoted most of his attention to species other than man. But in a chapter devoted to the implications of their research for man, he spoke freely of distinct human races, such as the Chinese, Japanese, Irish, and British races. Furthermore, he insisted, "the negro is inferior to the white. The negro has given the world no original contribution of high merit."[73] In 1913 Herbert E. Walter of Brown University, who had taken his Ph.D. at Harvard's Bussey Institution with Castle and East, declared in his genetics textbook that all human traits were inherited, although they "may each be developed by exercise or repressed by want of opportunity, nevertheless they are fundamentally germinal gifts."[74] With the possible exception of Davenport, all of these men had sound professional reputations and taught at major American universities; their statements conferred legitimacy and prestige upon eugenics. Other professional biologists published favorable statements about eugenics, including Alfred F. Blakeslee, Samuel Jackson Holmes, Michael F. Guyer, Leon F. Cole, Charles W. Hargitt, Edwin G. Conklin, William E. Kellicott, Henry Fairfield Osborn, and Vernon L. Kellogg.[75] They left the impression, as Walter did, that environment was of little consequence, and heredity was all powerful. Listen to what Conklin said in a series of invited lectures later published as *Heredity and Environment in the Development of Man* (1915), a book that went through several editions and was widely cited by eugenists as scientific gospel: "All the general laws of life which apply to animals and plants also apply to man," Conklin insisted.[76] Since heredity, through the germ plasm, governed plants and animals, it also controlled man:

The general tendency of recent work on heredity is unmistakable, whether it concerns man or lower animals. The entire organism, consisting of structures and functions, body and mind, develops out of the germ, and the organization of the germ determines all the *possibilities* of development of the mind no less than of the body, though the actual realization of any possibility is dependent also upon environmental stimuli.[77]

Michael F. Guyer of the University of Wisconsin and William E. Kellicott of Goucher College put it even more bluntly in their popular books on eugenics. Human progress came through the propagation of the germ plasm of the better classes and the elimination of the germ plasm of the "inferior stocks."[78]

Of course it can be said that the proportion of professional biologists who made such public statements was low in comparison to the total number of professional biologists in early twentieth-century America. But these were men who held good appointments in major universities and who also had distinguished reputations as scientists; consequently their actions enhanced the eugenics movement's prestige with the American public. Nor did professional biologists, as a rule, openly criticize eugenics—before the 1920s, only Herbert W. Conn, an able Wesleyan University bacteriologist, attacked eugenics in print.[79] If most professional biologists had reservations about eugenics, if they often winced at the political uses eugenists made of scientific ideas, or if they even had hesitations about the new hereditarian dogma, such feelings were scarcely evident in their publications or in their private correspondence. Even professional biologists who might have been expected to criticize eugenics, if not the more general hereditarian dogma, did not. For example, physiologist Jacques Loeb and geneticist H. J. Muller, as immigrants of Jewish background and socialists, might well have taken exception to the "Nordic" theories of the immigration restrictionists active in the eugenics movement. Yet before World War I, such doubts, if and when they existed, did not surface in public. Professional biologists seemed to endorse, by their actions, eugenics and the hereditarian dogma.

It was within this context in the scientific community and the larger society, in a time of growing racial and ethnic tensions, of the resegregation of the federal government, of the formation of inner city ghettoes, of a rising nativist backlash against the "new" immigrants from Southern and Eastern Europe, of a renewed martial temper, especially among those Americans who identified with Anglo-Saxon culture, and of new labor unrest, that the new hereditarian interpretation of human nature arose in American culture and the American eugenics movement began to win some of its goals. For example, eugenics became a part of the college curriculum; the number of colleges and universities offering courses in eugenics increased from forty-four in 1914 to three hundred and seventy-six in 1928, when, according to one estimate, some 20,000 students were enrolled in these courses.[80] Clearly a growing proportion of the next generation's business and professional leaders were being exposed to eugenics. Furthermore, eugenists influenced public law and public policy. Between 1907 and 1931, eugenists persuaded the thirty state legislatures to pass compulsory sterilization laws.[81] And eugenists played an important role in the enactment of America's most ambitious program of biological engineering, the National Origins Act of 1924, which imposed immigration quotas based squarely on eugenists' ideas of Nordic racial superiority and non-Aryan racial inferiority.[82]

## V

Why the biology profession permitted the impression that nature mattered far more than nurture to persist is a complex question, especially from the standpoint of the motivations of particular individuals. If instead we consider both the context and the social roles these men played, we can piece together a reasonable hypothesis. Quite obviously much happened quickly in experimental biology. Men who attended college in the 1880s, when Neo-Lamarckism was still the dominant orthodoxy, witnessed the rapid rise of experimental biology, the overthrow of Neo-Lamarckian ideas of heredity in the 1890s, and the electrifying new discoveries of the early 1900s. Many members of that first generation of experimental biologists had been nurtured on Neo-Lamarckism as young boys or men. They had broken with the acquired characters theory. And they no longer believed in purposeful schemes of nature. But like the Neo-Lamarckians they assumed that heredity was most probably the vehicle of evolutionary development, an assumption the new biology solidified. And quite a few of these men simply had unrealistic expectations about what any experimental discipline could achieve in a short period of time. Furthermore, they had been trained as biologists. Naturally they thought of the larger questions of life from the standpoint of biology. The importance of heredity was obvious to them. Hence, circumstances and prior intellectual traditions helped reinforce—quite subtly, to be sure—a strongly hereditarian point of view among many professional biologists.

But the explanation goes further. By definition, professional American biologists were members of and participants in the new, white-collar, professional classes. Even more importantly, virtually all of them were native-born Americans of Anglo-Saxon Protestant heritage. They were, of course, a particular segment of that culture. They were neither Kentucky mountaineers nor Mississippi sharecroppers. They comprised that element of the Anglo-Saxon culture experiencing important gains, individually and collectively, in social mobility, thanks to the opportunities offered by science and higher education at the close of the nineteenth century. Most of them came from solid middle-class backgrounds. Charles Davenport counted Puritan divines in his ancestry; Thomas H. Morgan was the son of a Kentucky gentleman-entrepreneur, and Conklin came from a comfortable merchant family in small town Ohio. Jacques Loeb—a Jew, a radical, an immigrant—was a social anomaly among his professional peers. Indeed, well over 90 percent of American professional botanists and zoologists came from Anglo-Saxon, Protestant, middle class backgrounds. They came from homes that were remarkably similar, if not always in objective economic fact, then certainly in parental ambition for them. Most had fathers who were merchants, professional men, or farmers. Few were the sons and daughters of immigrants or workers. Perhaps two professional biologists in these years had parents or relatives who had been slaves. In the kind of homes in which most professional biologists were reared, then, certain social values were taken for granted: the worth of the individual, the importance of upward social mobility, the significance of an individual's innate gifts, and a fuzzy but pervasive sense that the institutions of traditional American society, the institutions of social control such as the family, the church, the family firm, and local government, were being distorted or threatened by urbanization, industrialization, and immigration. Those raised in such a milieu were at least sensitized to the importance of individualism, or, in the parlance of professional biology, inheritance.[83]

It would be pointless to separate the cultural from the scientific in understanding why so many American biologists embraced this general hereditarian mood. For, indeed, the important conclusion is not which was more important, but rather that they interacted and reinforced each other. For many working American biologists, both their science and their culture taught them that the starting point for social thought and scientific explanation was the individual, that the processes of man and animals were wholly natural, and that a science of man, based on evolutionary science, was a possibility. The dramatic events of the 1890s and early 1900s in politics and the larger society—the agrarian revolt, the struggles between labor and capital, the agitation for reform, the problems of the cities—suggested that the time was ripe to create a science of man that could lead to social control, and to social stability.

# 2 THE NEW PSYCHOLOGY

*I*

D URING the Christmas holidays of 1918 two former
Harvard men, one the eighty-four-year-old president
emeritus, the other a forty-two-year-old assistant
professor of psychology, exchanged the season's
felicities, and shared their optimism for the future of the
younger man's science. "I congratulate you on the success of
the Army Mental Tests," wrote Charles W. Eliot to his one-time
protégé Robert M. Yerkes. "May one hope that these tests, with
suitable modifications, would be helpful in the public schools
and in the employment bureaus of factories which employ
people of several grades of intelligence? I am sure that a better
classification of pupils in respect to their natural and acquired
characters is urgently needed."[1] Yerkes vigorously agreed. As
chief of the Army's gigantic psychological testing program during
the recent war, indeed, as the man who virtually sold the program to dubious government and military officials and then ran
it with skill and verve, Yerkes took understandable pride in his
accomplishments. And he was persuaded now that the Army
tests showed his science could contribute to solving the nation's
domestic problems. "It is my hope and expectation that the application of psychological methods to employment may substantially help in the solution of our great labor problem. . . .
emphasis upon the relations of individual intelligence, tempera-

ment, and physique to occupation will at once increase the contentment of the individual and his industrial efficiency."[2] Eliot and Yerkes emphasized the exhilarating prospect that the precise and objective methods of modern science could fulfill an old American dream: the distribution of rewards in society on the basis of individual merit. This was a compelling idea for men and women who sprang from the same cultural and ethnic heritage as Eliot and Yerkes and were brought up to respect both the values of old-time Protestant America and the contributions science and technology could make to the commonweal.

This exchange of letters between Eliot and Yerkes underlined what a whole generation of Americans perceived as the electrifying promise of the new experimental psychology for resolving, not merely the basic issues of the origin and development of mind, but through the application of psychology, the social problems of a disorderly industrial society. In effect this exchange of letters suggested that American psychology had taken on its own identity and focus. In many respects American and European psychologists were heirs to the same intellectual traditions of nineteenth-century science. Physiological psychology taught them that mental life had a material basis in the nervous system and the brain. Darwinian biology instructed them that mind as well as body had evolved. The example of the physical sciences, and of psychophysics, inspired them to use the most rigorous and precise methods of inquiry they could imagine. Yet by the time Eliot and Yerkes corresponded, American psychologists had become far more didactic, far more concerned with questions of social application than their European counterparts. The American psychologists departed quite early from the static models of mental processes used by the Germans. The Americans wanted a science that would probe beneath the surface of mental activities, that was functional, evolutionary, and developmental, a science that would explain the causes of mental life, identify the statistical distributions of mental gifts among individuals and among groups, and suggest practical means of quieting social unrest and promoting social harmony in a nation of nations undergoing urbanization and industrialization. Although European psychologists had some interest in these matters, American psychologists had much more. The American psychologists concentrated their efforts on two general fields of research and explanation: the psychology of conduct, or the study of instincts, and the psychology of capacity, or the comparative measurement of intelligence. Both the psychology of conduct and the psychology of capacity contributed, in these psychologists' views, to the creation of an evolutionary science of man that might lead to the prediction and the control of human behavior.

The thrust of American psychological research and explanation to the 1920s attributed far more explanatory power to nature than to nurture. Heredity, instinct, individual differences, the evolution of mind, the sufficiency of original endowment—all these conceptions were given theoretical categories of the new American science of psychology. And the hereditarian emphases of the new psychology received powerful (if indirect) reinforcement from the new biology, which focused attention upon the power of inheritance. But there were other reasons why so many American psychologists emphasized heredity over environment. They shared the same general ethnocultural and class background as did the biologists. Like the biologists, they grew up in middle class, white Anglo-Saxon Protestant homes, in a subculture where the individual was the focal point of social explanation and comment, where the importance of original nature in human affairs was taken for granted, and where the importance of estimating the talents of individuals for the proper functioning of a self-balancing, stable social order was not seriously questioned.[3] The scientific and the cultural traditions of American psychologists reinforced one another, deepened the hereditarian perspectives of American psychology, thus shaping the psychologists' commitment to a science of social control.

## II

In 1890 William James aptly described psychology's state: "It seems to me that Psychology is like Physics before Galileo's time. Not a single *elementary* law yet caught glimpse of. A great chance for some future psychologue to make a greater name than Newton's but who then will read the books of this generation? . . . Meanwhile, they must be written."[4] Although experimental psychology grew out of contemporary European philosophy, its more important formative intellectual influences were European neurology and physiology, which posited the existence of neurophysiological structures for mental life, and Darwinian biology, which asserted that mind as well as body had evolved. European proponents of the new psychology brought the great issues of mind into the laboratory; they assumed that the proper study of psychology was the normal, adult, human individual defined as a psychobiological datum. Yet Continental psychologists quite naturally approached their discipline with prior philosophical assumptions and habits of thought. For example, they assumed consciousness existed in mind. They debated the mind-body issue furiously. And they used the highly personalized technique of introspection to study mental states.[5]

James' generation of psychologists in America was truly a transitional one. James and his contemporaries imported European psychology, borrowing the ideas and techniques that most interested them. In addition to being a bridge between the science of the Old World and the New, James' generation was transitional in another sense as well. They had commitments to both traditional philosophy and the new science of psychology. They sincerely believed in making the study of mind as scientific as possible, which meant, of course, learning the language, the methods, and the conceptions of modern science. Yet these pioneers had come to the study of psychology from other intellectual backgrounds, most commonly philosophy or theology, often in middle age. Perhaps inevitably they did not surrender the more subtle philosophical assumptions of nineteenth-century philosophy and theology. They perceived no contradiction in espousing experimental methods and mixing them with philosophical categories of thought. It never occurred to them, for example, to question many of their fundamental assumptions, notably the existence of consciousness or of instincts. Nor did they perceive any tension between their new scientific affiliations and their shared belief that mind fulfilled larger teleological goals in the grand design of nature. In other words, on a profound level (which they did not realize) they believed that the data of their science were composed of elements of differing moral and cultural values; they had not broken with Aristotlean categories of thought. Yet they convinced themselves they had done so by taking up the language and the concepts of contemporary natural science and grafting them onto their deeper philosophical and theological assumptions. Such is often the fate of transitional generations in the history of culture and of ideas.

Furthermore, James and his colleagues were institution builders, promoters of their nascent science in American higher education and science. They founded programs, journals, and professional organizations for American psychology. They established the major departments of psychology in a handful of prestigious universities. They started the first laboratories, organized journals, and the American Psychological Association, the major professional organization for their science and their profession. By the early 1900s these pioneers of American psychology were succeeding in their campaign to persuade other Americans of psychology's relevance to life—school administrators, professional educators, social workers, mental health professionals, business leaders, foundation executives, and even government officials. These psychologists were busy, active men, thoroughly caught up in promoting their science, in pointing to its potential for social policy, and in creating institutions that would advance it in the worlds of higher education and science. Perhaps their circumstances as pioneers and their roles as promoters blinded them to intellectual inconsistencies that later generations of psychologists would find in their ideas.

And they unleashed an era of phenomenal expansion for American psychology. Before the early 1900s, when James and his generation were just starting their careers, American psychologists comprised a small, intimate group whose members knew one another—perhaps too well. After the early 1900s, as the pioneers trained a second generation, American psychology began to develop into a modern, impersonal scientific profession. By the end of the World War this had run its course. Psychology's growth was illustrated by the increase of American doctorates in psychology. Between 1884 and 1898 American universities granted 54 doctorates in psychology. Between 1899 and 1908 the number of doctorates granted was 139, somewhat less than a threefold increase. An even more dramatic increase occurred between 1909 and 1918, when American universities awarded 234 doctorates in psychology, more than half of the 427 doctorates granted from 1884 to 1918. And it became increasingly difficult for any one group or faction in the profession to dominate the profession at large. An indication of this was the number of universities granting the Ph.D. in psychology. To 1899 only nine American universities granted a doctorate; over nine-tenths of these degrees were granted by five leading departments. Between 1899 and 1908 the number of universities granting doctorates in psychology increased to fifteen, and in the next decade this number grew to twenty-five universities. It is true that there always were five leading universities granting the majority of psychology doctorates, but their share of American Ph.D.s dropped from 90 percent before 1899 to 66 percent in the years 1909 to 1918.[6]

These pioneers and promoters pushed psychology's institutional growth in other ways. For example, they helped establish (or encouraged the founding of) over eighty psychological laboratories in colleges and universities. They made the psychological laboratory central to undergraduate as well as graduate instruction in psychology. Thus when an institution established a laboratory, no matter how rudimentary by later standards, this suggested that the institution could at least tolerate the new psychology, if not give it preferential treatment at the expense of traditional moral philosophy.[7] The very act of setting up a psychological laboratory signified a willingness to accept the idea that mind could be studied with the instruments (and the assumptions) of modern evolutionary natural science. James and his colleagues did not realize that the psychological laboratory in time would leave little room for the philosophical questions they so cherished. Change accompanied growth in other respects, too. It was this pioneer generation that founded the American Psychological Association, which in turn grew from thirty-two members in 1892 to more than three hundred in 1920 (even though membership standards became progressively more rigorous with the passage of time).[8]

This first generation of American psychologists did much to establish the new psychology in American culture as a science and a profession. They dominated the profession as a group of competitive, jealous, often cantankerous individuals, and they sought leadership in their science by publishing essentially personal interpretations of psychological phenomena. Their actions as promoters of their profession undercut their position in the long run, for by bringing into the profession and the science of psychology a much larger group of men and women who had no philosophical background and who fervently believed psychology was a natural science, they created a situation in which the ultimate symbols of scientific authority would be impersonal and experimental, not personalistic and philosophical.

Experimental psychology's growth in several leading universities after the mid-1870s was rapid and spectacular. And it demonstrated the tension in the minds of psychology's founders between their philosophical orientation and their definition of psychology as an experimental natural science, between their individualistic ambitions and the impersonal implications of building a modern scientific profession. Experimental psychology began at Harvard, thanks to James' efforts. While a medical student in the 1860s James discovered the new psychology; soon he was publishing notices about it in leading middle class periodicals such as *The Nation*.[9] At the Harvard Medical School in the 1870s he performed some psychological experiments, taught courses in the new psychology, trained the first American Ph.D., G. Stanley Hall, and set aside rooms for psychological experiments. Throughout the 1880s his courses grew, and in 1890 he decided he needed newer and larger facilities for the laboratory. He raised $4,300 from local philanthropists and opened the new laboratory in 1891. Yet he also realized he did not prefer to direct the psychological laboratory. He was interested in philosophical, not psychological, questions, and his elegant *The Principles of Psychology*, finished in the summer of 1890, represented his last substantial contribution to the ideas of psychology. James took steps to find a man suitable to run the laboratory and guide intensive experimental work, securing funds from Boston patrons for the position, and offering the three-year "trial" appointment to Hugo Münsterberg of the University of Freiburg. Münsterberg would direct the laboratory and graduate research, and teach undergraduate psychology. Münsterberg accepted,[10] and James changed his own title to professor of philosophy.

Münsterberg arrived in Cambridge in August 1892. A man of considerable energy, charm, brilliance, and teaching ability, he soon made psychology flourish. After 1897 his appointment was permanent. In the early years he was most impressed with Harvard graduate students, and they flocked to study with the

German professor.[11] James had trained but one Ph.D. before Münsterberg's arrival; Münsterberg or his junior faculty trained all the twenty-eight Harvard doctors who graduated between 1894 and 1908. Münsterberg was an exciting, stimulating mentor; he encouraged his students to venture into new areas, such as mental measurement, comparative psychology, industrial psychology, juristic psychology, and the psychology of advertising. He often advised them to take courses in experimental biology and in statistics, where they learned of the new biology and the new measurement techniques and formulae.[12] Münsterberg was also a popular undergraduate teacher; in the late 1890s, for example, over 360 students enrolled in his elementary Harvard College course every year. So many Radcliffe College students attended his lectures that university officials installed a laboratory there even though they initially opposed such instruction merely for women. Psychology's phenomenal popularity often worried Münsterberg. "What will the country do with all of these psychologists?" he wrote to his friend Cattell in 1898.[13]

Ironically Münsterberg the promoter of scientific, experimental psychology became increasingly interested in philosophical questions. He published his chief work in experimental psychology, *Beiträge zur experimentellen Psychologie* (1889–92) while still in Germany. In none of his subsequent publications, including *Grundzüge der Psychologie* (1900), did he reveal much interest in furthering experimentalism; in some he took up theoretical issues and advocated applied psychology, but increasingly his focus was philosophical. This fuzziness between philosophy and psychology was characteristic of Harvard's philosophers and psychologists in those years. Often philosophers, such as Josiah Royce, dabbled in psychology, and psychologists occasionally worked in philosophy. Finally in 1913 the department of philosophy (in which Harvard philosophers and psychologists taught) gave courses with psychological content psychological titles, and became a department of philosophy *and* psychology. The younger men at Harvard, the instructors and assistant professors, did the actual experimental work.[14]

An ambitious New Englander, G. Stanley Hall, created a major center of the new psychology at Clark University after

1889. Clark granted 108 of the 427 doctorates between 1884 and 1918, more than any other American university in these years.[15] In many respects Hall was the father of the new psychology: he founded the first journal, the first professional society, and either trained or strongly influenced more psychologists than any other American psychologist between the mid-1880s and the early 1920s. Yet Hall did not plan his career in psychology. He was born in Ashfield, Massachusetts, in 1844, and as a young man he decided to leave the farm for intellectual pursuits. Initially his choice was the ministry, but he discovered the delights of secular philosophy while a student at Williams College during the Civil War. It was when he studied philosophy in Germany that he discovered the new physiological psychology. He returned to America and in 1878 took his doctorate with William James at Harvard; he wrote his dissertation as a physiological psychological interpretation of the relationship between mind and matter, an essay in both the old philosophy and the new psychology. Finding professional opportunities in America extremely limited, he returned to Germany to take postgraduate work with Wilhelm Wundt at the University of Leipzig, one of the most distinguished psychologists of the day. When Hall returned to America he found positions still scarce, but luckily he won a professorship of psychology at Johns Hopkins University in 1881. In his seven years at Johns Hopkins, Hall did much to launch his career and to promote his science. He founded a psychological laboratory— the first one in America with a continuous history. He also trained a number of men later famous in the science, notably James McKeen Cattell, John Dewey, Joseph Jastrow, and Jujiro Motora, a founder of Japanese psychology. Hall's efforts in launching the *American Journal of Psychology,* the first American psychology journal, illustrated his boldness as an entrepreneur of science. Psychology journals were necessarily private ventures in those days, when there were not enough psychologists to make them self-supporting. The initial money for the journal came from a donor interested in psychical research. Hall invested his own funds, and collected subscriptions for more than a year. When the *Journal* finally appeared in October 1887, Hall published chiefly empirical studies and gave psychic research scant recognition as hypnotism. In 1888 Hall accepted the presidency of Clark University, in Worcester, Massachusetts, founded by Jonas Clark, a wealthy merchant. Hall hoped to pattern Clark after the example of Johns Hopkins: a purely graduate university in the sciences. With lavish promises of much time for original research (not always kept) and canny bargaining skill, he lured a brilliant faculty in many scientific fields. Hall turned out to be autocratic and somewhat devious in his dealings with his faculty and with Jonas Clark. Finally, Jonas Clark forced Hall to economize, which alienated most of the faculty. In April 1892, most of his better faculty resigned in protest.[16]

Clark University survived the crisis of 1892, chiefly as a strong center of the new psychology. Hall provided imaginative leadership by expanding psychology's scope (and not incidentally its utility) to include educational, child, abnormal, and animal psychology. He stressed that work in all of these fields would lead to understanding the causes of mind. Hall's genetic psychology included all these areas because it would study the development of minds from animals and children to normal adults. He claimed that genetic psychology, by shedding light on the causes of mind, could lead to social control. In 1891 Hall founded another journal, *Pedagogical Seminary* (later known as the *Journal of Genetic Psychology*), to publish work in applied psychology, especially educational and child psychology. This journal gave him important prestige with school administrators and with professors at teachers' colleges across the nation. Hall also encouraged Clark students to experiment with mental tests. He permitted his faculty to offer courses in such new and exotic fields as the evolution and inheritance of mental traits, the mental states of imbeciles, and the development of instinct in animals.[17] Hall promoted psychology in other ways, too—by organizing the American Psychological Association in 1892 and by hosting the conference at Clark in 1909 at which Sigmund Freud was introduced to American scientists.[18]

Ever the busy, imaginative, and, upon occasion, selfish entrepreneur of his science, Hall was a prophet and promoter more than he was a researcher. Of course his administrative responsibilities made research difficult. But he showed little consistent interest after the early 1900s in furthering his own research. His intellectual contribution to his science was his interpretation of genetic psychology and his assessment of its ramifications for social policy. His work was essentially speculative and impressionistic. He had discovered the new psychology in middle age, and his contributions to psychology were those of the promoter, the prophet, and the creator of professional institutions. His lasting intellectual contribution was in directly relating the new psychology to social problems.[19]

In the 1890s Edward Bradford Titchener, an Englishman, made Cornell University into another flourishing center of the new psychology. Yet Titchener did not share his American colleagues' vision of a didactic functional psychology. He fervently espoused Continental structural psychology and refused to adopt his colleagues' perspectives. Titchener believed psychology should be a pure science, unsullied by presentist cultural considerations or distortions of scientific method. As a philosophy student at Oxford in the 1880s Titchener discovered the new psychology. After taking a year to study physiology, Titchener left Oxford for doctoral work with Wundt at Leipzig. Titchener finished in the spring of 1892, and his friend Frank Angell, an American who had studied with Wundt, helped him win appointment at Cornell; Angell was leaving Cornell for another position at Stanford University.

Titchener found Cornell a congenial environment for the new psychology. From Cornell's founding in 1867 the University had welcomed empirical approaches to the study of mind; since the 1880s the philosophy department offered a senior course in physiological psychology.[20] Titchener worked hard at Cornell to launch both the curriculum and his research. He later remembered those early years as busy and important; when he "came into the field, the important thing for experimental psychology was to make it an academic subject, to assure its collegiate and university status. Laboratories had to be organized, and textbooks had to be written."[21] He fashioned the psychological laboratory into a sophisticated instrument of undergraduate and graduate instruction. He wrote laboratory manuals and textbooks. A popular if formidable lecturer, he rapidly attracted hundreds of undergraduates to his elementary psychology lectures. And he was an effective mentor of graduate students; he supervised fifty-four Ph.D.s in his thirty-five years at Cornell.[22]

In a profound sense Titchener was in but not of American psychology or American culture. He always felt himself an alien in his profession and his adopted land. He stayed chiefly because of his commitment to  scientific psychology as he defined it; his beloved England would not permit the new experimental psychology.[23] Titchener participated little in the doings of his American colleagues aside from cooperating as an editor of the *American Journal of Psychology* with Hall. He found the meetings of the American Psychological Association too didactic, casual, and miscellaneous for his taste. He organized his own personal group, the Experimental Psychologists, in 1905; he dominated its intense annual meetings. He influenced his peers most through the brilliant correspondence he maintained with them, constantly scolding them for deviations from pure science and clear logic. Titchener stood between England and America. His structuralist physiological psychology, his model of the normal adult mind, was too closely tied with natural science for English scientific culture and not sufficiently developmental for American psychologists. The philosophical parallelism implicit in his structuralist psychology underlined a deeper affiliation with contemporary philosophy—and separated him from American psychological theory.

James McKeen Cattell did more than any other American psychologist save for possibly Hall in pushing American psychology toward genetic or developmental perspectives. Unlike most other pioneers of American psychology, Cattell made a clean break with philosophy. He embraced experimental and statistical methods; he eschewed introspection. He urged both his colleagues and his students to follow his example. Born in 1860, Cattell grew up in a comfortable middle class professional milieu, for his father was president of Lafayette College. As a student at Lafayette, Cattell was converted to Baconian and Comptean Positivism; the idea of science as a powerful instrument to serve human welfare, and its concomitant panscientific and rationalistic implications, guided him for the rest of his life. After graduating in 1880, he studied the new psychology at Göttingen, Johns Hopkins, and, finally, Leipzig, taking his degree with Wundt in 1886. At Leipzig, Cattell was interested in genetic psychology, in individual differences, in mental measurement, in what might be called the psychology of capacity. Wundt's structuralism, however, left the young American cold. Cattell then went to England, where he was influenced by Sir Francis Galton, a pioneer of anthropometry and individual differences, and a founder of the English eugenics movement.

In 1889 Cattell became professor and director of the laboratory at the University of Pennsylvania. He remained three years, long enough to launch his career as an experimental psychologist interested in development of *minds* (not the static mind of German psychology) and in mental measurement. He published several important papers, inaugurated new courses, and trained a few Ph.D.s. In 1891 he resigned for a more attractive post as professor and director of the psychological laboratory at Columbia. There he promoted his interpretations of the new psychology—in his work, and in that of his colleagues and students. He trained a number of psychologists who later won distinction, including Edward Lee Thorndike, Arthur I. Gates, J. F. Dashiell, Harry Levi Hollingworth, Walter Dearborn, and E. K. Strong. These men worked in fields congenial to Cattell's interests, such as mental measurement, genetic psychology, and educational and industrial psychology. But his informal influence on Columbia's psychologists was even greater: as a colleague, as a discussant of projected research projects, as a critic of completed work. And Cattell was an able promoter; he did much to guarantee that his perspectives would become institutionalized. He persuaded the University to expand the department's space and personnel budgets, and John D. Rockefeller to endow a professorship—all the time pointing to the promise of the new psychology for a science of social control. He arranged Thorndike's appointment at Teachers' College, where Thorndike exposed the new genetic psychology to thousands of students for half a century. Cattell even influenced appointments outside his discipline; he helped recruit such statistically-oriented faculty as anthropologist Franz Boas. And Cattell was an important man in national scientific organizations. He edited several important journals. He took a leading role in the affairs of the American Association for the Advancement of Science, the American Psychological Association, and the American Society of Naturalists. Informally he was a major figure in American scientific institutions, often negotiating ably among generations of psychologists and advocating psychology to other scientists. He also was a spokesman for the whole scientific community in American culture, consistently speaking up for academic freedom, for faculty control over universities and other scientific institutions, and for improving the quality of scientific research, for making it relevant to the needs of the larger society and culture.

In many respects his impact was more impersonal than personal, for his was a personality that did not wear well with many contemporaries. He was a symbol of objective and statistical science, in particular of genetic psychology and of the possibilities in it for social control. His professional connections made him influential with his pioneering generation of psychologists. His methodological commitments enabled him to serve as a viable model of scientific professionalism to the younger generation of psychologists. He pointed the direction of American psychology's future to this new generation—toward the objective, statistical, and behaviorial study of minds which in turn would contribute to social stability and order.[24]

The pioneers of American psychology dominated their science and their profession long after the early 1890s. The sources of their power were many. They held the best professorships at the leading universities. They directed the most important laboratories and graduate programs. They dominated and edited all the journals; indeed, a small group of these pioneers necessarily owned the journals as nonprofit private ventures. The pioneers controlled the American Psychological Association's affairs, creating for many years more of a country club atmosphere than that of a national professional society. They published the leading textbooks, which gave them the license to define the ultimate boundaries of acceptable psychological theory. They were proud, ambitious, sometimes vain, men who squabbled among themselves over the symbols and the substances of professional authority. And most were philosopher-psychologists who told all who would listen of the wonderful possibilities of the new experimental psychology as a basis for a new science of social control. Their professional standing and their persistent advocacy of experimental psychology blinded them and their younger colleagues and students to the fact that they were, in reality, not genuine practitioners of the new experimental psychology.

Inevitably there would be changes, ineluctably the pioneers' power and status would melt away, given the coming of age of younger generations of psychologists committed to psychology as an experimental natural science who would seek professional careers in American academic institutions. With the exception of Cattell, the pioneers were, as one witness put it, the sponsors of psychology, not the psychologists who founded it as an experimental natural science. The philosopher-psychologists were "systematizers, thoroughly drilled in the arts of logic and scientific and philosophical criticism; but they were not laboratory men. . . . The incoming psychologist had a dominating goal— first-hand experiment; whereas the philosophers had an unparalleled art of logical insight and analysis in generalization. . . . It was the profound and irresistible devotion to the new idea of controlled, first-hand experimentation which rebelled against the systematic armchair organizers of experimental facts at secondhand."[25]

As psychology expanded in the early 1900s the young experimentalists who comprised the first large crop of American-trained Ph.D.s took their science with them to their new teaching appointments. Unlike the pioneers, this younger generation had no exposure or commitment to philosophy or theology. Their professors, after all, told them psychology was an experimental natural science. They believed their teachers. They even reaffirmed their faith in psychology as a natural science by making the laboratory the center of undergraduate instruction in colleges where there was no hope of a graduate program. Such teaching was not merely for future professionals, they knew, but for every undergraduate student.[26]

The post-1900 younger generation constituted an assertive followership within the profession. They soon began to complain that matters in the profession were not as they should be if psychology was an experimental natural science for specialists and professionals only. They pressured the Association's governing Council in the early 1900s to exclude philosophers, educators, and persons in other related fields from membership. Titchener joined the younger generation in calling for exclusion, but when the Association did not limit itself to pure psychology, he withdrew.[27]

Furthermore, the Association grew enormously in membership; and growth had important consequences. In 1892 the Association had but 32 members, in 1905 it had 105 members, and in 1920 slightly more than 300.[28] This growth changed the character of the profession. In the 1890s the number of psychologists was so few that the profession was a highly personalized, intense group in which everyone knew everyone else on a first-name basis, and in which personal symbols of authority were in fact important in the decision-making process. In that sense the profession of psychology was not an impersonal subculture. But especially in the 1910s, the Association—and the profession it was supposed to serve—became so large and impersonal that no coterie of individuals could control it for long, if at all. One symptom of the change was the consistent redefinition of membership, which excluded all but research-oriented experimental psychologists; another was a generational struggle for control of the Association and the journals. This did not mean there was suddenly a surfeit of personal and career jealousies, intrigue, and infighting, but that such tensions and conflicts had to be fought out and resolved in the context of impersonal professional and disciplinary symbols of authority.[29] Yet to those psychologists who lived through that period these changes were imperceptible, perhaps because they were impersonal. It would be a long time before men would look back and rub their eyes in amazement at the transformation of their profession—and their science.

## III

In 1912, University of Michigan psychologist Walter B. Pillsbury stated American psychology's central thesis: "The very general acceptance in recent times of the doctrine of evolution has forced us to read the story of mind in light of the development of the human organism from the lower forms of life."[30] American psychologists interested in the psychology of conduct between the 1890s and the World War I commonly embraced the human instinct doctrine, for instincts were commonly defined in post-Darwinian evolutionary science as a major psychic link between man and the animals. Quite apart from American psychologists' strong didactic interests in an evolutionary science of human conduct, post-Darwinian natural science taught them that instincts were powerful determinants of thought and action in men and brutes. In post-Darwinian natural science, in fact, instincts were a given category of description, analysis, and explanation. Charles Darwin outlined a theory of psychic evolution. Arguing that man's mind could be understood by comparing it with the minds of animals, he declared that man's mind had evolved from animal minds. In *Descent of Man* (1871), he defined instincts as the phylogenetic instruments of survival and adaptation in men and brutes. In other works he insisted that some of man's emotions had been inherited from the animals.[31] In the 1870s and 1880s, Darwin's friend, the British natural historian George John Romanes, transformed Darwin's suggestions into an elaborate theory of psychic evolution. Thus while Romanes was one of the first systematic students of animal psychology, he used the anecdotal method, compiling examples of animal intelligence from literary sources. He was not an experimental psychologist. He fervently wanted to reconcile evolution with contemporary theology and philosophy; he wanted to show that the universe had meaning and purpose. Accordingly Romanes interpreted animal minds anthropomorphically. That is, he tried to demonstrate the continuity of mind from the lowest species to man. He argued that consciousness increased in complexity and sophistication up the evolutionary scale to man. Not surprisingly he found many examples of clever, almost human-like actions among the brutes. Like the Neo-Lamarckians (who were also natural historians) Romanes believed in the acquired characters principle and the doctrine of innate ideas, which held that perfectly formed ideas were biologically inherited and came into play with little or no environmental modification. Romanes believed, therefore, that consciousness and all other mental "faculties" were innate and psychobiological.[32]

In his widely-read *The Principles of Psychology* (1890), William James combined the ideas of Darwin, Romanes, and contemporary physiology and neurology into a human instinct theory that two generations of American psychologists and other educated Americans followed. He mixed Darwin's teleological definition of instinct and doctrine of psychological evolution, neurology and physiology's assumption that instincts were given categories of the new psychology, Romanes' anthropomorphic interpretation of the evolution of consciousness, and his own belief in indeterminancy into an unstable but, for the time, appealing instinct theory and psychology of conduct. James believed his work would further a new evolutionary science of social control.

James' ideas were complex and transitional. They lacked inner consistency. He fashioned a "social psychology" from the determinism of post-Darwinian evolutionary science, in which the individual, defined as a biopsychological datum, was the central focus. He had no intelligible modern concept of a cultural group. Man had all the instincts that the animals did, and many more besides, he claimed, thus implicitly accepting Romanes' anthropomorphic interpretation of the evolution of consciousness. He defined instincts teleologically, as the purposeful instruments of the species enabling the species to survive and adapt. He argued that all human behavior had biopsychological roots. He listed over three dozen human instincts, ranging from winking and sneezing to love and modesty, from, that is, relatively simple reflexes to complex social behavior. Yet James also believed in free choice and indeterminism. In practice, James' instincts, at least in man, could be modified extensively by the individual to meet new, unanticipated conditions in the environment, for, since man had the most complex neurophysiological structure, his was the most plastic and intelligent in nature. Furthermore, although James publicly stated during the Neo-Lamarckian controversy that Weismann's critique of the acquired characters theory was probably correct, he did not understand that Weismann's rigid distinction between heredity and environment as distinct and disjointed processes in evolution ruled out James' teleological definition of instincts, for according to Weismann, evolution was only biological happenstance and natural process. Nor did he catch Weismann's statement that the doctrine of innate ideas was wrong, for if the theory of innate ideas was invalid, then how could "instincts," as anything more than mere biopsychological potential, exist, let alone be inherited?

James offered no experimental proof that instincts existed in man. He assumed the theory of evolution, and the science of physiology, made such substantiation superfluous. And, in any case, as he freely conceded, he was much more the philosopher than the experimental natural scientist. He simply listed what he thought were man's instincts. For proof he appealed to what neurology and physiology had "established." And James virtually undercut the significance, if not exactly the validity, of his instinct theory, when he used the concepts of free will and habit in his system of psychology. He added structure to human adaptability and plasticity by invoking the idea of habit. If he had been a consistent Lamarckian, of course, there would have been no logical problem, for the idea of acquired characters would have permitted him to shift back and forth between nature and culture, instinct and habit, and heredity and environment, with no sense of intellectual crisis. By accepting part of Weismann's position, James was forced to distinguish between heredity and environment, and inevitably, to contrast them, since, according to Weismann, they were entirely different types of factors in organic development. In a sense, the idea of habit was James' functional equivalent for Lamarckism. He defined habits as actions learned in the course of adaptation that had no definite inherited neurological basis save the general ability to adapt. By adding free will James made his inner logical problem worse, for if man had free will, or something approximating it, if he could solve new problems as they arose, if he could learn new behavior patterns and if he could modify his instincts, then his nature was hardly original or permanent. James tried to save his argument by insisting that habits soon became ingrained. This had some meaning when he referred to the individual, but little when applied to the group. What lent his argument an air of logical consistency was James' emphasis on the individual defined as a biological and psychological datum.[33]

Virtually no contemporary scientist noticed the difficulties in James' argument. After all, James appealed to scientific psychology, to the whole thrust of post-Darwinian evolutionary science, and to implicit cultural values, which understandably blinded him and his generation to his theory's inner problems. Grounding social psychology on the biopsychological individual made perfectly good sense, from the standpoint of American science, and of American culture, particularly the middle class Anglo-Saxon culture to which so many scientists in general and psychologists in particular belonged. James was propounding a psychology of human conduct, a psychology in which one could explain, and thus perhaps predict, human behavior. In the context of the age, that was indeed an electrifying and intoxicating prospect, at least to members of and participants in the dominant white culture. James always insisted that all responsible for the train-

ing of the young—parents, teachers, social workers, and others—
should understand both the cultural and the scientific importance
of inculcating the right habits in the young, that is, of recognizing and channeling the instincts into wholesome directions.[34]

Most American psychologists accepted James' formulation of
the instinct theory. This was true whether they belonged to his
pioneer generation or entered the profession at a later date. Indeed, only one psychologist, James Mark Baldwin, grasped the
implications of Weismann's doctrines for the instinct theory.[35]
Obviously American psychologists paid attention to other scientific ideas that reinforced belief in instincts: the tradition of
"faculty psychology" in pre-Darwinian times, the implications of
modern psychology, and the lessons of post-Mendelian biology,
all of which implied that man's nature was more original than
acquired. All of the leading psychology textbooks of the post-
1890 era devoted one or more chapters to human instincts; although each author compiled somewhat different lists of instincts, all followed James' general formulation of the instinct
theory. These psychologists had considerable power and prestige
in their profession. It occurred to virtually no one who read their
work to question what they said about human instincts.[36]

G. Stanley Hall strengthened the belief in human instincts in
the course of his championing genetic psychology. In technical
terms, Hall and James did not completely agree on the sources
of human thought and conduct. Hall placed great emphasis
upon the physiological causes of behavior. In 1904 he published
his magnum opus, *Adolescence: Its Psychology and Its Relations
to Physiology, Anthropology, Sociology, Sex, Crime, Religion,
and Education.* Hall drew his psychology from a different part
of nineteenth-century evolutionism than James. He used the
recapitulation theory, which held that the individual as embryo,
child, and, finally, adult, passed through the stages of evolution,
from one called animal to ape to normal adult human, the apex
of the evolutionary process. In Hall's hands developmental or
genetic psychology differed from James' psychology of conduct
chiefly in its particulars, not in its larger implicit themes. Hall
declared the causes of human conduct were biopsychological
and innate. Like James he stressed the importance of childhood
nurture, of inculcating the young with the proper and appropriate lessons. Hall's novelty was not his emphasis on proper child-

hood nurture, as such, but his advocacy of genetic psychology as the scientific key to social control. He stressed adolescence as a stage of life that should be studied scientifically. That had much vogue in a day when middle class Americans worried more and more about crime, delinquency, education, assimilation of immigrants, about, in the largest sense, adjustment of Americans to an urban, industrial society. And Hall trained a large number of students. Many became important members of the new human relations "semi-professions" of the age, such as education, mental deficiency, and criminology.[37]

William McDougall, an English physician and self-taught psychologist, went beyond James and Hall. He applied the instinct theory to social groups. His book, *An Introduction to Social Psychology* (1908), was an important vehicle for popularizing the instinct theory beyond a few hundred psychologists, sociologists, biologists, and other scientists, chiefly with ties to the natural sciences, to a larger and more heterogenous educated public of thousands of social scientists, intellectuals, commentators, journalists, and even literary figures. He argued that innate human instincts were the basis of social institutions. And he was one of the first authors in English to use the term "social psychology," and, indeed, to attempt to construct a viable "social psychology" based on the new psychology. McDougall completely reduced culture to nature. According to McDougall, instincts had an innate emotional core. They were the purposeful instruments of adaptation and evolution in all species, including man. He made blunt comparisons between human and animal instincts. Indeed, he argued that psychologists should examine animal minds at roughly the same time that American psychologists were shifting from human to animal subjects, which gave his theory an experimental aura, even though he advanced no experimental proof of the existence of instincts in man or the animals. Ultimately what interested McDougall was not experimental science but teleology. He believed that the universe had meaning, and that evolution was a purposeful process directed toward a final metaphysical goal. But most of his American readers probably did not notice that aspect of his work. More fascinating to them, presumably, was his frank appeal to the study of animal minds at precisely the time younger American psychologists were becoming convinced that comparative psychology was the key to developmental psychology. And for those interested in a science of social control, McDougall had more appeal than even James, for he devoted most of his space to discussing human instincts and social evolution and cultural institutions within the context of his theory. Thus he traced back human institutions to seven instincts he insisted were basic in man. He attributed the rise of modern warfare to the presumed instinct of pugnacity in man, and the rise of modern humanitarianism to the working out of the parental instinct.[38]

McDougall's impact among educated Americans was greatest, not among experimental psychologists, but among writers, commentators, and professionals in other fields interested psychology's social applications. Between 1900 and 1920 at least six hundred books and articles published in America and England advanced the instinct theory, and most published after 1908 cited McDougall.[39] Eminent American sociologists such as Charles H. Cooley of the University of Michigan and Charles A. Ellwood of the University of Missouri propounded the instinct theory.[40] A school of economists in theoretical revolt against laissez-faire economics' assumptions of the totally rationalistic individual seized upon the instinct theory as a handy instrument with which to reshape their own discipline. The instinct economists included Harvard's Frank W. Taussig, Columbia's Wesley Clair Mitchell, the University of Washington's Carleton H. Parker, Colgate's Lionel D. Edie, Thorstein Veblen, and labor relations consultant Ordway Tead. McDougall obviously inspired all save probably Veblen. (Veblen worked out his own particular psychosocial explanations.) The instinct economists accepted at face value psychologists' claims that human instincts existed; in fact, several instinct economists corresponded with psychologists on the instinct theory's technical aspects before publishing their work. The instinct economists' central argument was that modern industrial society was frustrating workers' constructive instincts, thus making them alienated and ready for rebellion. For the instinct economists, psychology obviously promised a science of social control. Their remedy was simple. Change social conditions, they cried, and permit the workers' constructive impulses to achieve satisfactory and socially acceptable resolutions. Otherwise chaos and perhaps even revolution were possible, the instinct economists assumed—

an understandable sentiment in an era of increasing social tension and labor strife. "If we supply a stimulus which is calculated to evoke only the more socially beneficial impulses of human beings," Tead declared in 1918, "we can rely upon the desired reactions taking place."[41]

## IV

In the second decade of the twentieth century, American professors of psychology and education seized upon intelligence testing. As the liberal educator Harold Rugg observed later, this era was, for professors of psychology and education, "an orgy of tabulation."[42] Following the publication of the first American version of the Simon-Binet scaled mental test in 1908, intelligence testing became the fastest growing professional specialty among psychologists and educators.[43] Most testers were younger men and women who entered the professions of psychology and education after 1900. The emerging psychology of capacity appealed to them for scientific and cultural reasons. Their commitments to natural science and to Anglo-Saxon culture convinced them that the tests measured innate intelligence undiluted by social circumstances and milieu. Thus they assumed from post-Darwinian natural science that it was legitimate to emphasize innate endowment. And the prospect of measuring the intelligence of individuals and groups was, to put it mildly, an exciting one for them. Quite obviously mental tests had social applications. The testers gave examinations to criminals, delinquents, and other social deviants, or to members of white and non-white minority groups. Their common standard of comparison was the middle class, white American. It was hardly an accident of history that the administration of mental tests to these groups occurred at a time of social unrest and polarization, or that the racial mental testing movement took place when there was growing a considerable backlash against immigrants from Southern and Eastern Europe and nonwhites. So dazzled were the testers by the new developments in biology and psychology, by their commitments to WASP culture, and by contemporary events, that they did not pause to wonder whether the tests actually measured innate intelligence apart from environmental influences.

The creation of methods for the measurement of differences in physical traits encouraged the belief among scientists on both sides of the Atlantic that innate mental differences could be measured as well. In the later nineteenth century, Sir Francis Galton and Karl Pearson founded the science of biometrics, or the statistical study of inheritance, in England. Their interest in biometrics was not merely technical; Galton and Pearson were also leaders of the English eugenics movement. They hoped that biometrics would provide a firm scientific foundation for eugenics, and would enable eugenists to predict the inheritance of traits through the generations. Galton and Pearson worked out some early statistical measurements of individual differences in mental and physical traits. In retrospect, it is obvious that their studies were often based on dubious, second-hand information. Perhaps they did not grasp the defectiveness of their evidence because they were so interested in promoting their new science. The statistical measurements of raw data they worked out did appear impressive on their own terms, and they seemed to Galton and Pearson convincing evidence for believing, as Pearson put it, that "the mental characters in man are inherited in precisely the same manner as the physical. Our mental and moral nature is, quite as much as our physical nature, the outcome of hereditary factors."[44] They gathered descriptions of physical and mental traits of unrelated and related individuals, and attempted to show, through the use of the coefficient of correlation, that mind and body were inherited since there was a higher positive correlation among related than unrelated individuals. The flaw was in depending upon impressionistic estimates of mental traits from third parties such as teachers.

Many American psychologists were impressed by the work of Galton and Pearson; there were in fact, some close ties between Galton and a number of American psychologists later influential in America. And the Galton-Pearson statistical methods appeared to be of unimpeachable scientific rigor and authority. Galton was probably the first scientist to use mental tests. He also argued that mental tests could be given to individuals and to groups as measurements of innate differences. Pearson refined Galton's coefficient of correlation, thus giving the infant field of individual psychological differences an instrument with which its practitioners could demonstrate the assumed inheritability of mental and physical traits. By the early 1900s, probably most American psychologists were gravitating toward the Mendelian, rather than the biometric, description of inheritance. But American psychologists still believed that Galton and Pearson had pointed the way to a science of human capacity. In the 1890s and the early 1900s a number of eminent American psychologists advocated mental measurement. At Columbia, for example, Cattell encouraged his students to learn statistical correlations with other Columbia faculty and thus promoted the study of human mental capacity. At Clark, Hall also recognized the promise of mental measurement and promoted it. So did Münsterberg at Harvard, who encouraged his young protégé, Robert M. Yerkes, to apply mental measurement techniques to animals, so as to expand knowledge of the development of mind from protozoa to man. Yerkes had taken as much work in experimental zoology as in psychology, and he was an early convert to Mendelism and to statistical measurement in psychology and biology. When Yerkes began teaching at Harvard in 1902, he encouraged his graduate students and the secondary teachers in his summer school classes to work with mental measurement. Overall the chief difficulty with these and other early American attempts at mental measurement is they focused on elementary human sensations, not the higher mental functions. They could not measure intelligence in action.[45]

In 1905 the French psychologists Alfred Binet and Theodore Simon published their famous scaled mental test, which became the model of the intelligence test. The tests were sequences of questions and problems of graded difficulty. Each cluster of questions and problems corresponded to what Binet and Simon judged as the norm or standard achievement for an age group. Inevitably the tests were "culture-bound." They measured the minds of persons at least six years old. Ability to pass the tests depended on such factors as recognition of verbal, mathematical, and cultural symbols, and, in particular, on scholastic skills. The tests were standardized on particular ethnocultural groups, whose achievements were thus the norm; then the tests were given to other ethno-cultural groups, whose experiences were different. The tests did not separate heredity from environment; they could only measure the mind in action. That is probably why the differences they revealed had some correlation to the differences in intelligence the testers could perceive in their own subjects, and gave the tests far more of an aura of "success" than the earlier, pre-Simon and Binet tests had. The measuring scales, in other words, confirmed what seemed to be common-sense observation, and, in the bargain, promised precise scientific measurement.

The Binet-Simon test became the model used by American psychological testers. Most American testers simply assumed that the tests were fair measurements of innate intelligence.[46] Henry H. Goddard and Lewis M. Terman, who had both studied with Hall, worked out the standard American adaptations of the new test for American examinees. Born in a small town in Maine, Goddard took his doctorate in 1899, at the age of thirty-three. After seven years teaching in a Pennsylvania normal school, Goddard became director of psychological research of the Vineland Training School, a private institution for the mentally retarded, in 1908. He discovered the Binet test the same year and began to work out a test appropriate for American children. Even before he published his results in 1910 he was speaking on the Binet test to such groups as the National Education Association and the American Association for the Study of the Feeble-Minded. Goddard did much to spread the idea that the feeble-minded constituted a threat to society and that mental defect was inherited. Good conduct, therefore, was innate. Goddard standardized the tests on "normal" and "subnormal" children by trial and error, by adding and eliminating questions until the right proportion of children of each age passed each age level, according to the expectations of the Gaussian probability curve.[47] In a stream of books, articles, and speeches, Goddard emphasized the menace he insisted the feeble-minded presented to society. Feeble-mindedness was inherited, he declared; the only solution was to segregate the subnormal from society. His famous studies, *The Kallikak Family* (1912) and *The Criminal Imbecile* (1915), impressed numerous educated, literate Americans. Goddard's central message was that modern psychological science held out the promise of a science of social control.[48]

Lewis Terman's revision of the Binet, commonly known as

the Stanford-Binet, eventually became the accepted American version, not Goddard's, largely for technical reasons. Born in Johnson County, Indiana, in 1877, and a graduate of Indiana University, Terman finished his doctorate at Clark in 1905, and began teaching in a California normal school. He learned of the Binet test and spoke at professional meetings about it, at least on one occasion with Goddard. Terman, too, recognized the tests' implications for both educational psychology and for social policy. In 1910 Terman accepted appointment as professor of educational psychology at Stanford University, where he spent the rest of his career. Like Goddard (and most psychologists) Terman assumed that nature prevailed over nurture, a belief that shaped his interpretation of his work in standardizing the Binet. He used school children in Palo Alto, all of them from middle class, native born, Protestant Anglo-Saxon homes as the subjects of his tests. He was satisfied his version was successful when he had the mixture of questions and problems for each age level that permitted the "right" number of children, on the Gaussian expectation, to pass. His scientific and his cultural commitments convinced him his adaptation was an accurate measurement of innate intelligence undiluted by social milieu.[49]

Many young psychologists joined the mental testing movement in the 1910s. Many took their doctorates with Hall at Clark, with Cattell or Thorndike at Columbia, with Titchener at Cornell, or with Münsterberg or Yerkes at Harvard, who encouraged them to work in this exciting new field. Opportunities galore beckoned to those with freshly-minted doctorates, for testing was a new field, not yet dominated by a few eminent persons and programs; tests were relatively easy to administer, and the results were publishable as reports of experimental investigations. The prospect of precise, scientific mental measurement had its own allure. Many young testers, such as Daniel La Rue, one of Yerkes' students, began their careers at humble normal schools; the more prolific and fortunate ones, such as Terman, soon moved to appointments in education or psychology in major universities.[50] The testers' cultural commitments were important in helping shape the questions they asked and the assumptions they made. In an ultimate sense theirs was an ethnocentric, middle-class bias, as was indicated in the very kinds of subjects they selected for measurement and comparison.

Rather quickly two groups of specialists in mental testing emerged. The members of the first studied social deviants. They gave mental tests to delinquents, dependents, and criminals. Following Goddard's suggestion, who proclaimed how menacing the feeble-minded were to society, these testers assumed that social worth and status depended upon innate intelligence as measured by the tests. Conspicuous among this group were former students and associates of Goddard and Hall, notably J. E. W. Wallin, E. B. Huey, F. Kuhlmann, and S. C. Kohs. Uniformly they discovered that their subjects had subnormal intelligence as defined by the tests. They claimed that intelligence was innate, and that sub-normal intelligence was the cause of anti-social behavior because it occurred overwhelmingly among social deviants. Their implication was that social remedies would not uplift these people; they were born to their fate.[51] The other group of testers measured "racial" intelligence. They compared the performance of white, native-born children with that of immigrant and nonwhite children on the tests. Since the tests they used had been standardized on middle class WASP children, it was to be expected that children of different backgrounds would have different scores on the average. Some of the testers who studied blacks, notably Marion J. Mayo, H. E. Jordan, Josiah Morse, and George O. Ferguson, were Southerners by birth or adult residence or both; others, such as S. L. Pressey and T. H. Haines, had taken their graduate training with mentors such as Robert Yerkes who were committed to racial interpretations of testing programs.[52] To understand why these testers took the positions they did, we must remember that several factors operated with varying degrees of effectiveness, depending upon the particular individual—the intellectual traditions of modern natural science, the influence of a specific graduate professor, the ethnocultural backgrounds of the testers themselves, and the patterns of current research in theirs and related disciplines.

The mental testing movement came of age, as it were, during World War I, thanks in large measure to the gigantic mental testing program the United States Army conducted. The Army was interested in employing some kind of yardstick for classifying the millions of new recruits for the many jobs in the military, and the psychologists provided the initiative and the proposal for such a program. A number of American psychologists saw in the War a marvellous opportunity for the expansion of applied psychology.[53] Robert M. Yerkes, of Harvard University, who was president of the American Psychological Association in 1917 when Congress declared war, offered the Association's services to the War Department, including the devising of suitable tests to separate the wheat from the chaff among the recruits. Through the offices of the National Research Council and its chairman, astrophysicist George Ellery Hale, the Wilson administration encouraged the American Psychological Association to proceed. As president of the Association, Yerkes then appointed a committee of nine Association members, including himself and such champions of mental testing as Goddard and Terman. In the spring of 1917 the committee members met at the Vineland Training School and hammered out over several weeks a test for group application (the first such examination in America), which became known as Army Alpha.[54] The men simply assumed that the tests measured innate intelligence. Indeed, it did not occur to them until several months later that many recruits were illiterate and would require a different examination. They standardized the prototype of Army Alpha (for literates) on the Stanford-Binet and assigned mental ages from the Stanford-Binet to Army Alpha, thus assuming that scales derived from suburban Palo Alto school children of middle-class, WASP background could be applied to adults from many backgrounds, adults who had been out of school for many years, and, in many cases, who knew English as a second language. They included in the test problems in arithmetic, grammar, definition, numerical comparison, and recognition of the artifacts of an urban, industrial, technological society. The test measured "practical" judgment in various social and industrial situations. And the circumstances in which the tests were administered were not always ideal. Recruits took the tests in groups; quite often this occurred under crowded conditions, shortly after arriving for basic training. This led to pressure, confusion, and anxiety, to be sure. For example, black recruits took the examinations in segregated groups—with southern white officers usually administering the tests.

After the War, Yerkes began to compile information for his massive report to the National Academy of Sciences and the

Army. From the more than two million examinations, Yerkes extracted a random sample of 162,526 for detailed analysis and interpretation. One generalization he discovered was that the various ethnic and racial groups had scores that could be arranged in a hierarchy. Native-born recruits stood the highest of all groups. Of the foreign-born recruits, those from English-speaking countries and from northern and western Europe in general did the best; those from southern and eastern Europe were below them. American blacks stood below the immigrants from southern and eastern Europe, and northern blacks scored higher than southern blacks. Although Yerkes was careful in his statements, quite obviously he believed differences in inherited mental capacities explained the varying levels of performance for the "racial" groups in the sample. Yerkes also discovered a high correlation between the extent of an individual's formal education and his test score, and between the level of occupation and test standing. Those groups with the most formal education, and those who worked in professional, administrative, entreprenurial, and even clerical occupations, stood higher than those with little or no formal schooling and those with semi-skilled or unskilled occupations. As with nativity and race he chose to explain these results as the consequence of innate endowment. Inherited intelligence, he intoned, determined the amount of education an individual or a group could receive, and the occupational status to which an individual could aspire.[55] In 1921, he wrote his young friend Carl C. Brigham, who had served with Yerkes in the Army's Psychological Division during the war, and who was finishing a semi-popular study of the army tests that would become powerful ammunition in the hands of the immigration restrictionists, that, "My own tendency is to emphasize on every occasion the importance of individual mental differences and also of racial peculiarities."[56]

By the early 1920's it appeared to men of science such as Yerkes and Brigham that modern natural science had resolved the nature-nurture issue. Mendelian genetics, the mutation theory, eugenics, the instinct theory, and intelligence testing all pointed, in their eyes, to the argument that in man as well as in the animals heredity prevailed over environment. Evolutionary natural science had now proven its worth, they believed, not merely for the greater comprehension of nature's laws, but also for the better understanding of how nature's laws directed the behavior of peoples and cultures. Yet the new scientific doctrines of hereditarianism owed at least as much to developments external to science—especially to contemporary events, and to the social policy commitments of American psychologists and biologists—as they did to the internal history of science. Rather quickly these scientific doctrines were seized upon by men in public life who possessed a direct stake in their application to American society. These ideas became the justification for the assertion of ideas of racial superiority and inferiority. As ideas of racial superiority became associated with ideas of science, a reaction set in. Coincidentally the American social sciences were just emerging as coherent scientific disciplines and professions in the second decade of the twentieth century. Perhaps inevitably American social scientists regarded the new hereditarian doctrines as a threat to the future of social science in America and to the prospects for a decent society.

# 2 *The Discovery of Culture  1900-1920*

# 3  OMNIS CULTURA EX CULTURA

## I

URING the early months of 1917, as the United States drifted into the World War, Robert H. Lowie, associate curator of anthropology at the American Museum of Natural History, delivered a series of public lectures at the Museum in New York City. His lectures, published later that year as *Culture and Ethnology*, were symptomatic of a momentous intellectual revolution in American anthropology that would soon resonate throughout the national scientific community. Lowie addressed himself to what he believed were the proper principles of modern scientific ethnology. He loudly insisted that the discoveries of evolutionary natural science did not give natural scientists the professional license to explain and interpret man's cultural behavior. Aggressively Lowie challenged those natural scientists who claimed that their scientific ideas and methods alone could account for the histories of peoples and cultures. Lowie flatly rejected natural scientific hereditarian interpretations of human nature and conduct. "The ethnologist will do well to postulate the principle, *Omnis cultura ex cultura,*" Lowie declared. "This means that he will account for a given cultural fact by demonstrating some other cultural fact, by merging it into a group of cultural facts or by demonstrating some other cultural fact out of which it has developed," he continued. Lowie defined culture as all the capabilities and habits men acquired as the consequence of belonging to society. Cultural development in man, therefore, was an entirely different order of phenomena than biological or psychological evolution. A given culture operated autonomously from the determinants of the natural sciences; it was not subject to the dictates of psychology's individual mind, biology's heredity, variation, and evolution, or physical geography's physical environment. Human cultures could be explained by cultural antecedents such as language, myth, religion, social institutions, values, and attitudes, not by the unlike antecedents of the natural sciences, such as instincts, genes, and landforms.[1]

Lowie's lectures signified that no longer would American social scientists tolerate natural scientists' incursions into their intellectual domain. Clearly the heredity-environment controversy had begun. Yet temperamentally Lowie was ill-fitted for the roles of intellectual rebel and scientific challenger. This gentle, tolerant Viennese preferred the quiet world of the scholar to the hurly-burly of public strife. He had strong left-liberal views on social issues; yet he usually remained detached. In the contentious profession of American anthropology Lowie often called for moderation even when issues of considerable moment arose. When Lowie spoke out against the natural scientists' presumptions to interpret cultural phenomena, he obeyed the imperatives of his role as a professional anthropologist rather than his personal inclinations. Lowie's words suggested that academic professionalization had now reshaped anthropology as it had the natural sciences a quarter century earlier. He thought and behaved as a professional anthropologist trained in one of the new universities. After finishing his doctorate with Franz Boas at Columbia University in 1908, Lowie launched his own scientific career, and he also consistently supported Boas' campaign to transform American anthropology into an academic science and profession, to shift its institutional locus from the museum to the university, to create a professional identity and role for academic anthropologists, to win respect for anthropology in American science and higher education, and, necessarily, to liberate anthropological theory from the assumptions and models of the natural sciences by proclaiming the modern social scientific theory of culture.

The roles the Boasians played as professional anthropologists in American science obligated them to speak out as Lowie did.

Obviously their efforts to professionalize American anthropology and to win for it a place of respect and authority in American science and higher education would have little purpose if natural scientists could continue to presume to interpret anthropological phenomena according to the lights of natural science. The Boasians knew that the culture theory was an indispensable intellectual weapon, for it permitted them to preempt man's cultural behavior as their special area of expertise, to retrieve it from natural scientists, without falling into the dilemma of attacking the theory of organic evolution. The Boasians were probably the first group of American social scientists to use the culture idea as a shared professional ideology. They placed a particular pattern upon the culture idea. After offering a brief and positive definition of culture, quite obviously borrowed from Boas' friend, the English anthropologist E. B. Tylor, they then devoted most of their energies and attention to saying what culture was not, what kinds of phenomena were not cultural, why culture was different from nature, why natural scientists could not use natural science determinants to explain cultural phenomena, and, above all, why natural scientists should not invade the domain of the social sciences. Clearly this peculiar pattern of argument reflected the context and circumstances in which the Boasians operated. No such pattern of argument had existed in the various European statements of the culture idea.[2] Nor did it among nineteenth-century American anthropologists.[3] And the handful of American social scientists who published statements with recognizable elements of the modern culture idea did not employ such emphases—or warnings to natural scientists to remain on their own turf.[4] Finally, such emphases were absent from statements of the culture idea in American social science after the mid-1920s.[5] Indeed, the emphasis of post-1925 American statements of culture was on elaborating its many positive ramifications.[6] The fifty or so statements of the culture idea published by the Boasians in the second decade of the twentieth century were the intellectual and theoretical components of their larger campaign to professionalize American anthropology in American university culture. Boas and his students realized that they must create a body of autonomous theory for their science; their rebellion against the domination of ethnological and social science theory by the concepts, the metaphors, and assumptions of the natural sciences, then, was a logical consequence of their efforts to transform anthropology into a respected science and profession in America.

Yet there was more involved. The culture idea had obvious implications for the stances men took in the world of politics no less than the world of science. The culture idea was at odds with the notion of "Nordic" superiority, not to mention many other values of WASP culture or the assumptions of "mainstream" middle class progressive ideology. The Boasians were separated from WASP culture; several were immigrants, of Jewish background, or both. If the Boasians had a common political posture, probably it was to the left of mainstream, middle-class progessivism; they were not impressed by the progressives' vision of a clean, efficient, and mechanistically democratic society. They believed that ethnic tolerance and pluralism were indispensable prerequisites to American democracy.

Before American intervention in the World War, however, public policy issues were distinctly secondary to the Boasians, who were busily preoccupied with promoting their science. American intervention in the World War, and the resulting consequences of American intervention for American public politics and for the politics of American science, hit the Boasians like a thunderbolt and drastically altered the political and scientific context in which they lived. For several years following American entry into the war, defense of the culture idea meant to the Boasians not merely the autonomy of their science and its prestige in the scientific community but also the legitimacy of their status as "marginal" men and women in an increasingly polarized society.

## II

"I only hope that I may be able to gradually develop anthropology in such a way that it may become a strong department of the University, and that it may materially contribute to the advancement of scientific research and of the scientific spirit in this country," Franz Boas told Seth Low, the president of Columbia University, in accepting his professorship of anthropology in 1899.[7] Boas' words were prophetic. Under his leadership Columbia became a major center of anthropological research and training in North America. His contributions to his science were monumental. From his base at Columbia, he promoted anthropology's status as a scientific profession in American science and higher education. As a product of the German universities, he understood well the criteria of America's post-1890 system of graduate universities. To his adopted land, Boas brought the ideals of German scholarship: of specialized research, of universalistic canons of evidence and interpretation, of the specialist as an autonomous professional. For many years he labored to build scientific institutions that would reflect these values and promote the advancement of his science. It is in all of these roles that Boas must be understood.

Boas settled in the United States just as the new system of universities was taking shape in American culture. The coming of the new universities suggested the possibility of a new era in the history of American anthropology. In the nineteenth century, American anthropology developed as a relatively minor discipline within the natural history movement. Anthropological institutions did not have collegiate or higher educational connections. Most anthropologists were not trained in that science, for the very good reason that no such formal training could be obtained in America. In post-Appomattox years, anthropology developed institutionally most notably in museums in several large urban centers, the most important of which were Philadelphia, Washington, and Boston-Cambridge. In each center a powerful figure dominated the local ethnological scene. He defined the proper lines of inquiry, spoke for the local movement, built its supporting institutions, and recruited its clientele. Consequently standards of ethnological practice were essentially rooted in localistic, not cosmopolitan, perspectives. In most instances there was little chance of placing anthropology within the emerging university culture, or, at least, little interest in doing so.

Daniel G. Brinton dominated anthropology in Philadelphia. His career is suggestive of the informal, heterogeneous ways in which men of science became anthropologists in the nineteenth century. Brinton graduated from Yale in 1858 at the age of twenty-one. He took an M.D. from Jefferson Medical College in 1861 and subsequently practiced medicine in Philadelphia. After earning enough money to retire in middle age to pursue his hobbies of linguistics and folklore, he published actively and soon won fame locally as an eminent ethnologist. Soon the American Philosophical Society and the Academy of Natural Sciences of Philadelphia granted him membership. In 1886 he became professor of linguistics and folklore at the University of Pennsylvania—the first such chair in Philadelphia, and probably the first in any American university. Yet for Brinton his chair did not lead to the formation of a school of disciples who would carry on his work. He did not deem it necessary to do so, for, to Brinton, there was no apparent relationship between his university appointment and his science save perhaps lectures and a library. And given the pattern of his career there was no reason to expect that he would have thought that relationship as more than he did. His main concern was the furthering of his own interpretation of the history of man, which was a variant on the evolutionary point of view so common among later nineteenth-century American anthropologists. As George W. Stocking, Jr., and others have pointed out, the evolutionary or Neo-Lamarckian point of view reduced culture to nature and assumed certain fixed stages of human "progress" through which all peoples must develop if they were to rise to the high plateau of white civilization.[8] As Brinton's career suggests, there was nothing about his background before becoming an anthropologist that was specific to anthropology. In fundamental ways he approached ethnological questions from biological, medical, and natural science perspectives. Because he operated on a fundamentally local scale—largely within the Philadelphia learned and scientific community—his peers were mostly not specialists in anthropology. This context made it perfectly natural for him to function, intellectually and institutionally, as he did.[9]

In late nineteenth-century Washington, D.C., Major John Wesley Powell did much to promote anthropology. The details of his activities differed, of course, from those of Brinton, but the larger patterns did not. When Powell organized the Bureau of Ethnology in 1879, he thus became an important institution-builder in Washington anthropology, establishing important links between the Bureau, the Smithsonian, the United States National Museum, and the federal city's scientific and learned societies, so that Washington became a vibrant center of ethnological activity. He commissioned studies; he arranged for the collection of invaluable materials and artifacts; he pushed for increased appropriations; he won approval for part time and full time ethnological positions (not an inconsiderable achievement); and he dominated the intellectual atmosphere of Washington anthropology. Like Brinton, he too was committed to the evolutionary point of view in anthropology, and so were those who worked with him. In an age in which natural scientists, and many laymen as well, thought the doctrine of evolution was the ultimate explanation for all living things, this was probably an understandable assumption. Powell and his disciples were trained and educated as natural historians; like Brinton, they came to anthropology in mid or late career. Both Powell and his successor at the Bureau, W J McGee, were geologists prior to taking up anthropology. While they were full time scientists, as was Brinton, theirs was more of a commitment to general natural history than to a specific discipline. Neither Powell nor McGee had important links, intellectually or institutionally, with the emerging university culture; although they used local institutions to promote a school of disciples who carried on the work, their ultimate standards of judgment owed little to academic or university culture. Washington anthropologists thought of themselves as "locals." Their chief identifications were with the evolutionary point of view, with Washington (and government) science, and with museums.[10]

In Boston-Cambridge, anthropology followed a different configuration of institutional development, thanks in no small measure to the influence of Harvard University and of its considerable resources. The pattern here foreshadowed those that would characterize American science, and American anthropology, in the next century. In the late 1860s philanthropist George Peabody gave Harvard a $150,000 endowment for a Museum and Professorship of American Archeology. This provided the impetus for the development of anthropology and archeology at Harvard within the context of the new graduate university culture. In the mid-1870s, Frederick Ward Putnam, a former student of Louis Agassiz, became curator of the Museum, and, in 1886, professor of archeology and ethnology at Harvard. In time Putnam dominated anthropology in Boston-Cambridge, but, unlike Brinton and Powell, he directed it within the context of modern university culture and built the Peabody Museum into the first exclusively anthropological museum in North America, to serve as a base for advanced instruction at Harvard. In the late 1880s he began accepting graduate students at Harvard; in 1890 Harvard granted its first doctorate in anthropology, and in 1894 Putnam began offering undergraduate and graduate courses on a regular basis. But Putnam did more than give his science academic roots. In a fundamental sense he was a different kind of scientist than Brinton or Powell; unlike them, he thought of himself as operating on a national stage, not merely the local authority who would promote learning among the genteel citizens of the community. Putnam became an important mover and shaker in national scientific circles in the later nineteenth century, serving, for example, as permanent secretary of the American Association for the Advancement of Science between 1873 and 1898, and as a consultant in archeology and anthropology for the Field Museum in Chicago and the American Museum of Natural History in New York during the 1890s. Boston-Cambridge became an important center of anthropology, but, thanks to the confluence of many circumstances—Putnam's own inclinations, the direction in which President Charles W. Eliot was pushing Harvard, luck, and funding, to mention only a few— it became a different kind of center than either Philadelphia or Washington. It became oriented toward advanced academic training and a specialized professional identity with anthropology, not merely general science or natural history.[11]

What happened at Harvard occurred at several other universities. In the three decades following Putnam's appointment as professor at Harvard, several other universities established positions in anthropology. Clark University established a docentship in 1889, the University of Chicago a professorship in 1892, Columbia three professorships by 1899, and California a professorship in 1901. By 1919, six universities had separate anthropology departments, thus in some sense recognizing anthropology as an autonomous science, and at least thirty-nine other colleges and universities of appreciable size offered instruction above the sophomore level in one or more of anthropology's subdivisions.[12] Anthropology's academic expansion in these years was certainly not spectacular, as compared with chemistry or psychology, but academic appointments did appear in a few major universities and in a number of respectable colleges and universities. Probably less than one hundred anthropologists worked full time as anthropologists in academic or museum careers as late as 1920.[13] If that meant to the administrators of universities that anthropology was but a marginal discipline and profession, nevertheless anthropology was included as both a discipline and profession in the American academy. Anthropology's marginality even extended to graduate training. For all practical purposes, Harvard and Columbia dominated advanced teaching in anthropology, granting over 90 percent of American doctorates before the 1920s.[14]

Franz Boas made Columbia into a strong center of anthropology. His background and experiences in Europe helped prepare him for his American career. Born in Minden, Germany, in 1858, in a liberal, middle class Jewish family, Boas grew up in a cultural milieu in which the values of scientific truth, of learning, of the life of the mind, and of the liberating visions of the Revolutions of 1848 were taken for granted. Even before Boas turned to a life in science he believed firmly in the dignity of all men, in their equipotentiality, and in the benefits that science and reason could bestow upon humankind if properly developed. Doubtless his position as a Jew in German culture helped solidify his left-liberal world view.

As a young man he had much contact with Continental scientific and academic institutions. These experiences encouraged him to turn to science for his life's work, and gave him the cosmopolitan values so common of European scientists. He studied at Bonn, Heidelberg, and Kiel; initially he worked in physics, but late in his graduate studies he became fascinated with geography, so when he took his physics doctorate from Kiel in 1881, his minor was in cultural geography. In the 1880s his scientific interests shifted further to ethnology. The reasons were complex and mutually reinforcing. His cultural values played a formidable role in directing his scientific curiosity toward ethnology. So did his experiences as a scientist in that decade: with a German government polar expedition to Baffin Land, where he made his first contact with primitive peoples, with the Royal Ethnological Museum in Berlin, where he worked with the famous Rudolf Virchow and his brilliant collaborators, and with the British Association for the Advancement of Science's Committee on the Northwestern Tribes of Canada, for which he did his first distinctly ethnological field work.[15] Boas learned another lesson too, thanks to alternating periods of professional employment and enforced idleness: the researcher required a science-related livelihood to carry on a continuous program of inquiry.

Boas settled in New York City in 1887 to be geographical editor of *Science* at the then-handsome salary of $2,000 yearly; his real purpose was to launch his career as anthropologist in America. He chose America for several reasons. America seemingly offered more professional opportunities for anthropologists, as her universities were just beginning to expand, and field work was obviously much easier than from Europe. Bismarckian Germany impressed him as less tolerant and more conservative for someone of his background and political outlook than did America. Boas' editorship permitted him to make his debut in American science. He published furiously in geography and ethnology. He made many scientific contacts. He began attending the American Association for the Advancement of Science's annual sectional meeting for anthropologists, for example, and he even attempted to revive the moribund American Ethnological Society in New York.

In late 1888 Boas entered a new phase of his life, a decade of temporary positions punctuated by eighteen harrowing months of unemployment. First he lost his job with *Science*. Nimbly he persuaded G. Stanley Hall, president of Clark University, to appoint him docent of anthropology. In Boas' three years at Clark, he started most of the work he was to develop for the rest of his career. He taught courses in physical anthropology, statistics, and African and North American ethnology. He conducted an extensive investigation of the physical traits of three thousand Worcester school children, which later blossomed, with different subjects, into his famous report on racial characteristics for the United States Immigration Commission. He devoted two summers to intensive field work among the British Columbia tribes, which became the focus of most of his specialized work.[16]

In April 1892 Boas resigned his position at Clark in concert with other faculty critical of President Hall's behavior.[17] Next Boas became Frederick Ward Putnam's chief assistant at the World's Columbia Exposition in Chicago, largely because Putnam had come to respect Boas during his brief tenure in Worcester. But the position was temporary, lasting only until 1894; then Boas was unemployed for a year and a half. Putnam again came to Boas' rescue when, in 1894, Putnam became a part-time consultant to the American Museum of Natural History on its anthropology collection. He soon persuaded the Museum's trustees to appoint Boas assistant curator of anthropology. In turn this opened the door to a position at Columbia, for, since 1890, Columbia and the American Museum had cooperated in developing certain fields of science (including anthropology) through a system of joint appointments. Thanks to Putnam's influence, and to that of James McKeen Cattell, professor of psychology and director of the pyschological laboratory at Columbia, Boas was offered a lectureship in anthropology at the university in 1896. In 1899 President Low appointed him professor on a two year contract; Boas' appointment became indefinite two years later, thanks to a handsome offer of a chair of anthropology Boas received from the University of Vienna. Now Boas had a firm institutional base in both a rising university and a prospering museum.[18]

In his early years at Columbia and the Museum, Boas demonstrated his awareness of the importance of promoting and building institutions that would advance specialized training, research, and publication. As he told a friend in 1901, he hoped to use the general Columbia-American Museum policy of cooperation to advance his science at the university. He hoped to use his curatorship at the Museum to develop its collections in many specialized areas. As each area of the Museum came to require a specialist's care, that would be "the opportune moment for introducing instruction in each particular line in Columbia University." Through this strategy Boas believed he could build a well-rounded school of anthropology, systematic and rigorous, that could prepare future generations of specialists. Preparation of specialists was, in Boas' opinion, anthropology's most urgent need; "I am very anxious," he declared, "that those who do take up the work should not be as unprepared as most of our generation have been."[19]

Accordingly Boas placed much emphasis on establishing anthropology's academic roots at Columbia in as comprehensive a program as possible. He showered Presidents Seth Low and

Nicholas Murray Butler with countless proposals: the founding of an East Asiatic Department, the creation of positions in East Asian studies, world civilization, and geography, the acquisition of East Asian cultural artifacts, the establishment of a cooperative Columbia-Yale School of Consular Service and Foreign Trade and an International School of American Archeology in Mexico City.[20] Most of Boas' more ambitious plans never materialized, probably because of anthropology's marginal status in the eyes of the university officials, philanthropists, and scientists who were the patrons of science of the era. Boas never lost an opportunity to suggest a proposal's social application. Thus he told President Butler that the Columbia-Yale school could prepare young men for government and business careers related to Asia—perhaps an all too obvious appeal to Republican Butler's Asia-first foreign policy sentiments.[21] And he couched a similar appeal for funds for an African museum in part on the grounds that it might teach Americans healthy lessons in race relations.[22]

In 1905 Boas' grand plans for Columbia and the Museum suffered a severe setback. That spring he angrily resigned from the Museum because he could no longer tolerate the new director's policies. The Columbia-American Museum policy depended, from the Museum, upon Morris K. Jesup, the leader of the Museum's trustees. In 1903 Jesup became seriously ill and relinquished daily control over Museum affairs to a new director, invertebrate zoologist H. C. Bumpus. Bumpus had little use for cultural anthropology—an attitude typical of many biologists of that day—and thought the Museum should cultivate its lay constituency, even at the expense of permitting research programs to deteriorate. For example, he revived the Museum's Department of Public Instruction, an extension program in natural history for teachers and the general public, not for research specialists. In this way he diverted funds and personnel from research; several of Boas' assistants, for example, had to drop their work and participate in the extension program. That was apparently the final straw for Boas.[23] As he explained his resignation to President Butler, the new director thought the Museum was to serve "the most elementary educational needs—a view with which I absolutely disagree. . . . Personally I do not believe that the policy of subordinating all scientific work under 'kindergarten work' can be carried out for any length of time."[24]

Boas won some modest successes at Columbia, but only with much effort and political wizardry. From the first he struggled for an autonomous department, a broad-based program covering all fields of anthropology, adequate research funds, and the like. The Columbia trustees authorized a separate department of anthropology in 1899. Boas then tried to staff it. He had psychologist Livingston Farrand, who was teaching anthropology courses, shifted to the department from psychology in 1903. When his proposals required new funds, he ran into difficulties. He asked for a graduate assistant in 1903. President Butler demurred, pleading insufficient monies. Boas then asked to have Clark Wissler, a recent Ph.D., serve as assistant without pay, while Boas attempted to raise the endowment for the position from outside sources. Butler agreed. Boas then used the same strategy to engineer the appointments of East Asian specialist Berthold Laufer and Central American archeologist Adolf Bandelier. For one year Laufer and Bandelier lectured without pay at Columbia while Boas sought endowment funds for their salaries. Unfortunately Boas did not raise quite enough money, so that he had to take a twenty percent reduction in his own salary to fund the positions.[25] Boas also introduced new courses and suggested others as a gambit to add new personnel, but far more effective was raising endowments himself.[26] Boas' department was always understaffed, and his burdens were heavy. As late as 1916 he taught over thirteen contact hours a week, supervised all graduate work, and handled all matters of departmental administration.[27]

Boas took great pains to prepare his graduate students, knowing he would have to train most of the next generation of American anthropologists. He was a rigorous and austere mentor, partly because of his own brilliance, and partly because he feared that the North American Indian cultures would be totally contaminated by white contact before professional anthropologists could record them. From the graduate student's point of view, working with Boas was a taxing experience not designed for the dilettante or the faintly curious. Customarily he registered beginning students in highly specialized courses for which they had little background. He plunged tyros into the icy waters of his course in statistical measurements for which they were usually ill-prepared and in which they floundered and failed at least once. In his ethnology seminar, Boas assigned ethnological literature without regard for the students' linguistic competence. Such a Draconian regimen had its disorienting effects. Thus, Robert H. Lowie entered the program with an undergraduate degree in chemistry. Boas made him take advanced courses in statistics and North American Indian linguistics instead of the beginning course in North American Indian cultures. Lowie's first year in graduate school left him totally befuddled; it was only by repeating all the courses the second year that "I gained some glimmerings of understanding."[28] Boas' indifference to his students' problems was legendary. He forced them to become independent, self-motivated specialists and researchers, and to teach themselves the various logical, mathematical, and linguistic skills necessary for anthropological research. Nor could he be fairly accused of training mere disciples who followed the master's way. He did push most of his early students toward research on North American Indian tribes, but chiefly because he wanted those cultures preserved before they vanished entirely. He did not assign dissertation topics. Students submitted their proposals to him for criticism and approval. He did insist that they explain cultural phenomena on cultural terms, not on analogies with the natural sciences. And he gave them for a working definition of culture that of the great English ethnologist E. B. Tylor. But he went no further. He and his students often disagreed about interpretations of phenomena; for example, he and Edward Sapir argued over linguistic data for more than a decade, and Sapir emerged from the experience rather critical of Boas's abilities in linguistics.[29] In other words, Boas did not create a school of ardent disciples, except perhaps methodologically. By the early 1910s he had trained Alfred L. Kroeber, Robert H. Lowie, Edward Sapir, and Alexander A. Goldenweiser, and had influenced several working anthropologists, including Clark Wissler, Elsie Clews Parsons, Wilson D. Wallis, and Roland B. Dixon.[30]

Boas also participated in the affairs of anthropological societies, in the hope of imposing the standards of university culture upon his science. At first his efforts met with mixed results. Most members of anthropological societies had neither academic affiliations nor a commitment to a career in specialized research. Yet these individuals were vital to the financial health of anthropological societies and journals. The number of full time anthropologists (let alone those with genuine academic credentials in anthropology) was too limited to support even one society, let alone a journal. Boas wanted any national society restricted to "professionals," regardless of financial considerations.

Consider his role in the formation of the American Anthropological Association. W J McGee, director of the Bureau of American Ethnology, and Powell's successor in Washington anthropology, took the lead in organizing the Association by writing several leading, full time anthropologists in late 1901, including Boas. He even included a proposed constitution and by-laws. According to McGee's provisions, persons without a full time commitment to anthropology would be admitted as members on a par with the professionals.[31] Boas vigorously dissented. If "the national society is established on the proposed basis, we may endanger the permanent interests of science by yielding to our temporary financial needs. . . . what is most urgently needed is a national society of anthropologists in which the amateur element is rigidly excluded." Only persons who "have contributed to the advancement of anthropology either by publication or by high-grade teaching" should be eligible for membership, Boas insisted.[32] McGee outmaneuvered Boas and persuaded his other correspondents to accept his, not Boas', membership criteria.[33] For several years Boas remained cool to the new society.[34] After 1910 he could influence the Association's affairs more, because his students were becoming important figures in the Association. They came to control the Association's journal, the *American Anthropologist,* taking over the book reviews in 1912 and the editorship in 1916, and thus elevating the journal to what they thought were academic standards.[35]

Boas worked in many ways to promote universalistic standards of training, competence, and performance in anthropology through anthropological societies. As a corresponding member of virtually all important European anthropological academies and societies he thought it his duty to articulate cosmopolitan criteria among American anthropologists. Consider his influence on the American Folklore Society, exercised through Kroeber. Kroeber became president in 1906. Boas then worked with him to strengthen it organizationally and financially. Boas also urged Kroeber to upgrade the Society's *Journal of American Folklore,* especially to restrict it to longer and more substantial articles. "It may be that our constituency won't stand them," he conceded to Kroeber, "but it is worth trying."[36]

Boas also worked to create a predictable and impersonal institutional structure for the support of anthropological research, in particular, field work, translators, and publications. He hoped eventually anthropologists would control their own research resources. For many years, however, such research resources as were available Boas cultivated, often ingeniously, as the consequence of personal arrangements he made with potential patrons of anthropological work. For example, he turned his appointment as honorary philologist for the Bureau of American Ethnology into an instrument for graduate student support for almost two decades. As philologist he commissioned students to collect linguistic data on selected tribes; the students in turn gathered ethnological data on the same tribes for their dissertations. The Bureau then reimbursed the students' expenses. Boas even tried, on at least one occasion, to have the Bureau's funds increased. In 1903 he asked President Butler, a Republican influential with the Roosevelt administration, to have Roosevelt increase the Bureau's appropriations. He argued that the Bureau's personnel would benefit by the addition of academically trained anthropologists, and that the Bureau could become an important agency for suggesting solutions to current social problems. "I believe that a number of important practical questions might be solved by the extension of the work of the Bureau . . . particularly if. . . it could include in its investigations the white and negro races."[37] Nothing came of Boas' suggestion. Nor did such institutions as the Carnegie Institution of Washington nor the Eugenics Record Office accept his ambitious proposals to study ethnological and racial problems, even though he carefully outlined their practical social applications.[38] Before the 1920s his most spectacular triumph in securing outside funding was his study of recent immigrants for the United States Immigration Commission, conducted between 1909 and 1911. This gave him the opportunity to refine his work in physical anthropology, to keep several graduate students busy collecting anthropometric data, and to fashion a sophisticated critique of scientific racism. But it was a temporary investigation that led to no permanent research institute.[39] Through personal solicitations he pieced together an endowment fund for research at Columbia, which yielded approximately

$1900 annually in 1916 and over $25,000 yearly in the early 1930s. But he had no success outside the University, except as noted, until the Depression.[40]

By the late 1910s Boas had left his mark on American anthropology, or, more precisely, on American anthropological institutions and anthropologists. He had created an important center of graduate training. He had produced a small cadre of academic anthropologists. He and his students had begun to influence other anthropologists, and anthropological institutions, not always with positive results, as we shall see. Above all, Boas stood for a vigorous redefinition of the anthropologist as an academic scientist, a member of the new university culture of America, a specialized researcher. It was appropriate that Boas' students, as they were launching their careers after 1910, insisted that anthropology was an autonomous science with its own determinants, methods, and phenomena, which no natural scientist, no matter what his or her accomplishments, could reduce to the level of the modes of explanation of evolutionary natural science.

## III

"Admitting without reserve the influence of physical environment and biological factors, I do not however see how social or historical phenomena can be accounted for without consideration also of social and historical—in other words, purely human—causes," Kroeber wrote Ellsworth Huntington in 1913. Huntington, a Yale physical geographer, eugenist, and publicist of the evolutionary point of view in anthropology, was clearly perplexed by Kroeber's recent publications. In particular, Huntington was puzzled by Kroeber's seeming refusal to be "scientific," by his insistence that the determinants of the natural sciences not be used to explain human behavior. So he wrote Kroeber: Was this not a flagrant and unscientific denial of the great principle of evolution, he asked, to reject the assumption that man is an animal? Kroeber replied that he, too, believed in evolution and in man's animal ancestry. Kroeber argued that biological laws did not explain everything about man. Man's cultural life was not influenced by his biological or psychological make-up; it constituted a wholly different order of existence. Man's cultural life arose from his unique brain and the unique cultural life that the brain permitted the human species. Culture and nature were distinct and separable levels of phenomena—even in the same species, man.[41]

Thus Kroeber and Huntington defined the heredity-environment's central theoretical issue: was evolutionary natural science capable of explaining all human behavior? Kroeber did not, it should be understood, deny the evolution theory's validity within its own jurisdictional boundaries. Indeed, he affirmed it. But he did rule out its applicability to cultural phenomena. One of the reasons the culture idea became such an effective weapon for social scientists in the heredity-environment controversy was because they reconciled it with the theory of evolution.

The origins of the culture idea as the Boasians advanced it can be easily traced. They commonly used E. B. Tylor's simple, enumerative definition of culture as all the habits and capacities of men acquired as the result of living in society. Boas had met the famous English ethnologist in Europe before settling in the United States, was impressed by him, and quite obviously appropriated his definition of culture.[42] This sense of culture was evident in Boas' earliest anthropological work, and certainly in his distinction between physical and cultural anthropology.[43] Boas passed this definition of culture on to his graduate students; Kroeber used it in his earliest research proposals as a member of the faculty at California, for example, and so did Lowie in his comments about his field work as a graduate student.[44]

Why did Boas' students begin their campaign to define and disseminate the idea of culture around 1910? Why did they continue their campaign throughout the next decade? It was only after 1910 that enough of them had earned their doctorates, begun their careers, and entered into a new stage of professionalization. Most of their more than ninety works published before then had been empirical, descriptive pieces intended to gather information, not to enunciate grand theories. After 1910 the Boasians turned to making theoretical and methodological statements. They addressed their remarks but to scientists in other disciplines, particularly natural scientists. They were engaged in the kind of activity that the distinguished historian of science Thomas S. Kuhn has noted is often associated with establishing a profession for a particular discipline and with articulating its central "paradigm." Boas' students knew perfectly well that most natural scientists did not consider cultural anthropology a worthy or respectable science. As Kroeber put it rather bitterly in a letter to Boas in 1919, biologists and psychologists "simply do not see what culture is, or in other cases while they admit it logically they have never got [sic] a real feeling for its existence. Consequently the sense always crops up in their minds that we are doing something vain and unscientific, and that if only they could have our jobs they could do our work for us much better."[45] Kroeber described a prevailing attitude among natural and physical scientists. For example, Robert S. Woodward, a physicist and President of the Carnegie Institution of Washington, told one of Boas' best friends that the Institution could not support research in anthropology because American anthropological work was "trivial" and "commonplace." Anthropologists did not deserve support because of "the diversity of opinion, and the petulance and impatience displayed by the numerous parties and interests."[46]

The Boasians' post-1909 campaign to announce the culture theory was the consequence of anthropology's development as an academic science and the contempt many other scientists held toward it. The Boasians emphasized anthropology's jurisdictional boundaries in social explanation, and its legitimacy within the university and the scientific community at large. In their several dozen statements of the culture idea before 1920 they attacked those who would preempt cultural anthropology and reduce it to a rigid natural science level: the evolutionary anthropologists, the physical geographers, and the new post-Mendelian psychobiologists.

The evolutionary anthropologists adopted the Neo-Lamarckian unilinear pattern of cultural evolution, according to which certain peoples succeeded or failed at rising through the universal stages of primitivism, barbarism, and civilization. This perspective represented a most explicit attempt to base social explanation upon evolutionary analogies. After 1900 most natural scientists abandoned it for the newer models of experimental biology and psychology. But many of the older generation of anthropologists, particularly the nonacademic ones centered in Washington, still embraced it. So Boas' students attacked the evolutionary anthropologists' unilinear scheme, with its implicit distinction between "irrational" primitive peoples and "rational" civilized peoples. In a summary of his dissertation, for example, Goldenweiser insisted in 1910 that totemism could not be explained by so simple a scheme.[47] Kroeber criticized the evolutionary anthropologists' assumption that morals had evolved, from primitives to civilized man, and those of civilized man were superior to those further down the evolutionary scale. Such analogies were false, misleading, and inappropriate; society did not evolve according to the presumed laws of biology, he insisted. Biology and anthropology were different sciences with their distinct order of phenomena.[48] And Robert H. Lowie argued in a paper before the American Folklore Society's 1911 meeting that ethnological phenomena could not be interpreted with false analogies borrowed from the biological sciences, thus making the same point as Kroeber.[49]

The Boasians spent even more energy marking off anthropology's jurisdictional boundaries from physical geography. Commonly they argued that geographical factors could not directly account for the existance of many different cultural systems in the same physical setting. At most the physical environment was merely a passive factor permitting many different cultural systems to flourish. "The geographical factor seems to be important only in that it offers barriers of a broad sort to a given culture," Wissler declared before the 1911 meeting of the American Anthropological Association; in most respects it was a "negligible factor."[50] Sapir summarized his dissertation research on linguistics by saying that we must "postulate an absolute lack of correlation between physical and social environment and phonetic systems, either in their general acoustical aspect or in regard to the distribution of particular phonetic elements."[51] From slightly different perspectives Goldenweiser and Wissler argued that the physical environment was static, whereas cultures were dynamic, within historical epochs; different cultures made various uses of the same environment. "All cultures, finally, are historical complexes," Goldenweiser concluded. "Every culture combines with traits that have originated within its own borders, other traits that have come from without, from other cultures, and have become amalgamated with the recipient culture . . . as a historical complex, every culture is largely independent of its environment."[52]

The Boasians also differentiated ethnological and psychological phenomena. They argued that psychologists studied the individual's innate and acquired responses to his environment, whereas anthropologists dealt with the group's patterns of response to its environment. Cultural patterns were essentially historically determined, and each culture had a distinct history. Mental talents were most probably distributed equally among all "racial" populations; peoples worked out their own specific behavior patterns in response to their own historical experiences. Clark Wissler, who had taken his doctorate in psychology, drew a fundamental distinction between cultural and psychological levels in man in a paper he delivered before the Anthropological Society of Washington in 1913. Culture was external to the individual mind and was carried on by learning processes. Cultures developed and had a historical development of their own. Since cultures (and culture) were not biologically inherited, he insisted, they could not be considered psychobiological in character. The individual mind was in a psychophysical sense the common property of all human groups, and "it enables any individual to practice any culture he wishes—although not with equal facility."[53] No Boasians denied psychology's ability to explain the workings of the individual mind. But they insisted it could not account for cultural phenomena. Wissler consistently argued that anthropologists studied the acquired activity complexes of human groups, whereas psychologists studied the individual's innate psychological equipment. Neither should intrude upon the other's domain, he declared. "It is clear that when we are dealing with phenomena that belong to original nature we are quite right in using psychological and biological methods; but the moment we step over into cultural phenomena we must recognize its historical nature," he wrote.[54] Before a joint meeting of the American Psychological Association and the American Anthropological Association in 1919, Wissler again drew the distinction between psychology and anthropology, "The psychologist deals with what goes on within the individual when confronted by the group and the environment, while the anthropologist gives his attention to what goes on in the group when confronted by other groups and environments."[55]

The Boasians' differentiation between the cultural and the psychobiological levels of existence implied a further theoretical distinction between the individual and the group. Kroeber developed this point more systematically than any other Boasian. In so doing he suggested the distinction between culture and personality. In several papers he aggressively announced the divorce of biological and social theory and its theoretical implications. In 1915, for example, he declared the determinants and methods of the natural sciences had no cultural or historical significance, "just as the results and the manner of operation of history are disregarded by consistent biological practice."[56] In 1917 Kroeber went further: "The distinction between animal and man which counts," he declared, "is not that of the physical and the mental, which is one of relative degree, but that of the organic and the social, which is one of kind."[57] He insisted that anthropologists could ignore the individual in cultural processes and concentrate on the patterns of behavior generic to the whole culture. Several Boasians took Kroeber to task for this latter idea. He replied that there "is something in culture which can be viewed without reference to the individual, something that is left when the individual is totally subtracted. My whole point is, that culture history proper can begin only after this subtraction is made, and that is only after we can profitably begin to connect its findings with any study that brings in the individual."[58] Yet despite this particular disagreement, the Boasians agreed on the general outlines of the culture idea.[59] As we shall see later, it could be reconciled with new ideas in evolutionary theory.


## IV

Well before the United States intervened in the World War in the spring of 1917 the Boasians had worked out the culture idea, at least in its negative meanings, and has disseminated it in various scientific forums. Before intervention they were busily engaged in building a paradigm, the culture concept, as the consequence of establishing anthropology in American science and higher education. Intervention did not interrupt these activities, nor did it alter the intellectual and theoretical messages they sought to bring before the larger scientific community. But intervention, and the political and social changes following it, drastically transformed the political, social, and emotional context in which the Boasians operated as citizens and as scientists.

As the Boasians' correspondence suggests, intervention and its consequences put them on the defensive as scientists and as citizens. The anti-German, anti-immigrant, anti-radical, anti-pacifist emotions aroused by the war, the activities of the Wilson administration and its self-appointed vigilante crusaders against German culture and the left, created a new and supercharged political and emotional atmosphere, forcing the Boasians to re-think their commitments to American society and American science. They had chafed for years at being regarded by natural and physical scientists as "unscientific" dilettantes, and they had resented the fact that their proposals for racial research had been consistently turned down by governmental and philanthropic administrators. But now a new threat, of preemption, appeared. Quite possibly those natural and physical scientists who held their work in contempt might conduct "anthropological" studies, with government backing, from a "hard" biological and racial per-spective. The Boasians noted in particular that certain eugenists, who also supported a war the Boasians privately detested, seemed favored to receive large grants to conduct such questionable racial studies. The Boasians concluded that their very identity as citizens and as scientists was threatened by this new turn of events. Accordingly they responded with new emotional intensity to national politics and the politics of science.

It was the politics of science, not national politics, that goaded the Boasians to action. Among the Boasians only Boas made his stances on public affairs widely known, in scientific forums and in the newspapers. His actions aroused resentment among pro-Allied, propreparedness scientists, even those who respected his scientific work. For example, the National Academy of Sciences established the National Research Council in 1916 to mobilize science and technology for the war. Several of Boas' professional opponents wrote George Ellery Hale, chairman of the NRC, arguing that a "disloyal" person such as Boas should not be in-cluded in the NRC.[60] "Frantz [sic] Boas has a letter in this morn-ing's *Times* which carried independence of thought to the point of sedition," wrote Princeton biologist Edwin G. Conklin to Hale in February, 1917. "It will be really unfortunate if at this time he must be included in any Committee of the National Research Council. He ought to be in Germany and not in America and I should like to help send him there."[61] Boas was so upset by

Wilson's foreign policy and the prowar atmosphere at Columbia that less than two weeks before intervention he was ready to emigrate abroad.[62] Boas could not accept Kroeber's advice, tendered in August, 1917, that war "is coercion, which must strike in as out. You would only be swimming against an impossible torrent, with no good to yourself or anyone else."[63] Boas felt impelled to speak out.

The issues of intervention and academic freedom came home to Boas at Columbia. The trustees and President Butler began to inquire into the loyalty of professors. In his June 1917 commencement address, Butler announced the "suspension" of academic freedom at Columbia for the war's duration. That enraged Boas, and his old friend, the psychologist James McKeen Cattell, with serious consequences for both. In August, Cattell sent a petition on university stationary to three Congressmen asking they not vote for legislation authorizing the presence of conscripts in combat zones. The Congressmen notified Butler. Butler and Cattell had been on poisonous terms for years; Butler took the matter to the trustees in the fall. They summarily dismissed Cattell at their October meeting, denied him his pension, and erroneously stated the faculty supported the decision.[64] The Cattell case affected Boas profoundly. He threw himself into organizing a campaign to have Cattell reinstated, but to no avail, although he did manage to help Cattell retain his several scientific editorships.[65] "I have not the happy faculty that you have to detach myself entirely from the events of the day," Boas wrote Lowie in December. "I feel them keenly. The present prosecution craze has upset me completely. In the Cattell case it is aggravated by the cowardice of the attack. . . . What I object to is the despotic way of managing so-called scientific institutions and the outrageous dismissal of men without any hearing."[66] Cattell's actions cost him his professorship and, for some years, his pension.

Among the Boasians, then, Boas was singular in his *public* protests against American foreign policy and tests of loyalty imposed upon academics and scientists. The differences between Boas and his followers were largely temperamental. Kroeber, for example, was German-American. He did not speak out on the war, which he hated, because of his rather easy-going, phlegmatic personality.[67] Lowie was also appalled by the war, and he felt his socialist political views as keenly as did Boas. Yet Lowie was quite detached from the rush of contemporary affairs. He was chiefly concerned with his science and its politics.[68] In varying degrees this was true of the other Boasians. Sapir, a native of Germany, was then working in Canada and was almost totally immersed in his linguistic work. Goldenweiser, a Russian immigrant who taught part-time at the leftist Rand School for Social Science, shared Boas' general views on public affairs, but chose to concentrate his energies on science. Of the native-born Boasians—Pliny E. Goddard, Clark Wissler, Wilson D. Wallis, and Elsie Clews Parsons—only Parsons made her political views public, but she was more interested in feminism than foreign policy. Wissler quietly supported the American war effort; Goddard and Wallis, on the other hand, remained apolitical. With the exception of Wissler, the Boasians agreed with Boas on public issues, but they did not feel so strongly to make public statements or throw themselves in the midst of political controversies over foreign policy or academic freedom.[69]

The politics of science animated all the Boasians far more deeply than did contemporary affairs. For years they had known that many natural and physical scientists did not consider cultural anthropology a worthwhile science.[70] They realized these scientists distrusted both their methods and their emphasis on the cultural rather than the biological side of man, and that this cost them dearly in the stiff competition for support of scientific research. They also understood the work they had done in physical anthropology earned them little respect among those natural and physical scientists because of its political and racial implications.[71] Thus Boas' rigorous work on the inheritance of head-forms for the United States Immigration Commission unsettled many natural scientists because Boas concluded that environment influenced head-form, and, perhaps, general cultural capability. Boas' views on the equipotentiality of human races were well known, and his blistering attack on eugenics, published in 1916, probably alienated those natural and physical scientists who cherished eugenics and other hereditarian nostrums.[72] Before intervention, occasionally one or another of the Boasians complained about eugenics or the hereditarian interpretation of racial mental testing.[73] But generally speaking, prior to intervention they kept their counsel; their feelings of animosity were there, under the surface.

Intervention changed the politics of American science no less than national politics. Intervention held out the prospect, through the mobilization of scientists and engineers for the war effort, of increased federal and philanthropic support for scientific research and the promotion of favorable public attitudes toward scientific research. At the same time, many scientists believed that only scientists of unquestionable loyalty should participate in the new government-scientific alliance. For our purposes, the major clashes in the politics of science revolved around the National Research Council and the Army's mental classification program, because in both instances the ultimate issue was whether the Boasians or scientists sympathetic to "hard" racial anthropology would conduct important ethnological studies possessing the government's imprimatur and large-scale funding.

The National Research Council was organized after the *Sussex* incident in March 1916, as an agency of the National Academy of Sciences. For all practical purposes the brilliant astrophysicist George Ellery Hale founded the Council. Hale was more than a highly capable scientist; he also promoted scientific institutions, much like Boas, except far more successfully. He wanted support for scientific research to come from the federal government, with the Council operating as a kind of broker between taxpayers and scientists; hopefully, too, Council members would coordinate scientific research, point to areas where further work was necessary, and, in the largest sense, use scientific and technological research to aid economic growth and national power. In September 1916, the Council began operations with a small fund donated by the Engineering Foundation of New York for operational expenses. The Council was to evaluate research proposals and to attempt to raise funds for those deemed necessary to the war effort.[74] If industrial research chiefly interested Hale, he was nevertheless willing to entertain proposals for "anthropological" studies. But those studies, for Hale, were worthwhile most probably if they followed the "hard" biological approach. Hale supported the eugenics movement, and even belonged to the Save the Redwoods League, a small California-based eugenics and conservation group. His tutor in matters anthropological was the distinguished ancient historian James H. Breasted, a dear friend and former colleague at the University of Chicago, who embraced the evolutionary point of view in anthropology that the Boasians so firmly opposed. And Hale was a conservative, Anglophile, prointervention, nationalistic Republican.[75] Small wonder he had some doubts about the Boasians.

The Council began operations in the fall of 1916 with Hale as chairman. He received much advice over Council policies, sometimes unsolicited. The question arose immediately as to which scientists from each recognized discipline to appoint to the Council's various divisions. This was an important issue, for those men would evaluate research proposals and help determine funding priorities. A Washington anthropologist—and bitter enemy of Boas'—wrote Hale on several occasions urging that Boas and his followers not be appointed to the Council's division of anthropology and psychology. He claimed the Boasians ignored biological anthropology for unscientific cultural anthropology, and he implied they were disloyal to the war effort.[76] Hale appointed Boas anyway, probably because Boas had too much scientific stature to be excluded. But Hale also selected men whom Boas regarded as having little or no anthropological expertise. Boas found Council divisional meetings most frustrating; proposals to study racial characteristics from a narrowly biological point of view received serious, extended discussion, whereas his suggestions were quietly shelved. Hale was much more interested in biological than cultural studies. Thus the eugenist Charles B. Davenport received warm encouragement and definite support to study the body build of Army recruits as a way to discover the "weak germ plasm" among immigrants in the population, and therefore protect American culture from disorder.[77]

Shortly after the Armistice, in December 1918, the growing hostility against Boasian anthropology and Franz Boas in particular among racially-minded natural and physical scientists took on a new organizational form with the founding of the Galton Society in New York. Important among the Society's charter members were prominent eugenists such as Charles Davenport, New York society lawyer Madison Grant, the famous publicist of racist theories, and several well-known natural scientists, notably Princeton cytologist Edwin G. Conklin and California paleontologist John C. Merriam, who succeeded Hale as chairman of the Council; none were anthropologists. These men wanted to encourage and support racially oriented anthropological studies. They were deeply opposed to the Boasians' work, and the Boasians, for that matter. As Grant explained the Society to Henry Fairfield Osborn, distinguished paleontologist and president of the American Museum of Natural History, the new organization would be an anthropological society in New York "with a central governing body, self elected and self perpetuating, and very limited in members, and also confined to native Americans, who are anthropologically socially [sic] and politically sound, no Bolsheviki need apply."[78] Members of the Galton Society met monthly, at fashionable New York clubs, to hear various lectures on man's biological traits and evolution. And they conferred about possible research projects that the Research Council might presumably fund. In March 1919, the Carnegie Corporation, which supported Davenport's work in breeding and eugenics, gave the Council five million dollars for a building and a permanent endowment. This paved the way for a peacetime Council that would fund all sorts of "desirable" research projects, including research into racial differences. Boas soon found that new proposals for such studies surfaced in the Council's divisional meetings.

Nor was this all. Several Galton Society members began a fund-raising campaign for a new anthropological journal, because, they complained, the Boasians' work dominated current anthropological journals. They believed the time appropriate for a new journal that would publish biological studies. Washington anthropologist Aleš Hrdlička, curator of physical anthropology at the National Museum, was to be the editor. By March 1919, Boas realized that the Galton Society might be able to establish such a journal, which would be little more, in his opinion, than a means of propaganda for scientific racism and eugenics. He was not opposed to Hrdlička's journal, or even to Hrdlička as editor, but he feared that men such as Davenport would sit on the editorial board and determine policy. He wrote Conklin, who was influential in the Galton Society a long letter about the whole issue of biological versus cultural anthropology and the specific issues in the Council and Hrdlička's journal. He complained that Hale had pointedly excluded him from many important Council meetings, and had invited men to participate whose "relation to anthropology is exceedingly slight." It was not true, furthermore, that biological anthropology had been slighted. He had tried many times to have work in physical anthropology supported, only to be rebuffed. Now he worked in cultural anthropology, for which he could get support. He complained that those who criticized him wanted "to throttle the kind of work we are doing." Furthermore, "the blame does not rest on anthropologists, but on the indifference of those upon whose financial support we have to rely." Boas implored Conklin to have the Galton Society fund Hrdlička's journal, with no strings attached, until it became financially independent; he argued this would be a test of the Society's sincerity about scientific anthropology.[79] Two days later Conklin told Boas the Society would help underwrite Hrdlička's journal.[80] Subsequently Boas discovered Davenport was a candidate for the editorial board; he asked Conklin to serve instead. Both Davenport and Conklin were appointed.[81] Boas was not pleased.

By the summer of 1919 Boas was becoming embittered. Over the last decade his grandiose plans for the promotion of anthropology had yielded few concrete results.[82] Post-intervention politics, public and scientific, depressed him deeply; his hopes for a more liberal and pluralistic America, uplifted by cultural anthropology, seemed in ashes. And in particular the politics of science bothered all the Boasians, not merely Boas, for they felt threatened, not merely by hostility or contempt, as had been the case before intervention, but by preemption of their whole science by men whose scientific credentials and political views impressed them as highly questionable. In August 1919, Boas complained to Sapir about the politics of American science as he saw them. Scientific administrators did not understand what anthropology was. They wanted to be shown immediate practical results. "It is hopeless to try to show . . . them. . . the purposes of linguistic inquiry, as well as of ethnological inquiry, although I am of the opinion that a mastery of the principles of ethnology might furnish a most wholesome background for the reconstruction of the political and social life of our day." In the Council's division of anthropology and psychology, he continued, what was desirable was the study of the population from the point of view of public hygiene and general racial health. This division was under the strongest pressure to be "entirely a department of research in heredity and other demographic features." Boas did not oppose such work if well done, but even then he feared that it might divert support for cultural anthropology.[83]

Boas and his supporters also found cold comfort in the Army testing program begun in 1917. They knew that the man responsible for the program, Harvard psychologist Robert M. Yerkes, was an able scientist, but a man who supported the eugenics movement for years, and that most of the other scientists in the program were similarly inclined. The Boasians were not opposed to mental measurement as such. But they were profoundly disturbed by the racial interpretations of mental measurement data that many psychologists were currently publishing in psychology and education journals. By 1919 the Army's program had ended, and Yerkes had not yet disseminated its results. But the Boasians prepared themselves for an interpretation of the test results that would challenge the implications of all their work in cultural anthropology. By the late summer and early fall of 1919, then, there were complex pressures on the Boasians from several directions—political, cultural, and scientific—that made them feel as if their work was in serious danger of being preempted, and, furthermore, that the government's approval was being given, however indirectly, to precisely the type of "hard" biological anthropology they had devoted their scientific careers to attacking.

On 20 December 1919, *The Nation* published an angry letter from Boas that brought out into the open the accumulating tensions between the Boasians and the much larger group of propreparedness, probiological anthropology natural and physical scientists, together with those anthropologists who despised the Boasians' theories and their academic orientation. In his letter, Boas charged that the Wilson administration had persuaded four government employed anthropologists to spy for the United States while doing fieldwork in Mexico immediately prior to American intervention. Boas charged this was a flagrant prostitution of the scientist's role in society. The scientist should be above the behavior of the common spy or politician, he declared. He angrily insisted that President Wilson's war goals and rhetoric were blatantly hypocritical. Wilson spoke grandly of freedom and suppressed it at home to wage a savage war.

Ten days later, at the meeting of the American Anthropological Association in Cambridge, a furious debate broke out between Boas and his allies and those anthropologists who had opposed Boas for various personal, professional, and scientific reasons. Those who were threatened by the rise of Boasian anthropology, by its explicitly academic orientation, or who were offended by Boas' stands on public policy, all joined together to move for Boas' censure. After protracted and acrimonious debate, they passed a resolution in the Association's Council that reprimanded Boas for his letter and expelled him from the Council, thus threatening him with expulsion from the Association and forcing him to resign from the NRC.[84] One of his enemies, a Washington anthropologist, wrote after the incident that "Boas and his followers have sought to gain control of and direct anthropological researches in this country, especially those undertaken on behalf of the nation. There appears now the danger that the Research Council may give this group control of researches and measures affected the most intimate and vital interests of the nation—interests which should be entrusted to those, and those only, whose standards of citizenship is wholly above criticism."[85]

Several of Boas' enemies in the scientific community attempted to impose further punishment upon Boas in the wake of the flare-up at Cambridge, testimony, certainly, to the tensions of the moment. Charles D. Walcott, geologist and Secretary of the Smithsonian Institution, in particular moved against Boas. Boas and Walcott had been on unfriendly terms for years, largely over the direction of policy of the Bureau of American Ethnology (a part of the Smithsonian), over which Walcott had much influence. And Walcott was a scientist-patriot instrumental in establishing the Research Council; his views on public affairs were similar to Hale's. Walcott thus engaged in a letter writing campaign to discredit Boas further. He transmitted his discontents to a Columbia colleague of Boas', the Hungarian-American Michael Pupin, a physicist who shared his distaste for Boas. Pupin replied that Boas "attacks the United States for the purposes of defending Germany, and yet he is allowed to teach our youth and enjoy the honors of being member [sic] of the National Academy of Sciences. This thought makes me long for the good old days of absolutism where the means were always at hand for ridding one selfe [sic] of such a nuisance as Franz Boas."[86] Walcott also wrote President Butler about Boas. He implied that Boas should be dismissed on the grounds of disloyalty. Walcott even wrote United States Attorney General A. Mitchell Palmer about Boas' transgressions. Fortunately for Boas, Palmer had larger prizes to capture than a professor of anthropology. Palmer referred the matter to the Justice Department's Bureau of Investigation (the forerunner of the Federal Bureau of Investigation) where it was lost in the chaos of post-Armistice Bureau activities.[87]

This was the context in which the Boasians shifted from publishing their statements of the culture theory to making attacks upon biopsychological, hereditarian social theories. Doubtless the events of the post-intervention years angered them, and pushed them into action to cry out against theories and interpretations that were an affront to their science—its validity, its autonomy, its politics—and to their culture. Since history deals with interdependent, not independent, variables, it cannot be said that the Boasians' professionalizing and theorizing activities before intervention were more or less important than the politics of science and of America after intervention in goading them to action. The Boasians' attempts to define the culture idea, and to promote their science (all of which occurred before intervention) obligated them to criticize, at some point, the hereditarian explanations of human behavior natural scientists offered. And clearly, too, the events in the wake of intervention, which polarized the scientific community and American culture, lent a new emotional urgency to the issues of race, cultural evolution, and ethnicity, and thus helped precipitate the heredity-environment controversy.

# 4 SOCIOLOGY AND CULTURE

*I*

SOCIOLOGY is at last shaking itself free from biological dominance and is developing an objective and a method of its own," wrote Luther Lee Bernard in 1923. "Thus it promises to be a science, not merely a poorly organized and presumptuous branch of biology, as some biologists formerly seemed to regard it."[1] This young University of Minnesota sociologist correctly estimated recent changes in American sociology. American sociologists were discovering that they must purge sociology of all natural science determinants, analogies, metaphors, and levels of explanation, if they wanted their discipline to become autonomous in modern American university culture. By the time Bernard wrote, most American sociologists understood the full meaning of the Boasian dictum *omnis cultura ex cultura.* Of course, this intellectual revolution in American sociology had important consequences for the scientific discussion of nature and nurture; it drew American sociologists into alliance with the Boasians and against the extreme hereditarian doctrines of contemporary natural scientists. American sociologists did not surrender belief in the doctrine of evolution as applied to organic phenomena or in the dazzling potential of an interdisciplinary science of man whose practitioners might predict and control human behavior. Nor did American sociologists necessarily abandon "neo-evolutionary" perspectives in cultural evolution. But since about 1920 American sociologists have consistently recognized the distinction between the biological and the cultural levels of human existence. They have accepted the idea that there does exist a qualitative difference between the minds of men and of brutes. They have embraced the essentials of the idea of culture, as this statement by sociologist William F. Ogburn in the 1940s indicates:

We consider social evolution as not including, at least in the past several thousand years, any biological evolution. What is evolving then is society or the various social groupings such as government, and industry, and the state. . . . What is evolving is 'culture', that is to say, the environment which men have, but which the wild animals do not have. This culture, or our social heritage, is a composite of many different parts, such as cities, families, farms, philosophies, art, science, etc.[2]

Broadly speaking, American anthropologists and sociologists "discovered" the culture idea—or, more precisely, discovered its necessity and utility for their disciplines and their professions—in the context of their efforts to establish their disciplines in modern American university culture. Yet the circumstances in which they labored differed radically, and they worked in virtual ignorance of each other's work for some time. From the first, the Boasians possessed a precise sense of their discipline's scope, its intellectual boundaries, its phenomena, its methods, its conceptual problems. They also identified with anthropology as full-time researchers and advanced teachers. Their commitment was not merely to an academic or scientific career, but to a lifetime of specialized research and teaching in anthropology. They responded to, and articulated as well, national criteria of training, competence, and performance in anthropology. They drew as bold a distinction as they dared between themselves as professionals in university culture and lay "amateurs." They developed anthropology as a graduate-oriented science. Anthropology remained a rather small discipline in American higher education; undergraduates and laymen did not flock to courses and lectures on ethnology, archeology, linguistics, and physical anthropology. Few nonanthropologists considered anthropology a science profession, with "practical" or "commercial" applications. And few opportunities for dazzling professional careers in anthropology existed. Anthropologists could muster little popular, political, philanthropic, or university support. Finally, the Boasians themselves were outside the progressive movement, and had little interest in most of the issues animating the native-born, middle class leaders of American reform.

American sociology's early history was quite different, having emerged in the later nineteenth century as more of a social reform movement in quasi-academic dress than a discipline or a science; for decades there was no sociological research based on prior achievements that could be used as the basis for further practice.[3] In contrast to anthropology, sociology enjoyed a remarkable institutional expansion in American higher education between 1890 and 1920, even though the most eminent pioneers of academic sociology did little specialized research, could not agree on what sociology stood for as a discipline, and were, in fact, best understood as the promoters of a subject rather than practitioners of it. And sociology prospered chiefly as an undergraduate offering in colleges and universities. Only the University of Chicago and Columbia University offered large-scale graduate work in sociology before the 1920s. If anthropology was a marginal subject of interest only to several dozen academic professionals and several hundred laymen, sociology was a "science" of consuming interest to many educated Americans who demanded courses and lectures and books and articles on "sociology." So popular was sociology by the 1920s that many secondary schools offered sociology courses.

Sociology's popularity among growing numbers of educated Americans may be attributed primarily to their deep interest in seeking "scientific" solutions to social problems. Outside the walls of the academy, citizen pressure groups—urban reformers, settlement house workers, Social Gospel ministers, social workers, and others—pressed colleges and universities to offer courses in social science, social welfare, and sociology. And America's first generation of academic sociologists lived up to these expectations and demands, for along that route lay both career opportunities and the possibilities of generating social reform. America's pioneer academic sociologists identified, not with sociology as a well defined science, but with a commitment to an academic career and to the promotion of "applied" sociology. Pioneering academic sociologists were both an emerging professoriate and a cabal for reform; they thought of these two roles as congruent. Each sociologist stressed these two goals somewhat differently, depending on personality and circumstances. "We older men who have assumed risks and taken punishment in order to procure what measure of academic freedom exists today for sociology must not let the oncoming generation devise plausible excuses for side-stepping all responsibility," wrote Edward A. Ross, perhaps the most politically outspoken pioneer sociologist in the early 1930s. "Raised as I was on Lester F. Ward I don't give a snap of my finger for a sociologist who wouldn't have an attitude on any of the social problems of his time. . . . A lot of our younger men don't want to take risks by assuming an attitude toward contemporary evils which have powerful support; so they duck and justify their ducking by these conceptions of 'science'."[4] "Ducking" contemporary issues was precisely what most pioneers of sociology could not do.

The first generation of academic sociologists, then, must be understood as both champions of social reform and of sociology as an academic subject in American colleges and universities. This duality helps explain the contrast between their stunning institutional successes and their tortured intellectual groping for a theoretical framework for sociology. From the 1890s to the 1910s sociology grew so rapidly, and those who became sociologists had so little conception of what sociology was, that even the founding fathers did not at first fully grasp that they were obligated by their roles as academics to divorce sociology intellectually from the natural sciences. Just as the circumstances of American anthropology encouraged the Boasians to make anthropology into a research-oriented science and profession in university culture, and heighten their awareness of the culture idea's manifest necessity, so the context of American sociology kept pioneer academic sociologists from understanding the full importance of the dictum *omnis cultura ex cultura*. Eventually American sociologists grasped the point, but this was only after sociology had become a roaring institutional success in American higher education.

## II

Since the 1870s, a growing number of colleges and universities had offered courses in social science, social ethics, meliorist social science, philosophy of history, Herbert Spencer's political economy, or some other such course addressed to the possibility of scientific social reform. Persons with dissimilar backgrounds and occupations taught these courses: ministers, social workers, economists, political economists, and even philosophers. Starting in the late 1880s, social science courses began to yield to courses with the designation "sociology" in their titles. Between 1889 and 1900, ninety-seven colleges and universities inaugurated courses called sociology, and in the next decade at least another three hundred institutions followed suit.[5]

Most "sociology" courses inaugurated before World War I were undergraduate offerings emphasizing such social problems as crime, vice, delinquency, and dependency. And sociology grew chiefly at certain types of institutions, most spectacularly land-grant institutions, and commonly denominational colleges. Most prestigious private universities with strong scientific graduate programs were cool toward sociology. For example, Harvard did not start a separate department of sociology until 1931, and Cornell permitted only rural sociology, and that through its extension program. Other major private universities, such as Johns Hopkins, and Princeton, resisted sociology too.[6] The impetus for sociology's expansion came from outside the walls of the academy, from secular and religious "reform" groups. Reform may not have been central to the era of 1890–1920, but reform movements did grow and prosper in these decades and reformers did seek to influence many kinds of institutions, including colleges and universities. Many private graduate-oriented universities could resist such pressures, at least for some years. But land-grant colleges and universities often could not resist citizen demands to pay attention to social problems, especially if citizen groups were organized; these were years of growing enrollments and relatively stable budgets. Similarly, many denominational colleges had strong ties, including financial support, to their sponsoring denominations. Particularly after the rise of the Social Gospel movement in the 1880s and 1890s, liberal ministers pushed denominations, and thus colleges, to respond to social ills.[7] Often secular and religious reformers joined hands in a variety of groups—the American Social Science Association, the Chautauqua movement, and the National Conference of Charities and Corrections—to lobby for the expansion of "sociology" in institutions of higher learning. Especially after the late 1880s, they became quite successful in their efforts, partly because of their growing skills at lobbying, partly because the social tensions of the 1890s gave their arguments added urgency. And it was difficult for college officials to refuse. Quite aside from the compelling humanitarian and social justifications for such programs, the sociological enterprise struck many collegiate administrators as very promising from the standpoint of their own institutions' welfare; they could train new occupational groups and probably increase student enrollments. It should not be assumed, however, that all citizens supported the new-fangled sociology; for example, the townspeople of Wahoo, Nebraska, refused to permit Luther College officials to inaugurate a course in sociology for fear it was a course advocating socialism.[8] But this was the exception, not the rule; in most instances citizen initiative was responsible for sociology's remarkable expansion.[9]

Accordingly, the men who assumed highly visible positions among the ranks of the first generation of collegiate sociology teachers understood perfectly their dual role as academic promoters of sociology and as champions of scientific social reform. They were entirely comfortable with the assumptions of mainstream reform thought. They were native born Protestants, usually of Northwestern European ancestry, of middle class heritage. They commonly identified, if only vicariously, with contemporary reform movements and with Anglo-Saxon, middle class culture. Many were the sons of intensely religious homes; often, they were Christian evolutionists who believed science would liberate men from the superstitions and follies of the past and usher in a new era of progress. As sociologists, they carefully cultivated their various lay clientele groups outside the university. They established important links with citizen reform groups by joining, taking official positions in those organizations, and speaking before them. They promoted sociology as an academic program within the academy, but, with few exceptions, they did little original specialized research. They formulated "systems" of sociology in their many textbooks and articles, statements of the laws or principles of society. They perceived no conflict between their roles as academic men and public men.

The University of Chicago and Columbia University sponsored the two major graduate programs in American sociology before the Depression. Over 90 percent of all American Ph.D.s in sociology took his or her work at one of these two institutions. Even at these two citadels of academic sociology, it was the general resources and circumstances of a major university, not the leadership of the sociologists, that created a viable graduate program. Given sociology's newness this was understandable. Under President William Rainey Harper's aggressive leadership, the University of Chicago established the first graduate department of sociology in America. Harper intended to build a great university in as many fields as possible; with Rockefeller support, he came very close to doing so. He could afford to make a potentially risky investment in sociology. For head professor of social science he appointed Albion W. Small, who became perhaps the promoter of sociology par excellence of his generation. Born in rural Maine in 1854, Small grew up in a middle class milieu; his father was a respected Baptist minister. As a college student at Colby in the mid-1870s Small demonstrated deep interest in a science of ethics. Initially he studied for the ministry. After three years in a theological seminary, he decided on a career in social science. He studied for two years at Berlin and Leipzig, and then taught history at his alma mater for several years. Small took another year of work at Johns Hopkins, and returned to Colby as president. He taught a course in "sociology," which he considered the science of ethics, instead of the moral philosophy course liberal arts college presidents usually taught. In January 1892, Harper appointed Small. From the first, the Chicago department was to have strong vocational and meliorist emphases: it would train social workers, gather scientific information for the solution of social problems, and prepare teachers. Small strongly approved. He saw no contradiction between an "essentially Christian" sociology and promoting sociology within a university context.[10]

Small ably promoted sociology at Chicago, building within a few years a full-fledged department that included Charles Henderson, George Vincent, and William I. Thomas. In 1895, President Harper virtually dared Small to found and edit a sociological journal as a test of sociology's academic credibility. Small accepted the challenge, and *The American Journal of Sociology* became the first sociology journal in America.[11] The Chicago graduate program quickly attracted students. By 1910, Chicago's department had granted twenty-eight Ph.D.s and another forty by 1926.[12] The Chicago department also developed a particular style of research, thanks to interdisciplinary contacts with the psychologists, philosophers, and other social scientists. Chicago students and faculty (most notably Thomas) who did actual social research worked from the standpoint of social psychology, using the case study method rather than statistics. Doubtless President Harper created the graduate department; he handed it to Small and told him to perform. And clearly the tone of the department was toward practical sociology, or, more precisely, "Christian sociology"; Henderson was occupied solely with social welfare courses, Vincent almost entirely, and Small remained a member of the NCCC for many years. Small himself did no specialized social research. He wrote several treatises of sociological systems and histories of social and economic thought. After 1905 he became a dean as well as chairman and editor, and had little opportunity for original investigation. Of the initial Chicago faculty, only Thomas had much time or inclination for original research.[13]

The founding of the Chicago department apparently goaded Columbia into following suit.[14] Since the early 1880s Columbia, under the leadership of John W. Burgess, had a prominent School of Political Science, with its greatest strengths in political science and economics. But courses in sociology were occasionally taught at Columbia. In 1894 the trustees appointed Franklin H. Giddings professor of sociology. Even more than Small, Giddings was a self-made sociologist who came to sociology in middle age. Born in rural Connecticut in 1855, the son of a Congregational minister, Giddings grew up in a rigidly pietistic home. As a pupil in secondary school he discovered Darwinism, which made him abandon formal religion. He attended Union College in the early 1870s to prepare himself for an engineering career. Then he taught school for two years and took his degree in 1877. After graduation he was a newspaperman for several years. Then he wrote editorials, which led to invitations to write articles on contemporary problems for the *Political Science Quarterly* and the Massachusetts Bureau of Statistics of Labor. This work brought him to the attention of Woodrow Wilson, who recruited him for an academic post teaching political science at Bryn Mawr in 1888. At Bryn Mawr he taught a course in 1890 on theories of society, which led to a lecturership in sociology at Columbia in 1891, and, finally, the professorship in 1894.[15]

Like Chicago, Columbia was to become a great center of training in social welfare—or so university officials, including the reform-minded president, Seth Low, thought. Giddings was quite conservative politically, with little enthusiasm for applied sociology. The construction of grand theories of sociology was, for Giddings, the only appropriate task of a professor of sociology. He advocated the science of society more than he practiced it. He did almost no empirical social research himself. He championed the use of statistics in social research, and was mainly responsible for the establishment of a modern statistical laboratory, complete with the most modern machinery available. But he was unaware of recent British statistics, so that he and his students were engaged in simple tabulations, not complex calculations of correlations and regressions. Columbia students interested in modern statistics had to work with Franz Boas in anthropology or Edward Lee Thorndike in psychology. Giddings was apparently a phenomenally successful lecturer with undergraduates, but his anti-Semitic, pro-war, and nationalistic prejudices, together with his considerable vanity, kept him from permitting the appointment of first-rate colleagues in sociology and from interacting fruitfully with such other faculty as Franz Boas, James McKeen Cattell, John Dewey, and James Harvey Robinson. Especially after Sarajevo, Giddings involved himself in lecturing around the country in support of the preparedness movement, in clamoring about the superiority of the "Nordic" races over all others, and in not promoting sociology as an academic or graduate enterprise. Yet distinguished sociologists did pass through the department as graduate students and perceive the promise of a statistically-oriented sociology. At Columbia, as at Chicago, it was the general condition of a major university graduate program that provided whatever professional and academic focus the next generation of sociologists possessed.[16]

But Chicago and Columbia were the only universities developing significant and continuous graduate training in sociology before the 1920s. Outside those institutions far different conditions prevailed. Most collegiate "professors" of sociology before the Great Depression were part-time teachers of sociology with little or no sociological training. Most of these part-time "sociologists" had no professional identity with sociology, and many were not men and women with advanced credentials. As late as 1910 there were fewer than sixty full-time college teachers of sociology, and as late as 1920, perhaps no more than twice that number.[17] The small cadre of full-time sociologists were most responsible for expanding sociology and for organizing and sustaining its academic organizations. Sociology grew the most at certain Middle Western land-grant institutions. Usually the initial impetus came from citizen groups pressuring institutional officials to respond to social problems by training social workers and offering courses in "sociology" and social problems. Once the program had started, further expansion commonly depended upon the professor's teaching abilities and student interest in courses in social work and social ethics or social problems. Often administrators and other social scientists were hostile toward sociology and permitted it only a "service department" status.

Consider what happened at the University of Illinois, where it was sociology's popularity with undergraduates and the charismatic teaching abilities of the staff that explained its remarkable growth, if not precisely its origins. In 1893, the chairman of the economics department introduced two courses in sociology, chiefly in response to citizen pressures for social welfare and social problems courses. In 1907 the chairman of the economics department welcomed the creation of a separate sociology department and the appointment of Edward Cary Hayes as the first professor of sociology. Born in rural New England in 1868, Hayes had a brief ministerial career before switching to the academy, first as a dean of a liberal arts college, then as a graduate student at Chicago and Berlin, as a popular lecturer in sociology at Miami University in Ohio, and, finally, as professor of sociology at the University of Illinois. He stayed at Illinois until his death in 1928. By all accounts Hayes was a talented, indeed melodramatic, lecturer.[18] Student enrollments in the department doubled every few years, from seventy-five in his first year to almost seventeen hundred the year before his death. Yet his successes as an undergraduate teacher aroused hostility among other social scientists. Often academic advisors discouraged students from taking sociology courses. University officials treated the department poorly, permitting new appointments at the most junior rank and penurious salary, keeping faculty-student ratios consistently at 250 students per professor, and throwing obstacles in the path of the graduate program Hayes wanted. Hayes was far more interested in sociology as an academic discipline and profession than as an instrument of social and political reconstruction. But his situation made it most difficult for him to do any original investigation, supervise the work of significant numbers of graduate students, or be much more than an undergraduate teacher.[19]

When the University of Wisconsin appointed Edward Alsworth Ross as professor of sociology in 1906, it gained an irrepressible reformer in the Populist-Progressive tradition. Yet Ross was an academic man, too, with a university oriented career. Born in Vandalia, Illinois, in 1866 and orphaned at an early age, Ross was a self-made, self-proclaimed reformer and intellectual. From boyhood he read voraciously in science, literature, and philosophy. After graduation from Coe College, near Cedar Rapids, Iowa, he taught in a business college in Fort Dodge. His collegiate and teaching experiences pushed him from the ministry to college teaching, for an academic life would satisfy both his intellectual curiosity and his reformist impulses.[20] He told his stepmother in 1887 that Darwin's *Descent of Man* "has made me a thorough-going evolutionist. I have never read a book so convincing as it is."[21] He studied briefly in Germany in 1888–89, but he returned to study economics with Richard T. Ely at Johns Hopkins University, a famous "liberal" economist of the day. Ross breezed through the doctoral program in two years. In the 1890s he had a meteoric career, winning appointments at Indiana University, Cornell, and Stanford. Ely helped his colorful, outspoken protege with positions and publication of his dissertation. By 1894 Ross was deeply engrossed in sociology, thanks chiefly to Lester Frank Ward's example and friendship, and began publishing a brilliant series of articles for *The American Journal of Sociology* later published as a book, *Social Control* (1901).[22] In 1900, however, he was dismissed from Stanford for his liberal, pro-Populist views; he had continually taken public positions that alienated Mrs. Leland Stanford, who controlled the University's pursestrings. Ross was able to make his case before the liberal public and many academics as a martyr to academic freedom. This helped him land a post at the University of Nebraska,[23] and his mentor Ely hurried forth publication and review of *Social Control*.[24] Ross enjoyed Nebraska; he lectured undergraduates in his classes and downtown business-men at luncheon meetings on the evils of contemporary society from a Populist-Progressive stance. He identified with academic sociology, writing books and articles, teaching courses, and interacting with other academic sociologists. But his more important role, in his eyes, was as social critic and reformer, and his more meaningful audience was the public, not professional sociologists. As he told Ely in justifying publication of *Social Control*, teachers and students of sociology would purchase the book, but most of all "preachers will get the book. It throws floods of light on what they are dealing with, viz., the regeneration of men. Surely out of the 75000 [sic] preachers in this country there are some hundreds of progressives who look for light to sociology rather than theology and who will want my book."[25] In 1906 Ely, now at Wisconsin, engineered Ross' appointment as professor of sociology. Ely warned Ross that some faculty opposed this first appointment in sociology because the courses would be popular, not scientific, and Ross would encroach upon existing courses.[26] Ross was amused. He threw himself into his career at Madison with characteristic gusto. Rapidly he became one of the University's more popular undergraduate teachers and, as a reformer-academic with a national reputation, an important member of the progressive "brain trust" at Madison. He published *Sin and Society* (1907), an argument for including corporate as well as personal evils within the law's purview, which won him much fame as a reformer. Eagerly he spoke and wrote for reform, but was less interested in promoting academic programs.[27] Sociology was not an autonomous department until 1929. Faculty expansion was slow. In 1912 he imported John L. Gillin, an old friend and a Columbia Ph.D., who handled extension and social work programs, and departmental administration, so that Ross could give public lectures and travel. Gillin and Ross constituted sociology until the 1920s, with Ross concentrating his energies on writing textbooks, foreign travel, and public issues until 1910.[28] Ross' national fame as a reformer-sociologist and his dynamic teaching abilities with undergraduates carried sociology at Wisconsin until his retirement in the late 1930s.

If Ross' national reputation as a reformer-social scientist es-

tablished sociology at Wisconsin, the creation of a chair in
sociology at the University of Missouri in 1900 resulted from
the pressure the Missouri State Board of Charities exerted upon
university officials. They told the president and curators, or
trustees, that they wanted the university to live up to its land-
grant ideals by establishing a sociology and social work program
that could train social welfare workers who would deal with
contemporary social problems. The man appointed was twenty-
seven-year-old Charles A. Ellwood, who was, like Ross, a pioneer
of academic sociology and a reformer. This gentle Quaker from
upstate New York turned from the ministry and then the law
to college teaching in search for a reformer's career. Ellwood was
a liberal Christian evolutionist, and he wrote his undergraduate
thesis at Cornell University on the possibility of a science of
social ethics; doubtless Ross, from whom he took a course, influenced him strongly. Ellwood then took graduate work at the
University of Chicago with John Dewey, George Herbert Mead,
Charles Henderson, and Albion W. Small. Upon graduation in
1899, with a dissertation on social psychology, he taught for a
year at the University of Nebraska, and organized a local charity
group and taught Sunday school. He thought of all three roles
as congenial.[29] When he accepted the offer from Missouri in the
spring of 1900, the university president made him the University's delegate to the NCCC conference in Topeka. Upon
Ellwood's arrival in Columbia, a member of the University's
board of curators told the young sociologist that he hoped
Ellwood would apply Christian principles "scientifically" to
social problems. The president told the young scholar-reformer
that his career depended on his success in teaching, publications,
and, above all, in public service—that is, the promotion of a
social work program. In the first several years, Ellwood was very
busy at Columbia. He organized a local charity group in
Columbia and the Missouri Conference of Social Welfare. He
also participated in the NCCC. As a teacher he stressed social
welfare; his students made social surveys of the socially deprived
as a part of their class work. He proved an effective and respected
teacher, although perhaps not as dynamic as Ross or Hayes. In
1902 the university president told Ellwood to make a scientific
survey of conditions in Missouri county jails. Eagerly Ellwood
and his students went all over the state, interviewing reticient

county officials and inspecting jails personally. They wrote a severe report implying massive corruption and negligence of official responsibilities. A political donnybrook resulted. Legislators allied with the offended county politicians pressured the University to fire Ellwood. In the ensuing controversy, Ellwood kept his job—and the jails were not reformed. The whole experience reinforced Ellwood's conviction that the sociologist must be both an academic and a concerned citizen.[30]

Ellwood's belief in sociology and social reform provided the unity of his career at Missouri. Enrollments in his courses grew, from sixty a year in 1900 to over four hundred in 1906, when he persuaded university officials to appoint another sociologist to take charge of the social work program. He won a grant from the Russell Sage Foundation to maintain a School of Philanthrophy at Missouri for several years; in 1909, Washington University in St. Louis took over the School and made it the germ of its social work program. Ellwood was proud of Missouri's social work program, and delighted that he had helped another university inaugurate a social welfare program. Ellwood taught Sunday school, delivered lay sermons on the general theme of a science of social ethics, and spoke before many citizen-reform groups in the state on the necessity of a scientific reconstruction of social and political life. After 1906, however, he could devote more attention to sociology proper. In 1904 he had begun work on a major statement of social psychology along the lines of William James' psychology. In the following years a steady stream of articles, reviews, and books came from his typewriter, all dealing with either systematic social psychology or with the necessity for a science of social ethics. He also trained graduate students; the first sociology Ph.D. was granted, in 1905, to a minister, and by the late 1920s six Ph.D.s and twenty master's students had worked with Ellwood. Ellwood dispatched several promising students to Chicago for doctoral work, including Luther Lee Bernard, E. B. Reuter, and Herbert Blumer, men who later achieved as much or more distinction as Ellwood. More than Ross, certainly, Ellwood played the role of the academic program-builder and professional. Yet Missouri never became a major center of advanced research and instruction in Ellwood's thirty years there, even though Ellwood published extensively. Undergraduate teaching bulked far larger than graduate training always—as late as the 1920s, for example, there were ten graduate students and over a thousand undergraduates in sociology—and Ellwood had little opportunity to become a director of an advanced program of instruction and research.[31]

Generally speaking, then, those sociology programs that prospered at all—as measured in multiple full time faculty appointments, growing enrollments, separate departmental status, and even graduate programs—did so because they profited from local citizen demands for meliorist social science and social work programs. Only rarely did a sociology program emerge because administrators and faculty thought sociology was a discipline worthy of status and recognition in the university, as happened by chance at Indiana University.[32] More often sociology's utility to administrators as a "service" program won it a foothold in universities. Typical was the experience at the University of Iowa, where sociology existed as a program from the mid-1880s to the late 1920s chiefly as a service operation to train social workers. Iowa sociologists were treated as poor relations whose professionalism was doubted by university officials and other social science faculty. As late as 1928, over 70 percent of all sociology courses were in social work, student enrollments were crushing, and faculty had few opportunities for advanced instruction and research. The University of Iowa continued sociology for so many years because of pressures outside the institution.[33] At the University of Southern California, on the other hand, the university president and trustees wanted a sociology program, that is, a social work program with strong practical emphases. In 1907 social work courses were offered for the first time. In 1911, Emory S. Bogardus, a recent Chicago Ph.D. in sociology, came to the University of Southern California and made the social work program prosper. His enrollments increased enormously. He even started a night social work program, a graduate social work program and raised funds for the *Journal of Applied Sociology,* which he edited, and which reflected meliorist emphases. He also made speeches before various charity, prison reform, social welfare, and civic groups around the state, thus establishing important links with those groups—citizen and professional—concerned with the care and administration of dependents, criminals, and delinquents. Bogardus wanted to do more theoretical or academic work, but he was also interested in applied or practical sociology, which took up so much of his time.[34]

The academic pioneers' dual emphasis on theoretical and practical sociology characterized both the American Sociological Society and the American Journal of Sociology. Both the Society and the Journal mirrored sociology's heterogeneous constituency of full time college professors, civic reformers, social workers, ministers, and corrections officials. By the early 1900s, the time seemed almost ripe for a sociology organization; there were perhaps 110 college sociology teachers, about 25 full time, and, in addition, at least 18 college sociology textbooks in print by 1905.[35] In the summer of 1905, C. W. A. Veditz, a George Washington University political scientist interested in sociology received the encouragement of several full time professors of sociology to try to start a society. Veditz sent a circular letter to almost three hundred persons suggested to him as potentially interested in such a society. He asked for a caucus with the American Economic Association's December meeting in Baltimore. Approximately forty responded in favor of the proposed society if it included "those who are engaged in practical sociological work" as well as those "interested in sociology from a purely theoretical and academic point of view."[36] About another forty or so persons attended the caucus. These conferees quickly agreed to form the American Sociological Society, with two dissenting votes, because they believed practical and academic sociologists needed a group separate from other social science organizations to discuss purely sociological issues. The issue of membership arose the second day among the remaining twenty academic conferees and the fourteen "practical" sociologists. The constitution the professors proposed stipulated open membership to those interested in sociological discussion and research. Several "practical" sociologists queried whether the critical issue was interest in sociology (regardless of one's occupation) or the holdings of an academic position in sociology. The professors replied that anyone could join so long as that person was interested in the scientific and critical aspects of sociology, not the "popular or propagandist."[37] The professors clearly walked a tightrope. They wanted a reputable academic group; they knew they needed all the members possible, including "practical" sociologists, to make the Society viable, but they did not want its purposes diluted.[38] Hence the compromise on membership qualifications. The Society belonged to the academic sociologists'—little doubt about that. They sent out the circular, and they convened the caucus in conjunction with an academic, not a practical, organization. Professors chaired the sessions, wrote the constitution, and held all committee posts. And the issue of whether those with no interest in practical sociology, but only in theoretical or academic sociology, could be denied membership never arose. But they knew they could not alienate "practical" sociologists amenable to their purposes, for the Society might otherwise flounder.[39] Consequently they unanimously elected Lester Frank Ward the first president. A vast gulf separated the academic sociologists from Ward, a self-taught intellectual and career civil servant innocent of any contact with the emerging university culture until the end of his life. The academic sociologists, as a group, were products of American university culture; trained at home or in Germany, they worked in a university setting and understood, however imperfectly, American university culture—its emphasis on specialization, disciplinary identification, original investigation, and, above all, its implicit recognition of the scientific information explosion. Ward's contributions to the American sociological movement were considerable: in the 1880s and 1890s he did much to popularize the term "sociology," to demonstrate its potential for social reform, and to suggest sociology should be "scientific" in approach, spirit, and method. If Ward was no social worker like most of the "practical" sociologists, nevertheless his achievements were those of the preacademic reformer.[40] The academic sociologists were interested in theories specific to sociology and the social sciences, not the grand, metaphysical philosophical systems such as Ward's, which sought to unify all scientific knowledge. In electing Ward, the academic sociologists recognized the "practical" sociologists and capitalized on his considerable reputation among the educated public.[41]

Necessarily, the Society remained heterogeneous for years. Academic sociologists dominated its affairs, but they recognized the "practical" sociologists as much as they dared without diluting the Society's essentially professorial purposes, and for good reason. A third of the 115 founding members were "practical" sociologists, whereas half were professors. For some time the "practical" sociologists were important; they constituted 20 percent of the 322 members of 1910 and were a recognizable fraction as late as 1920. The professors dominant in the Society were far more interested in recruiting "practical" sociologists than part-time sociology teachers at humble colleges, probably because the "practical" sociologists were more amenable to their conception of the Society's purposes than the part-time teachers. (Few of the latter ever joined or read papers.[42]) The doors were always open for the "practical" sociologists to belong and to read papers.[43] As president Ross launched a membership drive aimed at the "practical" sociologists; membership jumped from 579 in 1914 to 808 in 1916.[44]

Similarly the *American Journal of Sociology* reflected pre-World War I sociology's heterogeneity. Small maintained close ties with progressive reformers, belonged to the NCCC, and tried to maintain the journal as an academic enterprise. Many of the articles before the 1920s were meliorist, "practical," and reformist, written as much for a progressive public as for academic men.[45] It was not until 1915 that the first major statistical paper appeared in the *Journal,* and almost another decade until articles utilizing modern British statistics were published regularly.[46] There were, indeed, academic articles in the *Journal* before 1915 or 1920, but far fewer proportionately than in, say, the *American Anthropologist,* the *Journal of American Folklore,* or the natural science journals.

### III

In the 1890s a small cadre of seventeen academic sociologists came to the fore of American social science. As the most active and visible agents of sociology's institutional professionalization, they led the theoretical discussions of sociology, defining, in the largest sense, what were the phenomena, assumptions, concepts, and methods of sociology. Because sociology was so new, and more of a movement than a discipline, they experienced considerable difficulties in their deliberations. What was sociology's status as a university science? Why was it different from economics? From biology? What were its distinctive intellectual characteristics? Such questions as these occupied the seventeen's attention for years. The seventeen were full time teachers of sociology at reputable or major universities.[47] As a group they were relatively young men just embarking on their academic careers in the 1890s, as sociology became popular in colleges. (Twelve were born between 1861 and 1874.) As a group they were academic, not pre-academic, sociologists; they had university training in a social science, most commonly history or economics. Fourteen earned doctorates in history, economics, or sociology; ten took their degrees between 1893 and 1902. And they were pioneers of American sociology in other respects too. They helped found and promote undergraduate, and, in some cases graduate, programs, thus establishing sociology's place in the college curriculum.

From the perspective of the years 1890 to 1920, furthermore, they constituted a recognizable elite among academic sociologists. They held sociology's most desirable university chairs. They were prominent in the American Sociological Society's affairs; sixteen served as president before the late 1920s, and all helped make important Society decisions. As a group they dominated the *American Journal of Sociology*'s pages, publishing all out of proportion to their numbers. To 1920 the *Journal* carried 1,086 original articles by 380 authors; the seventeen accounted for 20 percent of those articles. And they wrote more than 30 percent of the 1,112 book reviews in the *Journal* before 1920.[48] They also wrote all the influential textbooks of sociology in an age when the textbook of first principles had high prestige among their professional brethren.[49] As they grew older, they became recognized senior statesmen of sociology, at least so far as the public was concerned.

In understanding the seventeen we may not depict them as a purely homogeneous group of self-conscious professionals who grasped, after the fashion of the Boasians, the implications of academic professionalism for their thoughts and deeds as professors in modern university culture. Sociology was too new for that. In many ways, too, they were a diverse group, each mixing his professional and social policy commitments and cultural values in somewhat different proportions. Many, most conspicuously Ross and less notoriously William I. Thomas and Ellwood, identified with the progressive movement, or at least with progressive reforms; yet William Graham Sumner, his disciple Albert

Galloway Keller, and Edward Cary Hayes, did not, at least not publicly. A number sprang from Protestant ministerial backgrounds, and some—Ross and Ellwood, for example—embraced liberal Protestantism, believing in the existence of evolutionary laws of society that the new science of society could discover to further mankind's progress. Yet others, such as Thomas and Sumner in their mature years, exhibited wholly secular and empirical outlooks. Some, like Hayes, Ross, and Ellwood, were proud of their small town heritage and tried to sustain its cultural values, whereas others, such as Thomas and Small, had an urban outlook. Several—Ross, Small, and Sumner, for example—studied in Germany, respected German culture and scholarship, and wanted to infuse American higher education with what they understood to be the spirit of German university scholarship. Yet others, notably Thomas and Charles H. Cooley, made noteworthy contributions to sociological theory and method even though they had never attended a German university.[50]

What unified these men as a group was not so much their personal characteristics, values, and backgrounds, although there were some similarities, to be sure, but the essentially impersonal circumstances in which they found themselves as academic professionals charged with defending and developing academic programs of sociology in modern American university culture. The need to define sociology in university culture, to differentiate it from other disciplines, to win recognition for it in universities, to delineate its methods, assumptions and body of data, to articulate the national and cosmopolitan standards of university training, competence, and performance, to establish achievements in research and theory that could be used by later generations of practitioners—all these imperatives defined their situation, lent the circumstances in which they operated a certain uniformity, and made them into a rather cohesive group, despite their differing attitudes on many issues, including at first their definitions of sociology. They agreed rather easily on what sociology was not, but they struggled for years to work out a positive and uniform conception of what constituted sociology, which strongly suggests that the impersonal circumstances of university culture and the needs deriving from academic professionalization, not personal attitudes, pushed along their theoretical deliberations over the next two decades.

In the 1890s some of the seventeen sociologists made the first attempts to define their discipline and thus give it a place in university culture. They began by forthrightly rejecting Herbert Spencer's biological sociology, his analogy between the biological and the social organisms that had so dominated sociological theory to that point. They agreed that social forces were psychological and mental in character, that the proper focus of sociological theory was the individual, and that society might be defined as the mental interactions of individuals. In this they were obviously influenced by the new psychology of the 1890s, in particular that of William James and his instinct theory as outlined in *The Principles of Psychology* (1890). The sociologists perceived no contradiction whatsoever in borrowing assumptions and analogies from a wholly separate discipline (psychology) and at the same time insisting that sociology was an independent discipline whose autonomy in university culture should be recognized. If sociology was an independent discipline, presumably its practitioners did not have to depend upon the assumptions, methods, data, and point of view of other disciplines. In 1894, for example, in the first formal statement of sociology by any of the seventeen, Albion W. Small and George E. Vincent declared in *An Introduction to the Study of Society* that society was "a complex of activities and movements originated by the energy of those physical and psychic attributes which determine human motives." At the same time, they insisted that the methods and assumptions of the biological sciences were completely inappropriate for sociology. Sociologists should use historical and analytical methods to study man as he actually was, without reference to the viewpoints of the natural sciences. Sociology was concerned with the thoughts of individuals, and these were "in a very large degree, an acquisition from the resources of society. The individual believes not merely the results of his own sensations and cognitions, but accepts on faith a vast body of social knowledge." Small and Vincent also insisted sociology was a science which only persons formally trained in the discipline, and who shared its sense of autonomy from other disciplines, should practice. Permitting amateurs to study sociology "would be like setting an artist in oils to paint bridges, or allowing a boiler maker to take command of a navy." But that did not prevent them from concluding it "is the psychical potencies of society, knowledge, taste, and criteria of conduct, which persist and constitute the real life of the social organism."[51]

In a well-known textbook published in 1896, Giddings declared that sociology should be a psychological rather than a biological discipline, for rigid biological analogies were inappropriate for sociology. "Every distinct science," he explained, "must have its own classifications and its own names for phenomena which, however they resemble the phenomena studied by other sciences, are yet different, and are the subject matter of a separate science only because they are different."[52] Subsequently Giddings argued that sociology would have to work out its own distinctive assumptions and methods apart from other disciplines. Sociologists, he insisted, would have to understand that they dealt with phenomena not reducible to mere natural forces. Giddings believed that social association, not natural forces, had a decisive impact "on the natures of individuals, and adapts them to social life. It creates a social nature."[53]

The seventeen sociologists, then, saw no difficulty with depending on the naturalistic assumptions of contemporary physiological psychology and asserting that sociology should be autonomous from all other sciences. Ellwood, for example, wrote his doctoral dissertation for Small at Chicago defending the instinct theory as the proper basis for sociology; he considered William James the "Darwin of psychology." Obviously, he thought James sociology's Darwin too.[54] Ross also attacked the Spencerian biological sociology, in his famous *Social Control* (1901). He insisted that society exercised its control over the individual by informal social means, such as public opinion, law, belief, suggestions, education, custom, and ceremony, not the biological means Spencer upheld. Human differences had biological causes, whereas human uniformities had social causes, declared Ross, thus segregating social phenomena from biological (if not psychological) phenomena. Like the other sociologists, Ross defined sociology as the study of the individuals of society, rather than of groups, classes, and social institutions. Yet he believed in sociology's autonomy. Analogies from other sciences have suggested what to look for, he declared in another book in 1905, but it "is certain, however that no recognized science borrows its laws from other departments of knowledge. The lasting possession of sociology will be regularities which, instead of being imported from without, have been discovered by patiently comparing social facts among themselves."[55]

Ross argued that sociology had to move from detecting vague and superficial analogies among a small number of facts to discovering large numbers of facts explainable on their own terms. In almost the same breath, he declared sociology was a psychological or psychical science; no ultimate nonpsychic factors—i.e., including those derived from the social order—could be admitted into sociological discourse until it was shown how they influenced such psychical factors as individual motive and choice.[56] Ross distinguished between the biological and social levels of human behavior. He believed in eugenics, for example, but "I take up the matter solely as a sociologist, seeking to understand to what extent and how human institutions improve or deteriorate the human breed. There are many other factors than institutions that enter into race improvement."[57] At this point he had no sense of the distinction between the psychological and the sociological levels of human behavior.

In the early 1900s several of the seventeen exchanged their ideas on sociology by correspondence.[58] Sociology's newness hindered them considerably, making their sincere efforts to transform sociology into an authentic inductive science recognized in university culture all the more difficult. With no well established intellectual traditions, they had a formidable task. Giddings, for example, in responding to a review Ross wrote of one of his books, said "I have attempted to present a pure inductive method, from which a priori assumptions should be excluded as far as possible." He insisted that social research should be empirical, inductive, and statistical.[59] For Giddings, that was a lofty goal. Five years later Ross reviewed another of Giddings' books. Giddings had not yet achieved his ideal, as Ross gently noted. Giddings responded, "You are entirely right in saying that the discussion of evolution in general has no proper place in a treatise on Sociology, but I had to include it in this volume because I wished to base my whole scheme of social causation upon it, and I did not elsewhere publish it in sufficient detail. . . to be able to count on familiarity with it."[60] Such intellectual confusion between the ideal of an autonomous, inductive science of society and dependence on evolutionary science characterized, in varying degrees, the writings of the other seventeen sociologists, including Cooley, Small, Thomas, and Blackmar, especially before 1906 or 1907.[61]

Around 1906 it began to occur to some of the seventeen they could not insist that theirs was an autonomous science and continue to rely upon naturalistic analogies and determinants borrowed from the evolutionary natural sciences. Inevitably this was a slow process consuming years. In an address before the American Sociological Society in 1906, for example, Small declared that sociologists "agree to differentiate sociology from antecedent psychology or cosmology or metaphysics." He still believed that sociologists might learn much from natural scientists, yet in the same breath he declared that sociologists studied human relations, which potentially pointed sociology away from the natural sciences.[62] In *Folkways* (1906), Sumner offered a complex model of society notable for its aggressively anti-naturalistic, anti-hereditarian tone; it was more indebted to contemporary British cultural anthropology than Spencerian evolutionism, suggesting that Sumner had entered a new phase of his intellectual career.[63] *Folkways* was a synthesis of other scholars' research, not a specialized monograph.

In *Social Psychology* (1908) Ross paid homage to contemporary naturalistic psychology by invoking the instinct theory as an explanation for much human behavior. But he also said sociologists were chiefly interested in the interaction of human minds, which occurred only in a social environment.[64] Ross still defined society as a collection of sentient individuals, akin to a psychological model, but his recognition of social environment signalled a new theoretical departure toward the social group and social tradition, or culture.

Cooley made the first decisive break from psychology's metaphors and determinants in *Social Organization* (1909). He argued that the social group, not the individual, was sociology's true point of departure. He insisted that human nature could develop only in face-to-face primary social groups such as the family, the playground, and the nursery; obviously for Cooley human nature was something more than, and probably different from, mere instinct (a change from his 1902 position) and something less formal than social institutions. Cooley defined human nature as a group nature that developed after birth in association with other humans, and which atrophied in social isolation. As Cooley followed the intellectual professionalization of sociology to its logical conclusion and insisted that man's behavior should be explained on social and cultural grounds, he came very close to recognizing that it was not necessary for sociologists to depend on contemporary naturalistic psychology and its assumptions, meta-phors, and concepts for their science of society.[65] In 1909 Hayes made a semantic and conceptual break with contemporary psychology when he declared sociology and psychology were wholly independent sciences whose practitioners studied entirely different orders of phenomena, neither of which could be explained in terms of the other.[66] George Herbert Mead, the brilliant philosopher and psychologist at Chicago, who had a profound influence on many academic sociologists, informed the seventeen sociologists in several sparkling articles that there was a qualitative difference between animal and human intelligence, that social psychologists and sociologists studied an entirely different order of phenomena than did the contemporary physiological and comparative or animal psychologists, and that social behavior—the proper concern of sociology—had social, not psychological, physiological, or biological, determinants.[67] His close friend, Thomas, argued in 1909 that the social sciences could learn far more from each other about human behavior than they could from the natural sciences. Thomas specifically recommended the work of the Boasians in cultural anthropology as required reading for all academic sociologists.[68]

In the 1910s several of the seventeen sociologists began to recognize clearly that they could not continue to insist on the autonomy of their discipline and still depend on the assumptions of contemporary psychology. Thus they abandoned explicit evolutionary analogies and led the way toward the divorce of biological and social theory. The full implications of the Boasian dictum *omnis cultura ex cultura* were now clear to them, and they made appropriate, if not uniform, changes in their posture. Consider the example of Charles A. Ellwood. Until the mid-1910s, Ellwood tried to base sociological theory upon natural science determinants and analogies; he believed firmly in James' instinct psychology, and in the autonomy of sociology as a science with no sense of intellectual crisis. So persuaded was Ellwood of man's biopsychological nature that he supported the eugenics movement for several years.[69] But in 1914–15 he took a year's leave and studied cultural anthropology with R. R. Marett in England, who was a cultural evolutionist. From Marett, Ellwood learned that cultural and social phenomena operated on a different level than natural phenomena—and the larger lesson that sociology would become an autonomous science only when sociologists abandoned natural science concepts, assumptions, analogies, and explanatory principles. He returned to America determined to incorporate this insight in a new book.[70] In 1915 he broke with the assumptions of eugenics, and in 1916 he criticized racial interpretations of human behavior, thus signifying his new cultural point of view.[71] Yet he did not completely emancipate himself from prior positions. In *An Introduction to Social Psychology* (1917), he did examine cultural and social phenomena from a cultural, not a psychological, perspective. But he still thought of cultural development in terms of steps of cultural evolution, after the fashion of British anthropology, probably because he could reconcile that with his progressive ideological commitments to liberal Protestantism and civic reform. And he could not shift from the individual to the group. But he continued to insist on the divorce of biological and social theory; he argued this was essential to sociology's autonomy in university culture. He even repudiated, in 1919, his earlier endorsement of Jamesian instinct theory.[72]

Several of the other seventeen sociologists made a roughly similar transition. In 1915, for example, Hayes insisted that cultural and social phenomena were the primary determinants of human behavior and that social scientists should stop looking to the natural sciences for the laws of society and of human behavior.[73] In *Social Process* (1919), Cooley analyzed society without recourse to evolutionary natural science conceptions and determinants.[74] His approach and terminology suggested that he was not familiar with the Boasians' statements of the culture idea. Whether he was or not, he had separated nature and culture; presumably if the impersonal circumstances in which he operated had not inclined him toward recognition of the necessity of divorcing biological and social theory, no Boasian statement would have convinced him.

William I. Thomas made perhaps a more profound transformation than any of the other seventeen, and largely because of his reactions to his experiences in original, specialized sociological inquiry. He went further than recognizing the necessity of divorcing biological and social theory, and even of distinguishing between the cultural and psychological levels of existence, what later generations of sociologists would call culture and personality. He also separated, at least as an ideal, the world of social facts from the world of ethical, cultural, and philosophical values in social explanation. Boasian cultural anthropology had influenced him deeply at a critical point in his own intellectual development, when he was beginning to question reliance on natural science determinants. In 1908 he began sustained ethnological research on Polish immigrants in America, using original source materials. He focused on the central problems of social disorganization and cultural adaptation in Polish immigrant culture. For years he worked assiduously on what was to be one of the earliest monographic research efforts in sociology by an American academic sociologist. In 1914 Thomas recruited, by chance, the Polish philosopher and emigration official Florian Znaniecki as collaborator. Their work appeared as *The Polish Peasant in Europe and America* in five volumes between 1918 and 1920. In the long methodological introduction, Thomas argued that social investigation must be based on social facts, not biological analogies, the researcher's values, or simplistic metaphysical schemes. Social facts had to be interpreted on their own terms, for social facts were of a wholly different character than the facts of nature, and were governed by different determinants than natural phenomena. Doubtless his experiences as a specialized researcher helped push him toward this insight, for to the specialized researcher, grand schemes or arguments from analogy would be of severely limited utility. Perhaps Thomas never achieved in this work his goal of actually separating the world of social facts from values, but he was probably the first American academic sociologist to make that argument in print. Not surprisingly, Thomas also insisted that social research and explanation were the social scientist's province, just as the investigation of nature was the responsibility of the physical or natural scientist.[75] Obviously by 1918 there were new currents of thought and new criteria of practice in American sociology.

## IV

The next several years demonstrated beyond any doubt that American sociology had undergone a major transformation. The clearest and most immediate sign of change was that after 1920

American sociologists caught up with the Boasians and announced the divorce of biological and social theory. They abandoned explicitly evolutionary or natural science models for sociological theory. In so doing they began to attack the human instinct theory, the eugenics credo, and the racial interpretation of mental testing. Throughout the 1920s, American sociologists participated in the heredity-environment controversy, sometimes explicitly by attacking hereditarian social explanations of human behavior, more often implicitly by doing the kind of specialized social inquiry that preempted or challenged the natural scientists' interpretations of social phenomena. The transformation constituted more than a change in point of view or pattern of social scientific explanation. In the post-1920 years sociologists, especially the new generation, approached the problems of social explanation on a new and more empirical, specialized level of discourse; they now understood perhaps the major contribution of sociology to modern thought, the insight that social structure influences human behavior, and that idea, together with the larger maxim of culture, guided their research and interpretation.

Why sociologists abandoned evolutionary explanations of the pre-1920 era was a complex story. To some extent the words and deeds of the pre-1920 generation of academic sociologists played a role in this transformation. All but five—Cooley, Hayes, Howard, Small, and Sumner—lived through the 1920s. Effectively the rest either made some accommodation with the next trends in sociological practice and theory or quietly lost influence among sociologists. Giddings became deeply involved in political and foreign policy issues from about 1914-on and faded from view. Dealey remained at Brown University for some years, attempting to carry on Lester Frank Ward's cosmology there, and refusing to adapt to the new currents of thought and practice; in 1928, he entered the newspaper business in Texas. Sumner died in 1910, and George E. Vincent left his Chicago professorship in 1911 for a university presidency; neither had much influence among sociologists in the 1920s. Blackmar faded away from sociology, as did Weatherly and Howard; all three concentrated on historical, not sociological, work. Before Small died in 1926 he recognized the new trends in sociology, but he was too involved in administrative, promotional, and editorial work to reorient his work as a sociologist. Keller actively disliked the new trends in

sociology. After 1916 he withdrew from the profession (but not
his Yale professorship), preferring to keep alive the memory of
his mentor, Sumner. Ross remained an elder statesman of sociology, a vigorous, outspoken man unto his death. But from the
1920s on he was an academic entrepreneur, not a sociologist,
writing high school textbooks, giving public lectures, and travel-
ling all over the world.[76]

Other members of that pioneer generation made various accommodations with the new trends. Thomas continued to do original
specialized inquiry; after 1919 he did so as a free-lancer, for
spurious charges that he violated the Mann Act forced his dismissal at Chicago. Gillin at Wisconsin did some important
research on criminology and penology that implicitly criticized
eugenic explanations of anti-social behavior. At North Dakota,
Gillette became interested in rural sociology, and did much to
launch that as a viable specialty. In his work he also recognized
the centrality of social determinants and the importance of specialized work. Cooley at Michigan and Hayes at Illinois did not
survive the 1920s, nor did they shift from enunciation of sociological "principles" to specialized sociological inquiry. But in their
work they did consistently recognize the importance, for sociology, of divorcing biological and social theory and of abandoning
natural science determinants. Probably Ellwood published the
most elaborate theoretical statements of any of the seventeen of
the importance of separating the biological from the social levels
of human existence. Throughout the 1920s at Missouri and the
1930s at Duke University he attempted to steer a middle course
in sociology, between the statistically-oriented behaviorists who
insisted on rigid environmental explanations, and those who
would not admit to the distinction between the biological and
the social planes of human behavior. In the 1920s in particular,
Ellwood attacked biopsychological models of human behavior
such as the instinct theory and embraced the culture theory. But
he did so in a context of a larger belief in progress and human
perfectability, so that unlike the Boasians his model of human
behavior was evolutionary. Especially after 1920 Ellwood became
even more interested in liberal Christianity, devoting more and
more of his energies to combatting fundamentalists. For example,
in answering an inquiry about a position at another university in
the 1920s, he wrote that "if I consented to leave my present posi-

tion, my principle [sic] reason would be to work in an institution where there was more of a religious atmosphere than there is in the State University and where the connections between Sociology and Religion could be made more vital."[77]

Yet the more important key to the transformation of sociological theory and practice was in the rise of a new generation of sociologists committed to specialized inquiry and to the imperatives of modern university culture. In turn, the change occurred in large measure because both the University of Chicago and Columbia University, the two major centers of graduate training, became different institutions from about 1915 or 1920 on. The appointment of Robert E. Park as professor of sociology at Chicago redirected the sociological enterprise there. Park was almost fifty years old when he began teaching at Chicago in 1914. He had a varied career: as a newspaperman, as a graduate student in Europe and at Harvard, and as an assistant to Booker T. Washington at Tuskegee. Park was always interested in facts, not so much grand systematic explanations; his contacts with southern black culture awakened his interest in race relations. And unlike most of the pioneer academic sociologists, Park made a rigid distinction between the science of society and humanitarian meliorism. At Chicago he encouraged colleagues and students to work in both the sociology of race relations and urban sociology. The work he and his colleagues did in race relations in the 1920s did much to diminish the credibility of scientific racism; their work in urban sociology, in the largest sense, demonstrated that the social disorganization of urban immigrant culture had social and cultural, not biological, roots. In 1921 Park and his colleague Ernest W. Burgess published an important textbook, *An Introduction to the Science of Sociology,* which effectively reoriented sociological theory from the vague "principles" and analogies of pre-1920 statements to the concepts, methods, assumptions, and definitions that thereafter characterized the interests of modern American sociologists.[78]

Columbia changed also. Columbia became an important center of quantitative method in sociology, just as Chicago encouraged the specialized case studies of social processes. After 1914, Giddings had precious little influence there. Several of his recent graduate students, most notably Stuart Chapin and William F. Ogburn, championed more specialized and monographic approaches to social investigation. Chapin, for example, began as a graduate student in 1909 completely enthralled by the prospect of using organic analogies in sociology. Yet once he was committed to specialized inquiry into a finite problem—which was the definition of a doctoral dissertation, after all—he became extremely uneasy with vague analogies. Eventually he shifted from broad analogical arguments to the use of modern British statistics in social investigation, such as regression analysis and correlations. In the decade following the publication of a seminal article in 1915, Chapin did much to disseminate the new techniques of quantification, and the more specialized perspectives they implied, among the newer generation of sociologists. Ogburn became also a pioneer of quantification in sociology, and in the period 1919–27, when he taught at Columbia, he was responsible for redirecting the focus of graduate research and training. Both Chapin and Ogburn became fellows of the American Statistical Association and (perhaps less importantly) the American Association for the Advancement of Science, symbols of their methodological and scientific reputations among scientists in more traditional disciplines. And Ogburn came into close contact with the Boasians in the late 1910s and helped diffuse Boasian cultural anthropology among American sociologists.[79]

It was within the context of these institutional, generational, and theoretical changes in American sociology that American sociology now became fully professionalized in academic culture and that American sociologists who identified with academic professionalism now sought to emancipate sociology from the natural sciences, and to insist upon cultural explanations upon a commitment to specific empirical research and upon investigation of discrete areas of society. As Bernard, a member of this new generation and an active critic of the instinct theory, noted in the late 1920s, in the last decade or so sociologists had abandoned the early twentieth-century philosophical and theoretical approach for a more dispassionate factual one, in which statistical and case study methods were used. In an obvious reference to the older generation of sociologists, Bernard remarked that there "are still some belated attempts in sociology to take seriously the making of concepts and of social laws, but most sociologists are now persuaded that if they get the facts the concepts and the laws will take care of themselves."[80] A few years earlier Albion W.

Small had recognized the changes that had swept over American sociology. Remarking on the tendency of the early twentieth-century sociologists to contrive theoretical statements rather than to bother themselves with hard empirical research into concrete social phenomena, Small concluded that "a humiliating proportion of the so-called 'sociology' of the last thirty years in America, both inside and outside of the goodly fellowship of scholars who were self-disciplining themselves and one another into the character of scientific specialists, has been simply old-fashioned opinionativeness under a new-fangled name."[81] "The true story of the American sociological movement," Small concluded, "would be a treatment of the theme: Up from Amateurism."[82] By the mid-1920s, clearly, sociologists considered themselves full members of American university culture—scientific specialists—or they effectively withdrew from sociology. John L. Gillin enunciated the new ethos in his presidential address before the American Sociological Society in 1926, an ethos that would not have been permissable a decade previously:

> In certain of our institutions it has unfortunately been true that sociology has been advocated by men who had no adequate understanding of scholarship. In their hands it was a mess of undigested, unsystematized, unscrutinized generalities which made a popular appeal to sophomore and attendants at Chautauquas. . . . While some of these, unfortunately, are still with us, the application of the scientific method and the increased emphasis on objective data have been acting as selective agents in consigning these enemies of sociology to a deserved innocuous desuetude. Doubtless we shall have to put with some of them longer, inasmuch as there is no sociological orthodoxy and no sociological inquisition or holy office by which these fellows can be eliminated. Emphasis upon rigidly scientific methods will attend to them.[83]

Gillin probably exaggerated the lack of a reigning "sociological orthodoxy" and a "sociological inquisition." The roles in which sociologists found it necessary to play if they were to participate in modern university culture, together with the larger institutional and attitudinal context in which they operated, performed those functions most effectively and, in the bargain, helped bring about a major transformation of American sociology as discipline and profession, thus resulting in the divorce of biological and social theory. In the post–World War years, American sociologists joined hands with the Boasians to shape a new science of man based on the dictum, *omnis cultura ex cultura,* a science whose practitioners considered themselves full members of American university culture. And now American sociologists confronted the nature-nurture issue.

# 3 The Heredity-Environment Controversy    1915-1941

# 5 EUGENICS, RACE, AND EVOLUTION

ITHIN two years of the Nazi collapse in Europe
*Science* published an article symptomatic of a
profound theoretical reorientation in the American natural and social sciences. In that article
Theodosius Dobzhansky, a geneticist, and M. F. Ashley-Montagu,
an anthropologist, summarized and synthesized what the last quarter century's work in their respective fields implied for extreme
hereditarian explanations of human nature and conduct. Their
overarching thesis was that man was the product of biological
and social evolution. Even though man in his biological aspects
was as subject to natural processes as any other species, in certain
critical respects he was unique in nature, for the specific system
of genes that created an identifiably human mentality also permitted man to experience cultural evolution. Man's unique brain
made it possible for him to organize complex political, socio-
economic, and technological systems and institutions that in
turn enabled him to master the earth's many environments on an
unprecedented scale. Man could learn, adapt, invent, and im-
provise as no other species in nature. "The effect of natural
selection in man has probably been to render genotypic differences in personality traits, as between individuals and particularly as between races, relatively unimportant compared to their
phenotypic plasticity," Dobzhansky and Ashley-Montagu con-

tinued. "Instead of having his responses genetically fixed as in other animal species, man is a species that invents its own responses, and it is out of this unique ability to invent . . . his responses that his cultures are born." Doctrines of racial superiority had no scientific justification, they concluded.[1]

Obviously the nature versus nurture dichotomy was in disrepute in American science. Scientists had stopped arguing which was the more important. They took it for granted that innate endowment and external environment were interdependent variables. As the fact of intellectual collaboration between Dobzhansky and Ashley-Montagu implied, natural and social scientists had resolved their theoretical and jurisdictional disputes quite handsomely. The new picture of heredity and environment made it increasingly difficult for those with extreme hereditarian views to maintain their scientific respectability. Furthermore, the new theory made it clear that the culture idea was not a criticism of the biologists' doctrine of evolution but, in a profound sense, an integral part of evolutionary theory properly conceived. Man's cultural evolution was not subject to the same determinants as his biological evolution, for these were different kinds of processes. Social scientists had always accepted the legitimacy of the natural sciences for the interpretation of man's biological aspects. Now natural scientists conceded the explanation of the cultural level of man to the social sciences.

The nature-nurture controversy was largely responsible for this reorientation of biological and social theory. In many respects the most important of the three debates constituting the larger heredity-environment controversy was the least dramatic one: the conflict over eugenics and race theories. To a considerable extent it was the labors of a new generation of geneticists which eroded eugenics and race theories. Since they were criticizing the work of men who in many instances had helped them launch their careers, they were understandably reluctant to speak out too loudly. And many of these younger geneticists sincerely believed in the ideals of eugenics. But the work they did in the laboratory was fundamental. It eroded whatever scientific legitimacy eugenics and scientific racism may have had by verifying a radically different model of evolution through mutation and natural selection, by painting a new picture of physical inheritance, and by working out statistically-oriented

at could be easily
new paradigm re-
stances beyond the
in the politics of
gh the specter of
eas of science, but
e of the complex
the politics of the

copy of. . . 'The
. . . I think it is
wledge of heredity
rtant advance since
tologist Edwin G.
September 1915.[2]
genetics changed
930s, and Morgan
rmation. In those
common fruit fly
the chromosomal
many traits with
iques and devices
for the breeding laboratory, and uncovered many complexities
of inheritance such as multiple factors, sex linkage, crossing-
over, and the influence of environment—all of which radically
modified scientific explanations of heredity.

Indeed, the two decades following 1910 witnessed major
changes in both genetics and evolutionary theory. Morgan and
his coworkers were not alone; many biologists in institutions
scattered across the United States and in Great Britain collaborated in building the new theory of the gene and the synthetic
theory of evolution. They exchanged ideas at conferences, summer institutes, laboratories, meetings of societies, in personal
correspondence, and, of course, in their publications. This was
largely an intellectual collaboration, a venture in the dissemination and exchange of ideas. But its institutional context can not

be ignored. Before 1910 few biologists did systematic research in genetics, although many more acclaimed Mendelism. And genetics workers were then located chiefly at Harvard's Bussey Institution, under William E. Castle, and the Station for Experimental Evolution, under Charles B. Davenport. After 1910 this changed. The number of geneticists grew dramatically, testimony to genetics' intellectual challenges and the doubling of botany and zoology doctoral graduates.[3] The emerging genetics profession became more impersonal and heterogeneous than ever before, thus lessening the influence of any particular individual or group on accepted genetics ideas. After World War I there existed a larger and more diverse array of institutions supporting genetics research than was the case in the 1900s; there were ongoing major genetics programs at Columbia, the University of California, the University of Wisconsin, Cornell University, the University of Illinois, Kansas State Agricultural College, Connecticut Agricultural College, Indiana University, and the University of Georgia by the early 1920s, for example; the impact of Davenport's Station for Experimental Evolution (and his Eugenics Record Office) was correspondingly diluted.[4] Many of the new workers did not share the early Mendelians' enthusiasm for eugenics, or their excitement over supposedly magical prospects of experimental biology. Many of these new workers represented a second rather than a first generation in the genetics profession. Their role was not to advertise their science's potentialities but to work out its problems in the laboratory. Many of these geneticists were specialists in agricultural hybridization, bound by intellectual and institutional commitments to be more concerned with the germ plasm of corn than that of men. And those laboratory genetics working in so-called basic problems, whether they could be considered in the first or second generation of geneticists—were obligated by their scientific value commitments and traditions to respect discoveries independently verified by qualified experts.

Thomas H. Morgan did not become fully committed to genetics research until he was in his early forties. Morgan was born in Lexington, Kentucky, in 1866, the son of an eminent hemp manufacturer and local gentleman. As a boy Morgan loved natural history, collecting specimens and making systematic observations. He graduated from the University of Kentucky in 1886 and took his doctorate from Johns Hopkins University in 1890, with emphasis in experimental embryology and morphology. Until 1904 he taught at Bryn Mawr College, then on to Columbia.[5] At bottom, Morgan was profoundly anti-religious; as a young man he took up science as an instrument with which to combat religion. Believing that religion fed on mystery, he devoted his career in science to showing that all biological phenomena were scientifically explicable.[6] Whether his family background, his educational experiences, or both, shaped his attitudes toward religion and science remains unclear; but the impact of his religious attitudes on his work was not. Consistently Morgan doubted any theory that struck him as mystical. As late as 1909, for example, he questioned whether Mendelian "units" actually existed in the germ plasm, or whether some scientists had invented them merely to explain their theories. And for many years he distrusted the theory of natural selection, even when the accumulated evidence for it was overwhelming, because as a young biologist in the 1890s he had decided it had mystical and teleological implications.[7] Morgan went about his work in genetics in a radically different way than did most early Mendelians, as, for example, his friend Davenport. Davenport promoted scientific institutions and, of course, the eugenics movement; he was too fascinated by eugenics to do careful work, and eventually his scientific reputation diminished.[8] Morgan was zealous, too—for a profoundly anti-mystical experimental biology, and for careful, rigorous, painstaking research. Their different interests led them to different careers and divergent interpretations of nature.

Morgan did his most important genetics work in the quarter century he spent at Columbia. In many respects this future Nobel prize winner was to profit from his milieu. The department of zoology in which he taught was one of America's most distinguished; he had able colleagues, most notably the cytologist Edmund Beecher Wilson, who helped him understand the important role the chromosomes played in inheritance, and the university attracted bright, enterprising graduate students who worked in his laboratory and who made important contributions to genetics theory. Morgan was more the inspirational leader of the Columbia genetics group than an original genius himself. Indeed, Morgan became interested in genetics in a circuitous fashion. An early supporter of the de Vriesian mutation theory, he began doing genetics research, not because he was a convinced Mendelian—for he was most emphatically not—but because he wanted to identify mutations. His interest in finding mutations led him to breeding experiments, first with a variety of species, eventually with *Drosophila melanogaster,* the fruit fly, which his friend William E. Castle told him about in 1907.[9] The flies could be bred quickly and maintained at minimal expense. After about two years of frustration, in 1910 he began to get results, and in time the appearance of mutations helped swing him over to Mendelism. By 1910 he was persuaded that Mendelian units or factors physically existed in the chromosomes, and functioned in inheritance and variation.[10]

In the fall of 1910, Morgan assembled a cadre of bright young workers in his so-called "fly-room," a room near his office in Schemerhorn Hall at Columbia that was the physical and social locus for the fruit-fly genetics work of the Morgan group. He brought two brilliant undergraduates, Alfred H. Sturtevant and Calvin B. Bridges, into the fly-room as the first of many such workers. The fly-room group functioned for another eighteen years at Columbia, and a decade thereafter at California Institute of Technology. The facilities of the Columbia fly-room were not impressive, not even by the standards of the day before "big science." The room was small and crowded, equipment was usually makeshift (flies were often bred in milk bottles) funds were barely adequate, and, even after the Carnegie Institution of Washington began supporting the fly-room group's work in 1915, there was no fat in the research budget. But the intellectual atmosphere was, by all accounts, intoxicating and exciting. Morgan did encourage students to work on their own problems, and, although he did a considerable amount of original work, several of his most able students, especially H. J. Muller, Alfred H. Sturtevant, and Calvin B. Bridges, accomplished at least as much if not more. There soon emerged from Morgan's fly-room a highly productive and creative researchers whose work had a major impact on the genetics profession. In 1909, Fernandus Payne, Morgan's first identifiable student of inheritance, graduated with a dissertation on chromosomal distribution and sex. Payne returned to Indiana University, his alma mater, where he founded a flourishing program in genetics research;

by the early 1920s he was nationally recognized for his research. In the 1910s and early 1920s other students left Morgan's fly-room and entered the genetics profession, including *Drosophila* workers such as Sturtevant and John S. Dexter in 1914, Roscoe R. Hyde and Herman J. Muller in 1915, Bridges in 1916, Robert S. McEwen, Harold H. Plough, and Alexander Wein-stein in 1917, John W. Gowen and Mary B. Stark in 1919, Shelley Safir in 1920, Donald E. Lancefield in 1921, and several others throughout the 1920s. All won recognition, and several—Sturtevant, Bridges, and Muller—became at least as distinguished as Morgan himself. By 1928, when Morgan left for California Institute of Technology, twenty-four of the Columbia department's eighty-four Ph.D.s had written genetics dissertations primarily with Morgan, and entered the profession between 1914 and 1928.[11]

Results came rapidly from the Columbia fly-room. Within a few years a mosaic of interpretation emerged that did much to undercut simplistic views of Mendelism and extreme hereditarian views of human development. Morgan's group made several important contributions to the new science of genetics. In his early papers, Morgan reported that sex was inherited in ways irreconcilable with the earlier idea of segregation of independent unit-characters, that the so-called unit characters were not always inherited independently, and that environment could influence the development of traits.[12] Edmund B. Wilson, Morgan's col-league in cytology, suggested in 1912 the unit-characters were not independent units. "The whole of this apparatus," he in-sisted, "the entire germinal complex, is directly or indirectly in-volved in the production of every character."[13] In 1913 Morgan criticized the independent unit-character idea again. In its place he advanced the multiple factor hypothesis, which the Swedish breeder, H. Nilsson-Ehle, first suggested in 1909. According to this hypothesis, many genetic factors might influence the de-velopment of a particular trait, and each genetic factor or gene could influence many different parts of the organism.[14] In 1913 also Morgan and H. H. Goodale, a former Columbia student working in hybridization experiments at Massachusetts Experi-ment Station, reported that the spotted coat in the guinea-pig was "a very complex affair, depending presumably on a number of factors."[15] While a graduate student, Sturtevant discovered even more complexity in Mendelian heredity. He claimed that certain genes in a particular species of *Drosophila* were linked together in inheritance; they did not behave as independent building blocks or unit-characters.[16] Hence the unit-characters of the early Mendelians and the genes of the Morgan school were not the same natural substances. In 1914 and 1915 Morgan and several students—Sturtevant, Bridges, and Charles W. Metz—extended both multiple factors and linkage to species other than *Drosophila*.[17] They also published papers demonstrating the dependency of the development of inherited traits on proper environmental conditions, noting, for example, the deleterious effects of unnatural environmental conditions.[18] The more they probed *Drosophila*'s heredity, the more complex the insect's heredity seemed to be, and the smaller the units of heredity, or genes, were—a finding that would have enormous consequences for evolutionary theory. Bridges reported in 1916 that at least seven different factors or genes modified the chief eye color of *Drosophila melanogaster*.[19]

In 1915 Morgan, with Sturtevant, Muller, and Bridges, as co-authors, published *The Mechanism of Mendelian Heredity*, a synthesis of the new gene theory. The portrait they painted of heredity differed from earlier interpretations of Mendelism on several counts. For the unit-character idea they substituted the gene, which was not necessarily an independent building block responsible for but one trait. Genes could and were influenced by the behavior of other genes, the cytoplasm, by their arrangement on the chromosomes, by various environmental factors. And the facts of genetic linkage, of crossing-over (or interchange of blocks of genes between homologous pairs of chromosomes) and multiple factors for traits, they insisted, meant that so-called unit-characters were not randomly and independently inherited. The development of traits was not a simple matter of a single unit-character leading ineluctably to the formation of a fixed trait. Development was highly complex. It depended on many genes, their interaction and interrelationships, and was, within limits, contingent. In other words, heredity was not the automatic, predetermined, simple process the early Mendelians implied it was. Here Morgan responded, not so much to his growing private distaste for eugenics propaganda, but primarily to his epigenetic views of biological development. In the nineteenth century, biologists had debated whether development was pre-formationist—meaning that somatic traits were preformed pre-natally—or epigenetic, which meant development occurred from complex interactions of organism and environment. Morgan subscribed to the epigenetic view because he thought the preformationist interpretation mystical and teleological—and lacking in experimentally demonstrable evidence.[20]

Morgan's other students also corroborated the new picture of heredity and genetics in their increasingly complex and precise investigations. H. H. Plough, for example, discussed the impact of temperature in the breeding culture on crossing-over, Payne and a graduate student the inheritance of orange eye color, Shelley Safir a new allelomorph of white eye color, Muller the mechanism of crossing-over and the question of whether genes were arranged on the chromosomes in linear order—thus open-ing the door to a new threshold of technical complexity, chromo-some maps.[21] Indeed, their activities and discoveries suggested that the science of genetics was rapidly changing in the 1910s; its pioneer era was ending. Genetics attracted many more workers in the 1910s than in the 1900s, as the founding of the journal *Genetics* in 1916 implied. And the field's criteria of scientific discourse had quietly, and radically, shifted to a far more technical, esoteric, and methodologically rigorous level. Concepts such as the unit character and dominance and re-cessiveness were either dropped or drastically changed. New kinds of instruments for the laboratory, new ideas, new defini-tions, new methods had to be mastered, not to mention the accompanying information explosion. American geneticists scrambled to keep up with each other and with their European colleagues.[22] Soon a scientific information gap between geneti-cists and eugenists appeared.

Other geneticists outside Morgan's circle verified the new interpretation of heredity. Naturally as specialists they accepted the new ideas, when verified, even if they conflicted with eugenics. Take the example of George H. Shull, a rising botani-cal geneticist and protégé of Davenport. Even though Shull took his doctorate with Davenport at Chicago in 1904 and was a resident investigator at Davenport's Station for Experimental Evolution, Davenport could not transfer his enthusiasm for eugenics to Shull. Shull was profoundly interested in basic genetic and evolutionary work. He accepted new ideas without regard for their implications for eugenics. For example, Shull embraced the multiple factor hypothesis in 1910. It took Davenport years to do so, and then reluctantly.[23] And unlike Davenport, Shull believed that many of the major issues of genetics were unsettled and current methods still too primitive. "The definition of unit-characters in terms of chemistry must gradually replace the methods now in vogue, of using external morphological characters," Shull wrote a distinguished botanist in 1911. "More minute analysis by the breeder will bring to light many features previously overlooked, which will give a fuller insight into the Mendelian behavior."[24] Emphasis on unit characters and external morphological characters were precisely what Davenport and other eugenists employed.

Another group of geneticists worked in land grant colleges of agriculture. These professors of agronomy, horticulture, plant breeding, and other specialities had as little interest in eugenics as the *Drosophila* workers or other basic investigators such as Shull, for they were under considerable political and institutional pressure from farmers, farm groups, and college officials to produce economically viable agricultural varieties.[25] Their work substantiated elements of the new genetics. In 1917, for example, a doctoral student in plant breeding at Cornell, E. W. Lindstrom, reported genetic linkage in corn.[26] And in some instances, these agriculturalists took a skeptical view of earlier interpretations. For example, Howard B. Frost, a plant breeder at the Citrus Experimental Station in Riverside, California, insisted in 1917 that too many persons had made careless statements about heredity; genetic factors were merely necessary biochemical elements that interacted with each other and the environment to produce somatic traits.[27]

In the 1910s, almost a score of land grant institutions established systematic programs in genetics research. Since both applied and basic geneticists were interested in the fundamental problem of hybridization, the work done at the agricultural colleges was often important. The University of California, for example, had a good program in genetics. Ernest B. Babcock led it as professor and department head after 1913. Like many professors of agriculture, Babcock was interested in the germ plasm of crops and livestock—not of men. In his teaching he kept up-to-date on the latest discoveries in genetics, and did not mention eugenics.[28] Babcock wanted a strong research-oriented department. In discussing the kind of person he wanted appointed in animal genetics, for example, Babcock said "My idea is to have him responsible for a minimum of instruction and also to give him time enough to conduct original research which will soon bring this department to the forefront of animal genetics."[29] Younger men in his department published contributions to the new genetics.[30] And Babcock himself enjoyed an estimable reputation among geneticists.[31] But California was not alone. The University of Wisconsin's tradition in biology stretched back to the 1890s. From 1905 on, Madison offered systematic instruction in genetics, first by Samuel J. Holmes, later by others, notably, M. F. Guyer and Leon J. Cole. A number of Wisconsin Ph.D.s published work and went to other colleges of agriculture to inaugurate programs, as for example H. L. Ibsen, who finished in 1916, published on linkage in rats, and became assistant professor of genetics at Kansas State Agricultural College. In 1918 Lindstrom finished his doctorate at Cornell, went to Wisconsin for a few years, and, in 1922, became chairman and professor of genetics at Iowa State College, starting a doctoral program. The University of Illinois Agricultural Experiment Station sponsored genetics research in the early 1900s, and within two decades the genetics department was regularly training productive researchers, as, for example, Joseph Krafka, who finished his work in 1919, published an important paper on the effect of temperature on Drosophila, and established a genetics program at the University of Georgia's college of agriculture.[32]

Yet before the mid-1920s or early 1930s, the only genetics program that could rival Columbia's was that of Harvard's Bussey Institution. The Columbia genetics program was highly focused on Drosophila; the Bussey's was far more diverse. The Bussey sponsored genetics research before becoming a graduate school of applied biology in 1909, but most of its major work in the field occurred between 1909 and 1936, when it was absorbed into the University.[33] William E. Castle supervised research in mammalian genetics for most of his career, as did Edward M. East in botanical genetics. Castle retired in 1936; East died in 1938. Each made important contributions in their own right to genetics. And each directed twenty successful doctoral students, a number of whom, notably C. C. Little, Sewall Wright, L. C. Dunn, L. H. Snyder, R. A. Emerson, E. Carleton MacDowell, D. F. Jones, and Karl Sax, became nationally recognized geneticists in their own right.[34]

The work of Bussey graduates did verify, generally speaking, the complex portrait of inheritance that Morgan's group was creating by extending it to species other than *Drosophila*. That the new picture of heredity was at considerable odds with the genetics "principles" eugenists were so fond of reciting did not deter the Bussey graduates from reporting their findings. Little, for example, published papers based on his work with Castle in which he questioned the conceptions of dominance and recessiveness in Mendelian inheritance and in which he claimed multiple factors were responsible for single traits in mice.[35] E. Carleton MacDowell published important papers on multiple factorial inheritance and on *Drosophila* that helped undercut the eugenists' simplistic interpretations of Mendelian heredity.[36] East's students contributed to the new interpretation of the gene as well. George F. Freeman, for example, was a breeder at Kansas State College and the Arizona Experimental Station before coming to the Bussey. In 1917 Freeman finished his degree and published an important paper on linkage in wheat.[37] R. A. Emerson taught horticulture at the University of Nebraska, finished his doctorate with East in 1913, became professor of plant breeding at Cornell University the next year, and published many technically elegant studies on corn, as, for example, a 1920 paper on the complexities of multiple factors.[38] L. C. Dunn completed his studies with Castle in 1920, taught at Connecticut Agricultural College, and published papers strongly supporting the still-developing theory of the gene.[39] Dunn succeeded Morgan at Columbia in 1928, and rapidly became a major figure in the genetics profession and the larger scientific community. Most of the Bussey graduates entered the profession between 1915 and 1928 (roughly the same period as Morgan's group) and about one-third of the Bussey graduates won national distinction for their research before 1940.[40]

East and Castle did important work as well. East contributed much to the development of hybrid corn. Born in 1879, the son of an engineer, East grew up in Illinois and took all his collegiate degrees from the University of Illinois, in chemistry. As assistant chemist at the Illinois Experiment Station in the early 1900s he became fascinated by genetics. From 1905 to 1909 he did important work on tobacco, potatoes, and corn at the Connecticut Agricultural Experiment Station in New Haven. This work helped him win his appointment at the Bussey. After 1909 he worked chiefly on corn, bringing the inheritance of multiple factors in corn under the general rules of Mendelism, essentially what the Morgan group was doing with *Drosophila*.[41] And East discovered that the interpretations so many eugenists placed on Mendelian heredity in their writings were, at best, oversimplifications.[42] This was ironic, for East remained a moderate eugenist well into the 1920s.[43] For East as for other geneticists, quite obviously science was more important than social policy. East's hybridization work led him at one point to a long-term project on inbreeding, a subject with certain ramifications for devotees of eugenics and race theories. East and a former student, Donald F. Jones, reported the results of their work on inbreeding in corn in *Inbreeding and Outbreeding: Their Genetic and Sociological Significance* (1919). Their central thesis was that inbreeding per se was not necessarily deleterious; the good or ill of inbreeding depended entirely upon the characteristics involved.[44] Such a conclusion could, in fact, be embraced by those who believed in superior and inferior races and those who did not. East, for example, was a eugenist, he wrote of human races as if they were distinct, and he even said that intermarriage among blacks and whites would be disastrous. But his good friend, the physiologist Jacques Loeb, who sternly opposed racism, wrote East that his work confounded those who believed in "pure races": "your work is extremely important for the public at large, since you know what the Germans have done with the. . . pseudo-science of 'pure races'."[45] Yet East believed in the existence of innate racial differences.[46] He replied to Loeb, perhaps to be politic, that he thought of his remarks about the social implications of his work as "more in the nature of suggestions than anything else."[47]

Castle's work had an important bearing on the new picture of nature, too, although largely in a negative sense. For many years he worked on the mutation-selection problem. In the early 1900s he accepted de Vries' mutation theory, but by 1906 his work had convinced him this was wrong, that natural selection was the effective agent of evolution, selecting among the many inherited variations that blended in inheritance. Starting in the fall of 1907, he initiated a long-range breeding program in which he and his students bred over fifty thousand rats.[48] Castle's assumption, that the genetic factors were unstable and thus blended in inheritance, came under increasing criticism after 1910 as new discoveries, especially the multiple factor theory, persuaded many geneticists that Castle's theory of blending inheritance was incorrect. Factors did not change during inheritance to produce new traits; they were stable and far more numerous than had been previously imagined. The *Drosophila* workers in particular subscribed to this latter view.[49] After 1910 Morgan and his coworkers were backing away from the de Vriesian mutation theory and the concept of macromutations. In their eyes an especially damaging discovery for the de Vriesian theory was that new mutations were often lethal and usually had no phylogenetic importance. The Morgan group, in other words, was drifting to natural selectionism, but at the same time criticizing Castle for assuming that the inherited variation on which selection operated was itself unstable rather than stable, and that the number of genetic factors was relatively small rather than relatively large. Morgan soon made his position public. His good friend E. G. Conklin, now at Princeton, arranged for him to deliver a series of lectures at Princeton in February and March 1916. Quite obviously his work with *Drosophila* influenced him powerfully. "Of course I shall work the flies into the story wherever possible," he told Conklin in January when he accepted the invitation.[50] In his lectures, Morgan argued that evolution proceeded through the interaction of single, small mutations and natural selection. Selection acted upon each; it preserved some. Morgan's theory was one of single gene replacements, what has been called a "bean-bag" theory of species formation. This was of course a most congenial and logical idea for Morgan, who was a major architect of the multiple factor hypothesis and who strongly believed, on the basis of the *Drosophila* work, that genes were very numerous, small, and highly stable.[51]

The controversy between Castle and the Morgan school continued for another few years. In those years Castle continued to doubt the Morgan hypothesis of multiple factors and to believe that natural selection, acting on blending genetic units, was "an agency capable of producing continuous and progressive racial changes."[52] But the controversy went against Castle. Several members of the Morgan group, most conspicuously Sturtevant, marshalled overwhelming evidence that Castle's blending inheritance theory was wrong. Castle's so-called "modifiers," which acted in blending inheritance upon Mendelian units, were actually small, highly stable Mendelian units; the Morgan group even demonstrated the positions of these units on the chromosomes with chromosome maps.[53] Sewall Wright, one of Castle's students, helped disprove the blending thesis in an ingenious experiment with the rats, conclusively demonstrating that Mendelian factors could not be modified, i.e., "blended." In 1919, Castle graciously capitulated.[54]

The implications of these developments were momentous for genetics, evolution, the nature-nurture issue, and eugenics. The complexities of Mendelism, which the Morgan group and others had conclusively demonstrated (multiple factors, crossing-over, linkage, and the like), made inheritance far more contingent and complex than implied by the genetics "principles" the eugenists fondly recited. Furthermore, eugenics suddenly became highly impractical; biological engineering involved far more than the elimination or the encouragement of particular "unit-characters." If the eye color of Drosophila depended upon the interaction of several genes, presumably social traits in man, to the extent they were influenced by inheritance, were necessarily far more complex. Even more important was the new model of evolution, for the model of evolution helped reorient genetic theory and the nature-nurture issue. Henceforth, most American geneticists reconciled a revised Mendelism with the natural selection theory. It took more than a decade for geneticists on both sides of the Atlantic to stitch together the new theory in all of its complexity; but from the start their assumption was that nature (micromutations) and nurture (natural selection) worked together in phylogeny and ontogeny. Extreme hereditarian interpretations were no longer accepted. The new model prepared American geneticists in the 1920s and the 1930s to embrace the brilliant statistical elaboration of population genetics and the "synthetic theory" of evolution created by Sewall Wright and two British scientists, Sir Ronald A. Fisher and J. B. S. Haldane. Although Wright, Fisher, and Haldane approached the reconciliation of Mendelism, mutation, and natural selection in different ways, and their arguments differed in some important particulars, for present purposes we may view their contributions as complementary.[55] By arguing that statistically-defined organic populations evolved (these were radically different conceptions of species and evolution than those of nineteenth-century biology) by, in other words, providing a logically and empirically defensible conception of a biological species, and by insisting that mutation and selection were interdependent yet distinct factors in evolution, population geneticists left open the door for an eventual synthesis of natural and social science and a resolution of the nature-nurture controversy. As culture could not be reduced to nature, nature could not be explained in cultural terms. The abandonment of the model of evolution popular in the early 1900s seriously eroded the scientific justification for extreme hereditarian interpretations of human nature and culture.

This was certainly evident after 1920, as the younger generation of American geneticists, the students of the field's pioneers, and some of the pioneers, too, backed the emerging new paradigm of evolution. And some took a further step and endorsed the culture idea, which lent the prestige and authority of modern evolutionary science to the social sciences. Thus in 1924 Thomas H. Morgan argued that most human traits were not transmitted as simple Mendelian units, and human behavior could not be reduced to mere biological traits. Man differed from the animals in that he had a social inheritance; he lived in a cultural world of ideas and symbolic language that shaped his behavior.[56] Herbert Spencer Jennings, a distinguished geneticist at Johns Hopkins, took up the nature-nurture issue directly in the mid-1920s. He listed the many discoveries of the past decade that so severely undercut extreme hereditarian ideas: the multiple factor hypothesis, the complexity of genes, the complexity, indeed, of the processes of inheritance and development, and the importance of environment. "Neither the material constitution alone, nor the conditions alone, will account for any event whatever, for it is always the combination that has to be considered," he argued.[57] Inheritance provided potentiality. Actuality came from the interaction of heredity and environment. Jennings, too, accepted the culture idea on its own terms. Man's culture could not be reduced to nature, nor could his heredity be facilely predicted, for he inherited many potentialities. "In man the number of diverse sets of genes that may be. . . produced is very great; although it is of course not unlimited. But what the limitations are can not be stated from general biological principles or from what we know of any other organisms," Jennings insisted. "They can be discovered only by concrete studies of man himself."[58] Why biologists of the stature of Morgan and Jennings spoke out in defense of the culture theory is not completely clear; both had been disillusioned by the eugenics movement, and by its appropriation by pro-Nordic publicists such as Madison Grant. And whether they were influenced by the inner logic of the culture idea, by the new institutional legitimacy of the social sciences in American university culture after 1920, or both, cannot be definitively answered.

Well into the 1930s American and British geneticists expanded the new ideas of genetics and evolution, and in the process dwelt at some length on the ramifications of the new synthesis of nature and culture, on man's uniqueness in nature, and on the absurdity of the nature versus nurture dichotomy.[59] The influences animating them were not limited to scientific considerations; but scientific doctrines were powerful and compelling in their own right. In 1938, the British scientist Julian Huxley published an important statement of the new view in the prestigious magazine *The Yale Review*. He denied the validity of the eugenists' analogies between man and the animals and the credibility of scientific theories of racial superiority and inferiority.[60] Huxley did not write with the same sense of urgency on race that Dobzhansky and Ashley-Montagu did nine years later; but otherwise, his article of 1938 and theirs of 1947 were virtually interchangeable. Obviously the reorientation of natural and social scientific theory was at hand.

## III

On 17 September 1941, Charles B. Davenport invited Nicholas Murray Butler, President of Columbia University, to serve as honorary president of the Fourth National Conference on Race Betterment, to meet in 1942 in Battle Creek, Michigan. Three days later, Butler asked Frank D. Fackenthal, his administrative secretary, to investigate whether it would be wise to accept. "This invitation is one which I think it would be useful for me to accept provided the organization is on a high plane and has a good record," Butler told Fackenthal. "I dare say it relates to the negro problem. If it relates to eugenics, however, I would not be interested."[61] Fackenthal checked into the matter, and several weeks later informed Davenport that President Butler decided he must decline the honor because of other commitments.[62]

This small episode suggested how serious the eugenics movement's image difficulties were with educated middle-class Americans by the early 1940s. For Butler was profoundly conservative, a man with political views congenial to eugenics nostrums. Many educated Americans turned away from eugenics in the interwar years because they learned in some fashion it had little scientific credibility. As president of Columbia, Butler naturally heard of the new genetics. Certainly many scientists and other educated Americans read about the new ideas in scientific and popular media. College students in many majors listened to lectures on multiple factors, linkage, crossing-over, mutation, selection, and the like. In such ways many educated Americans heard of the new ideas.

Yet the decline of the eugenics movement's reputation among educated middle-class Americans owed far more to the public criticisms of eugenics made by recognized scientists after the early 1920s. Those who did not hear of the new ideas, or saw no relationship between them and eugenics, could nevertheless follow a drama in their daily newspapers and in popular magazines: the withdrawal of geneticists' support for eugenics and scientific racism. This was devastating for eugenics and scientific racism, for their ultimate sanction lay in science; divested of that legitimacy of science, both were now perceived as elitist political ideologies.

In the 1910s and early 1920s a scientific gulf appeared between working biologists and geneticists on the one hand and eugenists on the other. Initially, most working biologists and geneticists were not, as a group, hostile toward eugenics. Their research simply crowded active participation in eugenics from their personal agendas. Thus Castle wrote Davenport as early as 1910 that he could not serve on a eugenics committee because: "My hands are quite full enough with the study of animals. I am quite willing to leave to others the study of human heredity."[63] Most working geneticists worked on infrahuman species; as Castle's statement implied, participation in eugenics would have meant shifting their research to human heredity, technically a more difficult experimental subject than rats or *Drosophila*. Of the leading genetics workers of the 1910s, only Morgan was doubtful of eugenics. Occasionally geneticists grumbled in private letters about the eugenists' outmoded ideas and sloppy methods.[64] But overall they did not criticize eugenics in public; the only recognized biologist who did so before the early 1920s was Wesleyan University bacteriologist Herbert W. Conn.[65] Far more typical for working geneticists who became disenchanted with eugenics was to withdraw quietly, or to ignore eugenics. Even Morgan kept his counsel; when he resigned from a committee of the American Genetic Association and an editorial position with the Association's *Journal of Heredity,* he explained to Davenport that "if they want to do this sort of thing, well and good; I have no objection. It may be they reach the kind of people they want in this way, but I think it is just as well for some of us to set a better standard, and not appear as participators in the show."[66] Most of the leading geneticists of the 1910s accepted the new research. The gulf between them and eugenics activists widened even more.

An even more obvious gap was between the eugenists and the younger geneticists who entered the profession after 1910. With a few exceptions, these geneticists were too involved in research to take an interest in eugenics. Only one of them joined a eugenics organization. As a group they did not permit eugenists to use their reputations, except for their own purposes.[67] To a far greater extent than was true of the early Mendelians, this second generation had little or no time for eugenics, because they were much more interested in solving the new riddles of genetics and evolution. And in any case human genetics was a far more difficult and frustrating field for an ambitious and productive researcher.

How much the gap between working geneticists and eugenists had widened by the 1920s was illustrated by the issue of whether geneticists should participate in the Second International Congress of Eugenics, scheduled for September 1921, in New York. Some geneticists, notably Morgan and the English geneticist

William Bateson, strongly opposed participation in the Congress, even though an important committee of the National Research Council supported participation, because, as Bateson put it to Morgan, "we might do ourselves great harm."[68] At least fifteen prominent working geneticists participated in the meeting by reading papers, but largely on their terms; all but two of them ignored eugenics issues entirely and gave detailed reports of their own genetics research.[69] The two who dealt with eugenics did so from the relatively noncontroversial perspectives of population and agricultural problems.[70] That these geneticists skirted matters of great moment to eugenics propagandists, such as race suicide, immigration restriction, hereditary incompetence, and the dangers of race-mixture, suggested they had little in common anymore with leaders of the eugenics movement. As Alfred Sturtevant, one of the most brilliant and creative *Drosophila* workers, wrote to a geneticist at Berkeley, saying, "Eugenics conference wasn't so worse [sic]. The long-haired guys weren't so numerous as expected, and flocked mostly by themselves."[71]

Presumably most working geneticists and biologists might have been willing to let well enough alone, and not make public criticisms of eugenics, had the tone of American politics not changed in the several years after the Eugenics Congress. At least they left that impression in their correspondence. But eugenics became entwined with the abrasive political issue of immigration restriction, and many prominent eugenics publicists, such as Madison Grant, tied eugenics and the Nordic theory of racial superiority together.[72] In 1921 Congress passed a temporary restriction law; apparently economic, not racial, considerations were chiefly responsible. Immigration restrictionists then clamoured for a permanent law in late 1921 and after. Now they made conspicuous use of racist arguments. These were years of intense WASP backlash against the "new" immigrants, as the flowering of the Ku Klux Klan and the reception of the Army intelligence tests, for example, suggested. And apparently there were some shifts within the pro and anti coalitions on restriction diminishing the interest of some business firms in restriction, and, thus, made the racial issue even more obvious.

And a number of highly visible proponents of restriction drew attention to the racial issue also. The House Committee on Immigration and Naturalization, which began to hold hearings on a permanent immigration restriction law in 1922, was largely dominated by rural and small town Congressmen from the middle west and the south who made no secret of their belief that "non-Nordics" were inferior and unacceptable material for American citizenship. These harsh, narrow attitudes offended the sensibilities of the many reputable biologists who, as men of science, considered themselves cosmopolites above such unsophisticated and crude cultural nationalism.[73] Nor did the attempts of eugenist propagandists such as Madison Grant and Lothrop Stoddard to make use of genetics for their thinly-veiled Nordic arguments sit well with many working geneticists, who, regardless of what private feelings they may have about minority groups, were committed as scientists to making careful, reasoned statements on the basis of the evidence. From the standpoint of working geneticists, perhaps even worse was the spectacle of Harry H. Laughlin, a protégé of Davenport's, making wild statements in the press and before the House committee as the latter's "expert" consultant in eugenics. Quite apart from his activities as a proponent of restriction, Laughlin's reputation as a scientist was not enviable. But when he repeatedly justified restriction in the most strident pro-Aryan terms in public forums, even geneticists who were politically conservative and had supported eugenics found Laughlin too much to swallow. Raymond Pearl, for example, a reputable statistical geneticist, intimate of Henry Louis Mencken, and sometime supporter of eugenics, became appalled by the likes of Laughlin and Grant. He wrote his friend Jennings in 1923, "I have a strong feeling that the reactionary group led by Madison Grant and with Laughlin as its chief spade worker were likely, in their zeal for the Nordic, to do a great deal of real harm."[74] Pearl put it well. A number of reputable biologists of varying political hues—the physiologist Jacques Loeb, a socialist, E. G. Conklin, a conservative Democrat, H. J. Muller, a radical, not to mention and old-tie Republican conservatives such as East and Little—were quite willing to tolerate the eugenics movement and to perhaps even support some of its goals so long as the movement appeared disinterested, scientifically credible, and dignified.[75] But the activities of Laughlin and others put science—genetics, after all, so far as the public could tell—in a distasteful and ugly light.

From January to May 1924 Congress debated a new restriction bill. The Nordic issue was highly visible. Congressmen in the House often made blatantly racist speeches, and cited Laughlin approvingly. Indeed, Laughlin figured so prominently that Jennings became outraged; he took time away from his own research to prepare rebuttals in the popular press and before the House committee.[76] In January 1924, Jennings published a heavy critique of much of Laughlin's testimony in *The Survey*. Several distinguished scientists wrote Jennings approving letters, including Hubert Lyman Clark, a Harvard zoologist, Edmund C. Sanford, a psychologist at Clark, E. Carleton MacDowell, a genetics investigator at the Station for Experimental Evolution, Samuel J. Holmes, a geneticist at the University of California, and Vernon Kellogg, a former Stanford zoologist now permanent secretary of the National Research Council. These gentlemen were obviously disturbed by what they regarded as the flagrant misuse of science by such persons as Laughlin and Grant, not to mention the Congressmen who so vehemently favored restriction.[77] And when Representative Emanuel Celler, one of the new anti-restrictionist members of the House Committee, arranged for Jennings to testify before the House committee, Jennings was treated badly by the majority; this shabby spectacle infuriated other reputable biologists and blackened eugenics' reputation among them even more.[78]

At first some reputable geneticists were reluctant to criticize eugenics in public in part because of Charles Davenport's dual role as champion of eugenics and head of a major genetics research institution. Davenport had befriended many a neophyte geneticist and had been provided support to many geneticists. And apparently Davenport was always hospitable and gracious. Hence personal considerations did, in part, keep some geneticists from speaking out.[79] But the pressure was too great; something had to be done. In 1922, Conklin, Castle, and Little—all working geneticists who had not shared Morgan's severe reservations about eugenics—published statements in which they warned eugenists to catch up with recent developments in genetics and to tone down their arguments.[80] Those reputable biologists still interested in eugenics in the 1920s were most circumspect and cautious.[81] Apparently, however, the public furore over the National Origins Act pushed a number of prominent geneticists into taking further steps. Between 1925 and 1930 Jennings,

Castle, Pearl, Morgan, and, finally, Conklin, published thorough criticisms of eugenics, emphasizing in particular the gap between eugenics propaganda and modern genetics.[82]

By the late 1920s, then, the gap between eugenics champions such as Grant and Laughlin and most reputable geneticists was ineluctably widening. Eugenics came into serious disrepute, in large measure because divested as it was of scientific legitimacy, it was now a political movement that deeply embarrassed many working geneticists.[83] Yet more geneticists kept their counsel then spoke out on eugenics. Some who remained silent simply believed it was not the responsibility of the geneticist to discuss political issues. In some other instances, quite probably constraints of a different order operated; geneticists were well aware that those laymen who controlled much of the research funds for genetics had favored eugenics in the past, and some were afraid of giving offense.[84] And at least some geneticists kept quiet about the current eugenics movement because they still believed in the ideals of eugenics, and hoped for the reconstruction of a new scientific eugenics once the current movement had collapsed. Silence would emphasize their distance from contemporary eugenics.[85]

By the late 1920s, then, the eugenics movement was in serious difficulties. Symptomatic of eugenics' declining fortunes after the late 1920s were certain changes in eugenics institutions. In 1929, John C. Merriam, president of the Carnegie Institution of Washington, appointed a committee of scientists to evaluate the Carnegie's Eugenics Record Office. Merriam had for many years supported eugenics, but, as he stringently insisted to eugenics popularizers Albert E. Wiggam and Madison Grant, eugenics had to develop scientifically and responsibly.[86] That Merriam felt it necessary to appoint a committee to review the work of the Eugenics Record Office was eloquent testimony to the eugenics movement's scientific reputation. Committee members included anthropologists A. V. Kidder and Clark Wissler, psychologists E. L. Thorndike and Carl C. Brigham, and geneticist L. C. Dunn, in addition to, of course, Davenport and Laughlin of the Eugenics Record Office. The committee met several times, but, as Kidder said later, its hands were tied, and no changes could be made until Davenport retired, which he did in 1934.[87] In 1935, President Merriam appointed a new committee: Kidder,

Dunn, two physical anthropologists, E. A. Hooten and A. H. Schultz, and an ethnologist, Robert Redfield.[88] This time Merriam excluded Laughlin, perhaps because he questioned Laughlin's scientific reputation.[89] The new committee recommended closing the Eugenics Record Office. This happened, and the egregious Laughlin was retired a year early.[90] A scarcely less important manifestation of the declining fortunes of eugenics was a change in the *Journal of Heredity,* the official organ of the American Genetic Association. Apparently for years a donor had supported the *Journal* with the understanding that it published articles favorable to eugenics.[91] Of course the *Journal* also carried technical papers. In the 1920s the discrepancy in the *Journal* between its enthusiastic editorials extolling eugenics and its sober, technically rigorous research reports became painfully apparent. In the early 1930s several recognized geneticists helped arrange the appointment of a new editor who was closer to their views.[92] By 1939 the *Journal of Heredity* had made its new policy of caution towards eugenics open and explicit.[93] In the wake of these developments, geneticists could develop human genetics as a respectable scientific field, and a moderate group of leaders took over the eugenics movement.[94] The old eugenics movement and its ideology had died out.


## IV


Social scientists influenced the ebbs and flows of the heredity-environment controversy in American science. While geneticists criticized eugenics, social scientists spoke out on the race issue. Some social scientists did criticize eugenics.[95] But they emphasized race—the Nordic hypothesis—more than eugenics, probably because the "Aryan" interpretation of human cultures was closer to their expertise than was eugenics. Beginning in 1918, a growing number of American sociologists began to produce detailed studies of the "new" immigrants, such as the classic *The Polish Peasant in Europe and America* (1918) by William I. Thomas and Florian Znanecki, and on black Americans, such as University of Iowa sociologist Edward B. Reuter's monograph on the American mulatto. The hallmark of these studies was their attention to cultural as well as biological influences, their insistence that cultural phenomena operated autonomously from the dictates of race and other biological determinants.[96] Throughout the 1920s and into the 1930s sociologists published many such studies, which, in their totality, affirmed the importance of the cultural level of human existence and which also implied that level of human existence to be the domain of the social sciences. American anthropologists often attacked the Nordic theory directly, especially in the years in which the restrictionist controversy waxed hottest.[97] Anthropologists and sociologists alike argued from the culture theory's assumptions; they never embraced a single-minded cultural environmentalist position that denied man's animal ancestry, or rejected the power of heredity. In a manner strikingly similar to some reputable biologists in the 1920s, they insisted that the cultural and the biological were separable levels of human reality that nevertheless interacted in development. Hence a consensus was emerging among natural and social scientists on man, and, one might add, an amicable division of disciplinary territory. Even when social scientists published semi-popular essays, with a definite political purpose—for example, in a famous series on race in *The Nation* in 1925—they debunked racism, but not evolution or the natural sciences.[98]

Yet more than sweet reasonableness, if that is what it was, figured in the disputes, hidden and public, between natural and social scientists on race in the 1920s. Part of the struggle they fought was within the secluded world of American scientific institutions. These were important conflicts because in large measure those individuals who controlled scientific institutions could influence future patterns of research on race.[99] After the World War, some powerful American scientists wanted to promote scientific studies of race—of the "new" immigrants in particular, but blacks as well—through the auspices of the National Research Council. Robert M. Yerkes had directed the Army's testing program, and Charles Davenport had done an anthropometric study of servicemen. These seemed promising beginnings for a long-range study of racial groups in the American population. John C. Merriam, former California paleontologist, supporter of eugenics causes, and, after 1920, president of the Carnegie Institution of Washington, actively pushed for such a research program. In 1921 and 1922, he had the Institution sponsor a series of conferences on fundamental research in biological problems. All invited participants shared Merriam's belief in eugenics, in heredity, in race, in biological anthropology, and many of the proposals of these conferences centered on race and immigration in one way or another.[100]

Robert Yerkes participated in the conferences, which helped inspire him to launch a major research program on race through the auspices of the Council's Division of Anthropology and Psychology. As the Council's director of information services, he sat on the Council's board. Yerkes persuaded his colleagues to appoint him chairman of a new committee on the scientific problems of human migrations. This meant racial migrations—of immigrants to America, and southern blacks to Northern cities. Yerkes was interested in the public policy ramifications of race because the World War had convinced him that racial differences mattered in human affairs. In 1922 Yerkes organized support for the migrations committee, and secured a small grant from the Russell Sage Foundation to defray the committee's administrative expenses. In November he presided over a special conference attended by a number of scientists, including some openly favorable to eugenics and the Nordic hypothesis. The purpose of the conference was to present ideas for new research to the committee.[101]

Yerkes then recruited the rest of his committee and laid the ground for its smooth operations. Matters went awry almost immediately. The Russell Sage Foundation persuaded Yerkes to appoint one of its officials, Miss Mary Van Kleeck, director of industrial research and a former social worker, to the migrations committee. She and Clark Wissler were the only social scientists on the committee. The rest were natural scientists who were probably sympathetic to eugenics and to racial interpretations of man. At the January 1923 meeting of the full committee, Yerkes proposed general areas of research, all skewed toward the biological sciences: better methods of racial testing, studies of race-mixture, studies of the immigrant problem using individual case studies, environment and race, and better methods of selecting aliens.[102] Miss Van Kleeck brought the committee members up short when she suggested that the committee should cooperate fully with the social scientists, who were then organizing what later became the Social Science Research Council. Yerkes and the other members accepted Miss Van Kleeck's proposal, because they saw no alternative; but now social scientists had their foot in the door.[103] John Merriam's brother, Charles Merriam, the Chicago political scientist most important in launching the Social Science Research Council, hurriedly appointed a subcommittee on migrations research in March, two months before the Social Science Research Council was officially founded. Now Yerkes' migrations committee was obligated to meet with the social scientists; indeed, Merriam appointed Yerkes to the Social Science Research Council's committee.[104]

On 29 March 1923, the Social Science Research Council group met in the National Research Council's Washington office. In his opening remarks, Yerkes declared that his committee was really not encroaching into fields in which it had no expertise; it was simply that migrations problems were, by definition, interdisciplinary. The social scientists took up that theme, and masterfully preempted much migrations research from Yerkes' committee. There was no bitterness, no hostility; all were polite, discrete, and indirect. But Yerkes had a long day. Chicago sociologist Robert E. Park, for example, deftly retrieved the whole field of immigrant communities, arguing that the valid way to define immigrants was as subcultures, not as biological individuals. Columbia sociologist William F. Ogburn insisted on the culture-nature division of labor among the social and natural scientists. Others remarked in passing on the importance of Franz Boas' work in physical anthropology, engaged in learned discourses on research methods and other technical issues, and insisted human groups could not be arranged in hierarchical rankings of superiority and inferiority. The social scientists also discussed employing black scholars and researchers, notably Carter G. Woodson and Charles S. Johnson, a conversation that simply did not occur in the National Research Council migrations committee's deliberations. Overall the social scientists left the clear impression that much detailed, sophisticated work remained to be done—by social scientists, or, at least, by investigators using social scientific methods and perspectives. Even Raymond Dodge, a psychologist and National Research Council committee member—and dear friend of Yerkes—declared in the afternoon session that natural scientists had no adequate way of studying cultural behavior and processes.[105]

It was a measure of Yerkes' integrity, not to mention his commitment to scientific specialization, that despite what must have been a disturbing meeting he cooperated fully with the Social Science Research Council, he accepted the social scientists' claims graciously, he tried to promote amity and cooperation between the two committees, and, so long as he dominated the National Research Council committee, he steered it in ways respectful of the social scientists. In the spring of 1923, Yerkes' committee generated several research projects. The most ambitious was Carl C. Brigham's proposal to develop culture-free intelligence tests; all proposals were oriented toward psychobiological issues. Then Yerkes secured a large grant from the Laura Spelman Memorial Fund of the Rockefeller Foundation, which financed migrations research for three years.[106]

The understandings between the two committees became unstuck in 1924, largely because the Social Science Research Council group had not yet launched a program, which suggests that the social scientists' initial efforts were probably organizationally premature, perhaps animated by the possibility of natural scientists preempting social scientific work. Discussion arose in 1924 in the National Research Council's committee of new projects, including studies of Negro intelligence, human racial blood groups, and immigration—at the time when Congress heatedly debated the immigration restriction bill.[107] Relations between the National Research Council and Social Science Research Council committees subtly deteriorated, in part because Yerkes was now absorbed in his new position at Yale and anxious to begin the primate work that became the capstone of his scientific career. In the spring of 1925, Yerkes asked to be replaced as chairman. At a tense meeting that April, Harry H. Laughlin was included as an investigator for the immigration project. Miss Van Kleeck and Columbia University economist Wesley C. Mitchell criticized the committee for slighting the social sciences and pushed through a resolution insisting on joint cooperation.[108] At the next month's meeting came the second turning point. Miss Van Kleeck secured agreement that all social scientific work would be turned over to the Social Science Research Council; Yerkes' resignation was accepted, and the committee's dissolution was openly discussed.[109]

Thereafter the Social Science Research Council was sufficiently organized to take over the social scientific migrations work.

At a Social Science Research Council conference held at Dartmouth College in the summer, social scientists discussed the areas of research most worthy of encouragement, including migrations work.[110] Subsequently the Social Science Research Council appointed committees to propose research on the American Negro, on labor, crime, mental defectives, and migrations—all topics of intense interest to eugenists as well as the National Research Council migrations committee.[111] Social Science Research Council officials persuaded Yerkes' successor in the National Research Council committee that the work of the two groups was complementary.[112] In 1926 the National Research Council migrations committee began to fade. No new proposals were accepted, and current projects were quietly shelved. The Social Science Research Council now moved aggressively into the field of race. Its Advisory Committee on Interracial Relations, with a genuinely integrated membership, did much to propose and support research in black-white relations in the later 1920s; work on Negro history and culture by Charles S. Johnson, Howard W. Odum, E. Franklin Frazier appeared, and so did work on race psychology recommended by Joseph Peterson, a white psychologist, Melville J. Herskovits, a student of Boas', and Ernest E. Just, a distinguished black zoologist at Howard University. The Advisory Committee was discharged in 1930, but not before it had established the scholarly study of black people, formerly consigned to black institutions and scholars, as a legitimate field of inquiry in the Social Science Research Council, the National Research Council, and in social science departments in many major universities.[113] Somehow the Social Science Research Council migrations committee never launched a systematic program; it was formally organized in May 1924, agreed upon a research program in December 1926, and was reconstituted as a committee on population problems in 1927. But since the National Research Council migrations committee was wrapping up its affairs in 1927, there was no need, as there had been three years before, for the Social Science Research Council committee on migrations.[114]

The influence of social scientists and of the Social Science Research Council on the study of race in the late 1920s was shown in other ways, too. In 1927, the National Research Council's Division of Anthropology and Psychology appointed a new committee on the American Negro. Perhaps symbolic of the delicate balance between the natural and social sciences was that Franz Boas *and* Charles B. Davenport were committee members. Indeed, Boas' stature with the National Research Council figures had improved dramatically since the early 1920s. Part of the reason was that Boas had interceded two years before, when the American Anthropological Association all but condemned the National Research Council's Division of Anthropology and Psychology for ignoring anthropology and favoring psychology. Boas acted the shrewd diplomat in the affair. In effect, he prevented the Association's censure of the National Research Council by obtaining a new "understanding" on the part of Research Council officials of the importance of cultural anthropology.[115] And both Boas and his students effectively answered the criticism, voiced by natural scientists, that anthropologists ignored physical and psychological aspects of man. In the 1920s, Wissler promoted physical and cultural race studies in the National Research Council; Margaret Mead began her investigations of Polynesia, for example, with Council backing. Melville J. Herskovits began intensive work on American and African blacks. And Boas himself did research on race. By the late 1920s, the Boasians' professional credentials were in much better order than in the late 1910s; and so were those of all social scientists to speak out on race. Thus in the National Research Council's new committee on the American Negro, Boas easily won his argument to cooperate with the Social Science Research Council's Advisory Committee on Race Relations.[116]

How much things had changed, and were changing, in the 1920s, as compared with the prewar years, was suggested by the publication of two major studies of blacks in the later 1920s. The first, *Race Crossing in Jamaica* (1929), was by Charles B. Davenport and Morris Steggerda. In 1926 an anonymous philanthropist gave the Carnegie Institution a gift to investigate "the problem of race crossing, with special reference to its significance for the future of any country containing a mixed population."[117] Davenport and Steggerda did work that essentially looked backwards, toward pre–World War hereditarian science. Davenport never really accepted the new genetics and evolutionary theory of the post-1910 era, although if pressed he would concede that the single unit character theory was wrong. Yet he firmly believed in polygenism. Man, he told Conklin in 1925, "is an imaginary being that does not exist in reality . . . made up of a series of biotypes which might be isolated like those of Shepherd's purse, if only breeding were controlled. It is just because of the existence of these biotypes that different races of mankind have different physical characters, different mental qualities and different mental values."[118] This notion of a hierarchy of races permeated the Jamaica study. Davenport and Steggerda worked up physical and mental measurements of "pure," "brown," and "light" blacks, to determine whether racial hybridization was beneficial, a question of little moment to most reputable biologists after the work of East and Jones in 1919. The anthropometric data was carefully done, but, as Castle pointed out, Davenport and Steggerda drew untenable conclusions from it.[119] Nor did the use of mental tests widely assumed by then to be culture-bound help. Davenport and Steggerda concluded, for example, the races did differ, and suggested, among other comparisons, that blacks were more musical and less intelligent than whites.

In 1928 the other study by Melville J. Herskovits appeared, entitled *The American Negro: A Study in Race Crossing.* His focus was on the physical traits of American and African blacks. Whether he and Boas intended this as an answer to the criticism that Boasian anthropology ignored the physical aspects of man is unclear. But that was its effect; and Herskovits took steps to conciliate natural scientists by, for example, corresponding regularly with Davenport about his research. In his book, Herskovits argued that American blacks constituted a homogeneous population, created by social circumstances (most notably slavery and segregation) as well as by biological forces. The interaction of culture and nature, then, was the book's leitmotif; perhaps its most salient conclusion, supported by a searching critique of contemporary genetics, was that neither fears of race-crossing nor categorizations of race were justified. We know nothing about race, he concluded; we cannot provide a systematic definition of it, thus illustrating "how much we take for granted in the field of genetic analysis of human populations."[120] The essential concept in Herskovits' study, then, was not race, biotype, or the like. It was *population,* a statistical concept signifying, in physical terms, the sharing of a large number of similar genes by a definable group of people. Human populations were capable of considerable interbreeding. Herskovits' implication was clear: man was a single, interbreeding species, influenced by his nature and culture. The concept of population as he used it was strikingly and remarkably similar to the one Wright, Fisher, Haldane, and other population geneticists were using to reorder and reshape the theory of evolution. Herskovits' study, in direct contrast to that of Davenport and Steggerda, had enduring value for American anthropologists and social scientists; it also spoke to the fundamental issues of genetics and evolutionary theory as geneticists perceived them then—and ever since.

## V

"What a world is facing us if Britain goes down as I dreadfully fear she will," wrote Lewis M. Terman to Robert Yerkes in September, 1940, at the height of the Battle of Britain. "Intellectual freedom and bills of rights gone except in this country and a few other small areas; brutal repression of the individual elsewhere."[121] Indeed Terman wrote in times distressing for many American scientists. The Depression's economic and psychological impact upon the personal and professional opportunities of American scientists had been harrowing enough. Far more upsetting to many American scientists was the rise of repressive military dictatorships in the world, and especially in Europe.

For American scientists, to a far greater extent than their fellow countrymen, what happened in Europe was of considerable importance. In many respects Europe was, for American scientists, the center and the source of science and culture, the continent from which the science of the western world had been elaborated in the past three centuries. Many American scientists were horrified by the rise of Nazism in Germany. The Third Reich had special meaning for those American scientists who concerned themselves with the nature-nurture issue. The rise of the Third Reich must be viewed against the broader context of the history of Europe between the wars. In the 1920s and 1930s it appeared to any intelligent person who read the international news (or heard it on the radio) that all Europe was progressively falling under the control of repressive dictatorships, starting with Admiral Horthy's regime in Hungary in 1920. Hitler was the ninth such dictator in Europe. Especially after the Nazi-Soviet Pact of 1939, it seemed the Old World had passed into a new age of darkness.[122]

Doubtless the Hitler regime represented to American scientists the most brutal and vicious example of contemporary political barbarism. They asked themselves why Germany, the center of so much of the world's culture and science, was now controlled by a government that dismissed famous scientists, purged the civil service, took over control of the universities and of cultural activities, and embarked upon a ferocious campaign to discriminate against Jews within the Reich and to implement "eugenic" policies. Much of the suppression of the German academic and intellectual communities had occurred by the summer of 1934. And by the summer of 1934, American scientists and academicians were doing what they could to rescue these refugees. In the spring of 1933, Robert M. Hutchins of the University of Chicago helped form the Emergency Committee in Aid of German Displaced Scholars in concert with other university and foundation executives; the Committee acted as a fund-raising operation that helped German (and later all refugee) scholars find appointments in American colleges and universities. Alvin Johnson founded the University in Exile, which helped other refugee scholars and scientists. And Abraham Flexner, founder and director of the Institute for Advanced Study in Princeton, appointed several famous refugee scientists, most notably Albert Einstein. Of course American higher education's financial resources were diminishing in the 1930s; naturally relatively few refugee scholars could be placed in America. Thus between 1933 and 1945, the Emergency Committee received over 6,000 appeals for assistance, and was able to help but 335 individuals for periods ranging from one to seven years.[123] The refugee problem hit American scientists hard; they formed political action committees, signed petitions, wrote memoranda and letters to university officials attempting to find places for the refugees, and raised funds, all to help resolve the problem.[124] In the 1930s many leading scientists in many disciplines worked together in such groups as the Emergency Committe or the American Committee for Democracy and Intellectual Freedom, founded in February 1939; men and women of science who a decade previously had taken diametrically opposed views on nature and nurture found new and common ground in political activism, precisely that kind of political activism that made the most sense to an American scientist: the battle for intellectual liberty.[125]

It would be inaccurate to say that the experiences and attitudes of American scientists in the 1930s and 1940s—in particular, their collective reaction to the rise of dictatorship and the implementation of racist policies by the Nazis—changed the technical descriptions that natural scientists made of the operation of nature. The most important elements of the new technical model of nature and nurture were worked out in the later 1920s, several years before the rise of Nazism. The tumultuous and horrifying events of the 1930s and 1940s persuaded American scientists of the centrality of intellectual liberty and freedom to the free and disinterested pursuit of science, and provided ample evidence of the vulnerability of science to unscrupulous politics.

What the 1930s and 1940s added, then, to the problem of nature and nurture was not a new inner revolution in the technical description of heredity, variation, and evolution, but a new sense of the importance of the relationship between science and democracy, of the dangers of the misuse of science by governments. That was the context in which American scientists joined their fellow countrymen in the years after Pearl Harbor in the common battle against fascism and dictatorship. International politics and domestic politics reinforced their commitments to descriptions and explanations of nature already known and accepted in the scientific community. It was from that perspective that American scientists stood up for the Four Freedoms of the Atlantic Charter. And it was in the spirit of this new awareness of the interdependency of liberal democracy and the pursuit of science that Dobzhansky and Ashley-Montagu wrote their article in *Science* in 1947.

# 6  INSTINCT

*I*

"THE old conception of *instinct* certainly is not useful in present-day biology," conceded psychologist Robert M. Yerkes to Zing-Yang Kuo, a Chinese graduate student in psychology at the University of California in 1922, "but the same is true of many other conceptions which originated in connection with philosophy or natural history."[1] "It is very unfortunate that many of the present-day psychological concepts are still inherited from Aristotle," Kuo replied a few days later. "I think that experimental work is the only cure for the metaphysical mind in psychology. . . . this is my main reason of [*sic*] denying instincts . . . the conception of instinct. . . is unexperimental and I therefore consider it a stumbling block in experimental psychology."[2] Indeed, the instinct controversy was all but over in January 1922. It had lasted barely half a dozen years. Before 1917 it was difficult to find any psychologist who questioned the instinct theory. By 1922 it was almost impossible to identify more than a handful of psychologists who still accepted the human instinct theory as a legitimate category of scientific explanation. A similar shift of opinion occurred among social scientists, who recognized after 1917 that they must attack the human instinct theory if they were to emancipate the social sciences from the intellectual thralldom of the evolutionary natural sciences.[3]

Why psychologists made such a rapid about-face on the human instinct theory was a complex story. Yerkes and Kuo suggested, quite accurately, that suddenly psychologists recognized that the instinct theory was a philosophical concept not congruent with psychology defined as an experimental natural science. Yet that explanation hardly suffices. The more penetrating question was why did psychologists abandon the instinct theory after supporting it, however superficially, for approximately thirty years? Changes within their discipline and their profession were largely responsible for the psychologists' abrupt shift on the human instinct theory. In the early 1900s, a new generation of psychologists came to the fore. As products of the newly established psychology programs, not mid-career recruits from some other discipline, these men and women were more closely attuned to the canons of experimental natural science than were the founders of academic psychology. These post-1900 psychologists considered themselves experimental psychologists, not merely advocates of experimental methods. More than their predecessors, they sought to establish psychology as an experimental natural science. They were far more conscious than their predecessors of the distinctions between philosophy and psychology. As young, ambitious, emerging professionals, many of their activities to establish themselves were attempts to purge psychology of all prior concepts and methods that had any relation to philosophy, natural history, or teleology, and to make psychology a behavioristic science. At the same time a number of the younger psychologists sought to transform the subculture of psychology from a small, highly personalized clique, in which a few dominant individuals controlled most of the honors and the power, into a more diverse and impersonal scientific profession in which the bases of scientific authority were impersonal, not personal, and which depended upon larger cosmopolitan and universalistic canons of training, competence, and performance. The younger generation did not consciously decide at a certain moment to throw the old rascals out. To a considerable extent they were the captives of the roles they played in university and scientific culture. They responded to underlying or implicit assumptions and normative proscriptions, often without directly recognizing the implications of their words and deeds. And in some instances purely personal circumstances and considerations mattered too:

individual resentments, jealousies, and conflicts. Yet the members of this new generation, in their espousal of experimental psychology, of a new evolutionary science of human social control, in their insistence psychology be defined as a full time scientific profession with an explicit identification with the science of psychology, were only attempting to respond to the rhetoric, the ideals, and the challenges that the senior statesmen and pioneers of the psychological enterprise in university culture had constantly urged upon them in many informal and formal situations. Appropriately, it was by accident that a leader of this new generation of behavioristic psychologists stumbled onto overwhelming evidence, most unintentionally, that the instinct theory was no more than a literary convention carried by the older generation of psychologists into the discipline of psychology from philosophy.

In both the attack on the instinct theory and in the formulation of newer explanations of the sources of human conduct in the two decades after 1922, psychologists and social scientists reconciled the cultural and the psychobiological levels of human existence. The psychologists' contribution was to construct experimentally viable models of the learning process and of human personality. The social scientists' achievement was to reconcile psychological models with cultural perspectives and models. Taken together these distinct contributions implied the balance of man's natural and cultural evolution.

## II

"Pity that the thoughts of the young men, up and down the country, run so much to status and positions! It wasn't so in England, in my time, I am sure," lamented Cornell psychologist Edward Bradford Titchener to Robert M. Yerkes, a rising animal psychologist at Harvard, in 1906.[4] Titchener's complaint possessed considerable accuracy. At least a visible proportion of the younger generation of American psychologists were indeed ambitious and restless. Indeed, the profession's inner structure was rapidly changing, with consequences none could foresee in 1906.

When Titchener launched his career in America in the early

1890s, psychology was a very small enterprise in American university culture. The psychologists with advanced training in their science and who earned their living from it numbered less than several dozen. Only thirty-two men enrolled as charter members of the American Psychological Association in 1892, and not all of these men possessed either advanced training in psychology or derived their income from teaching it; several taught in related fields, such as philosophy, religion, and education. It was not until the late 1890s and early 1900s that the pace of graduate training in psychology quickened. From 1884 to 1893, for example, American universities graduated about one Ph.D. in psychology a year. Between 1893 and 1897, American universities produced an average of five a year, a remarkable change, to be sure; but they turned out even more thereafter. From 1898 to 1907 American universities trained an average of about thirteen doctors of psychology a year, and from 1907 to 1919 they graduated twenty-three annually.[5] This rate of expansion was mirrored in the American Psychological Association's growth; the Association's membership trebled from 105 in 1905 to over 300 in 1920.[6] Hence the number of psychologists in the 1910s ran in the hundreds, whereas in the 1890s it was only several dozen.

And most psychologists before 1920 earned their living as college teachers. Even as late as 1916, when public agencies and business firms were beginning to hire psychologists, over 90 percent of America's psychologists taught in colleges and universities.[7] Most psychologists, especially the younger generation, did not win appointments in the leading departments, because many of those positions had been filled in the 1890s. The younger men and women took appointments at far more humble institutions— denominational colleges, normal schools, and land-grant colleges—where the opportunities for a professionally rewarding or an intellectually exciting career in research and teaching were minimal. Graduate education, and the professional advantages accruing to it, took place chiefly in a few leading departments, which were established by the early 1890s and which continued to dominate the profession for many years. Clark, Columbia, Harvard, Chicago, and Cornell were the major five departments; together they accounted for 71 percent of all psychology doctorates between 1884 and 1919. Only three other departments— Pennsylvania, Yale, and Hopkins—granted significant numbers of doctorates in these years. The eight departments together awarded 81 percent of all psychology Ph.D.s between 1884 and 1919. Yet they made most of their appointments in the 1890s, so that the younger psychologists commonly took their positions in humble institutions in which it was often difficult to obtain funds and space to equip a psychological laboratory for the purposes of undergraduate teaching, let alone their own research.[8] Furthermore, the younger psychologists' teaching burdens were crushing, salaries were not high, and often college officials were either unsympathetic to their research ambitions or simply had too little money for faculty research. To be sure, not all of the young psychologists wanted a career in research. But those who did often found themselves in situations in which it was extremely difficult for them to fulfill the research ideals they learned in graduate school. And it occurred to a number of these younger psychologists that many of the older generation in the leading departments appeared less interested in doing original research than in advocating that others do so.

Take the example of Thomas H. Haines, who eventually rose to a position of considerable distinction in psychology. He finished his doctorate at Harvard in 1901; his major professors were Münsterberg and Yerkes. He found his graduate education exciting and stimulating; he had worked on important experiments in animal psychology that, he believed, possessed significant implications for the causes of the evolution of mind. Yet Haines' appointment was as the only psychologist in the department of philosophy at Ohio State University. His colleagues did not always approve of his perspectives, or even of psychology. His research in animal psychology—with its monist philosophical implications—appalled them, for they embraced the dualist position on body and mind, according to which there was a wide gap between the minds of men and those of the brutes. Nor were university officials particularly sympathetic to Haines' requests. He could not get support for a modest undergraduate psychological laboratory, for teaching purposes, until 1908. And in contrast to Harvard, Ohio State was an impoverished institution for many years, which frustrated Haines. "We are horribly poor here," he wrote Yerkes in 1903. "Absolutely no new equipment here next year in any dept. [sic] save Bacteriology. And no raise in salaries."[9] Haines was strongly committed to his own research. Indeed, he even took an M.D. so that he could become a specialist in mental health. He found the local situation close to intolerable. But more than local conditions galled him. Within a few years he had become resentful, perhaps understandably, of the pioneers of academic psychology who held the best appointments, dominated the Association and the psychology journals, and possessed seemingly limitless opportunities for the very research he so earnestly desired to do. "The young fellows in this work need to stay by each other," he wrote to Yerkes. "I need the support of all my friends right here now."[10] Yet in the long run the prospects for the kind of career Haines wanted were far brighter at Ohio State University than at many other colleges and universities—thanks in large measure to Haines' efforts to upgrade psychology there and the kind of institution Ohio State became. In 1917 the newly organized department of psychology granted its first Ph.D., and by the 1930s that department held major stature in the profession—it ranked third in the production of doctorates, for example, in that decade.[11]

There were other changes as well. Increasingly the experimental ethos, and the evolving definition of psychology as an experimental natural science distinct from philosophy, became accepted, especially after 1900, and particularly among the younger, post-1900 psychologists. It was indeed the younger psychologists who did the most to foster this definition of psychology and to attempt to follow it. After all, they had not been recruited into psychology in mid-career from a background in theology or philosophy. They had directly entered psychology and could not think of psychology as anything other than an experimental natural science. In other words, increasingly, there was emerging a new definition of the psychologist as an experimental natural scientist who identified professionally and intellectually with psychology, and who drew a clear distinction between psychology on the one hand and philosophy and theology on the other. The younger psychologists, perhaps without always being conscious of it, drew a much sharper distinction between psychology and philosophy than did the 1890s generation of pioneers.

These changes were reflected also in the internal affairs of the American Psychological Association. At first, in the 1890s, any person interested in psychology, philosophy, education, or any related field could join. But the membership criteria underwent a succession of changes that excluded all but experimental or research oriented psychologists. In 1906 the Association's Council, chiefly because of pressure from younger psychologists, restricted membership to psychologists, that is, to persons who had a career commitment to psychology. In subsequent years the Council, again under pressure, required that prospective members submit a bibliography for approval, and, in 1911, prospective members had to submit reprints of their experimental papers in support of their applications. By 1916 the younger experimentalists were in full control of the Association, and they now decreed that all prospective members must possess an academic degree in psychology—which usually meant the doctorate. Subsequently the Council created two classes of memberships. Full members were senior psychologists who had already distinguished themselves in research and publication. Associate members were new Ph.D.s in psychology who had already shown promise of a research career.[12]

There were other indications in the profession of this growing sense of definition of the psychologist as an experimental natural scientist who possessed the proper academic credentials, who was not a philosopher, and who identified with the science of experimental psychology. Before 1902, the types of papers read at annual meetings ranged widely over psychology, philosophy, religion, education, and related fields, including the history of philosophy, thus reflecting the Association's heterogeneous constituency. Reports of experimental investigations were by no means absent, but they were not dominant. After 1902 the number of historical, philosophical, analytical, and introspective papers decreased, and the number of reports of experimental investigations increased correspondingly, especially in papers dealing with the objective study of individual differences, child psychology, animal psychology, and general physiological psychology. By 1913 at least three-fourths of the papers read at the Association's meetings were reports of experimental investigations, and the proportion increased as the older members in related fields left the Association or died. As James McKeen Cattell, one of the few pioneers of psychology who had broken with philosophy, remarked in an address before the twenty-fifth annual meeting of the Association, the decline of "philosophical papers at the American Psychological Association . . . does mean the establishment of psychology as a science completely independent of philosophy."[13] Another manifestation of the separation of the disciplines, and hence the professions, of philosophy and psychology was the philosophers' founding of the American Philosophical Association in the early 1900s, because the psychologists would not organize a philosophical section of the Psychological Association. The two Associations met jointly for several years, chiefly because of the close personal ties between the older philosophers and psychologists.

And there was a striking change in the type of individual elected president of the American Psychological Association. Of the fifteen persons elevated to the Association's presidency before 1908, all but four—James McKeen Cattell, E. C. Sanford, Joseph H. Jastrow, and James Rowland Angell—possessed strong intellectual commitments to philosophy as well as psychology and blurred the distinction between the two disciplines. In a few cases, the pre-1908 presidents identified with philosophy entirely. Most of the pre-1908 presidents, however, perceived no clear, definite break between the two disciplines; they taught psychology *and* philosophy. By contrast the fifteen persons who held the Association's presidency between 1908 and 1922 perceived many fundamental distinctions between psychology and philosophy. Even the two post-1908 presidents who displayed even a passing interest in philosophy, Howard C. Warren of Princeton and J. F. Baird of Clark, spent most of their time running experimental laboratories and publishing experimental papers. For them philosophy was at most a hobby.[14]

An even more dramatic indication of the changes sweeping the profession came when the younger psychologists forced a major "reform" in the governance of the Association. The episode clearly illustrated the transformation of American psychology into a modern scientific profession in which the bases of scientific authority would rest with the profession at large rather than a small group of dominant personalities. In the 1890s the Association was not yet an impersonal scientific subculture. A few famous individuals held the best appointments, owned and edited the journals, and, thanks to the Association's by-laws, constituted a self-perpetuating elite of officers within the Association. As late as 1900, attendance at meetings involved a few

dozen persons who knew one another well. In the decade of the
1900s, however, attendance rose dramatically at Association
meetings, and the bulk of the newcomers were the younger
psychologists who identified so strongly with experimental psychology. Here and there in the Association's meetings, the
younger men and women began to criticize this or that decision
that struck them as based upon the personal whims of the older
men—not the impersonal standards of modern science and university culture as the younger generation understood them. Until
1911, the retiring members of the Association's powerful Council
chose their successors and the new officers. The membership
was then asked to ratify the Council's nominations. But in 1911,
a group of the younger psychologists, dissatisfied with the behavior of the older generation, demanded that the Council find
a more democratic—that is, impersonal—method of selecting the
Association's officers, a way that would reflect the current research interests of the profession. In a sense this was beginning
to happen anyway, as in the nomination of presidential candidates from 1908 on. (But this hindsight was not necessarily
available to the young psychologists of 1911.) In 1912 the
Council began to implement the new plan; the entire membership nominated and elected the president in that year, and by
1915 the whole membership nominated and elected all officers
and council members.[15]

Another indication of these changes toward a more impersonal professional subculture was the issue of control of journals.
Initially individual psychologists owned journals as nonprofit
enterprises. Thus Stanley Hall owned *The American Journal of
Psychology* and the *Pedagogical Seminary and Journal of
Genetic Psychology*; perhaps inevitably the articles published in
their pages reflected his interests and those of his students and
disciples. In 1894, James McKeen Cattell of Columbia and
James Mark Baldwin of Princeton founded, with their own
capital, the *Psychological Review* and its satellite publications,
the *Psychological Index*, an annual bibliography, and the *Psychological Monographs,* a series in which their students could
publish their research. Eventually the experimentalist Cattell
and the philosopher Baldwin found it difficult to cooperate, and,
for a while, each took primary responsibility in alternating years.
In 1903, after much discussion, they agreed to separate; Baldwin

bought Cattell out, and Howard C. Warren, who succeeded Baldwin at Princeton in 1903, purchased the *Psychological Review* and its allied publications in 1908.[16] Initially, private ownership of journals was a necessity in the 1890s due to a lack of enough subscribers to sustain the journals. Many of the pioneering psychologists wanted to start, or did start, journals so they could publish the work of their students. Only Cattell wanted centralized publications, impersonally or corporately owned, edited, and managed because of his commitments to scientific professionalism. The phenomenon of private ownership and editorship of psychology journals began to take on a different complexion after the early 1900s, as the profession grew in size and changed in character. Apparently the editors did not engage in favoritism, at least not in any obvious manner. But as the profession changed, private ownership became less consonant with the profession's new ethos. The problem became all the more acute as more psychologists entered the profession and attempted to publish their work.[17] In 1909, for example, the young Johns Hopkins animal psychologist, John B. Watson, wrote his friend Robert M. Yerkes, "I believe personal ownership of journals is bad. I did not think so a few years ago but I see some rocks ahead for the *Psychological Review* I fear."[18] Within a few years the Association purchased the *Review* and its allied publications from their owner, Howard Warren, and put ownership and editorship on a corporate and impersonal basis. Ironically, there never seemed to be enough space to print all worthy material, yet there was often little alternative to private ownership, especially among the more specialized journals in their early years. A dozen journals by 1910 could not accommodate psychology's continuing information explosion. Thus Watson and Yerkes found it necessary to launch and edit the *Journal of Animal Behavior* privately, much to their regret, so that work in that mushrooming field could be published. There were not enough subscribers. Watson and Yerkes did obtain donations from their editorial board and some institutional subscriptions, but the *Journal* remained essentially a private, nonprofit venture until it folded in 1919 after eight years. And journal space was a persistent problem. "For some time the *Psychological Review* has been badly crowded. Lately we have had enough material offered to fill twice our space," wrote Watson, editor of the *Review,* to Yerkes in 1914. "The editors of the *Review* feel that something must be done to relieve the congestion and to afford authors a more speedy avenue for publication."[19]

By the decade of the 1910s, then, psychology as both a discipline and a profession was changing in mutually reinforcing ways. The post-1900 psychologists who defined psychology as an experimental natural science, not as a quasi-philosophical system, were now taking their place in the profession, and they were succeeding in imposing upon American psychology the mores and folkways of modern academic and scientific professionalism.

### III

In the decade before World War I, a small band of animal psychologists gradually undermined the human instinct theory. So powerful was the hold of the instinct theory over them that they did not perceive clearly that the path they chose in research would eventually lead them to repudiate the instinct theory. They took the first step when they questioned whether the conception of consciousness was appropriate for the science of psychology defined as a behavioristic, experimental natural science. Then they criticized the validity of the introspective method of psychological research, which the pioneer generation of philosopher-psychologists had utilized. It dawned on them that both were relics of traditional philosophical psychology. Correspondingly they abandoned both. Finally, their research on instincts made them realize the instinct theory, rooted as it was on the Romanes-James-McDougall anthropomorphic theory of the evolution of man's mind from its animal antecedents, had no basis in experimentally demonstrable fact.

From the start, the animal psychologists were uneasy with the concept of consciousness. It struck them as somewhat incongruent with their conception of an experimental natural science. "I quite agree with you that an 'objective criterion' of consciousness is impossible," declared Raymond Pearl, a zoologist interested in animal psychology, to Yerkes in 1903.[20] Pearl had difficulty reconciling the philosophical psychologists' assumption that consciousness existed in all sentient organisms and their use of introspection with the methods and assumptions of modern experimental natural science. Like other experimental animal psychologists, Pearl relied on testing animal reactions under controlled conditions, not on Romanes' method of compiling anecdotes of animal cleverness from literary accounts. There was, of course, no way for an animal psychologist to interview his or her subject, so that some kind of indirect means was necessary to investigate the animal's mental processes. In the early 1900s, Pearl and other researchers were shifting the focus of animal psychology from such conceptions as "mind," "mental states," and "introspection," to "behavior," "process," and "experimental manipulation of variables." Neither introspection nor the assumption of consciousness were useful to these investigators once they had decided to observe animal behavior directly rather than to depend upon anecdotes in literary sources. Of course psychologists interested in the human mind could interrogate their experimental subjects as extensively as they wished, so that the difficulties of the introspective method— and the conception of consciousness—did not occur as rapidly to them as they did to the animal psychologists.

And in retrospect it is clear that the new experimental animal psychology undercut the human instinct theory of Romanes, James, and McDougall. In a sense, what made the human instinct theory appear logical to James' generation (aside from general appeals to evolution) was the anthropomorphic interpretation of the evolution of mind that Romanes explicitly formulated and which James implicitly assumed. According to this interpretation, consciousness existed in all sentient organisms. The complexity of consciousness depended upon the complexity of the neurological equipment in particular species. There was an unbroken progression of neurological complexity, and thus complexity of mental life, from the lowest sentient species to man. In other words, the gap between the minds of men and of the higher animals was relatively narrow. Yet what the new experimental work on animals showed quite clearly was that the gulf between men and brutes was far greater than Romanes and James believed. Consciousness did not evolve; the physiological capacity for increasingly complex behavior did. Since most animal psychologists did not understand that James' definition of instincts was anthropomorphic, it did not dawn on them for some time that the deanthropomorphic conclusions of their work constituted a repudiation of the Romanes-James-McDougall instinct theory.

Consider the implications of Edward Lee Thorndike's doctoral dissertation on animal intelligence, which Cattell supervised and which was published in 1898—but nine years after James published his famous *The Principles of Psychology*, and ten years before McDougall brought out his *An Introduction to Social Psychology*. Thorndike's effort was the first systematic experimental research done in animal psychology in America. Thorndike did not follow Romanes' methods, assumptions, or conclusions. He constructed puzzle-boxes to test the intelligence of a number of different species, including chicks, dogs, cats, and monkeys. Then he placed the animals in the boxes, and recorded the number of times it took each species to learn how to open the door to the box and retrieve its reward of food. Here was objective, experimental measurement of animal behavior, according to the highest canons of modern science as Thorndike and Cattell understood them. Thorndike found that the animals, far from being almost as intelligent as humans, as Romanes and James' anthropomorphic interpretation of the evolution of mind assumed, were actually rather stupid as compared with humans. They showed no ability to reason, they did not imitate other members of their own species to learn new acts as man did, nor did they transfer learning from one concrete experience to another. They learned through trial and error, through their own experiences, through gradual association. Furthermore, the animals seemed to have many instincts that they gradually modified and perfected by experience, by association, by trial and error. Thorndike declared that Romanes' interpretation of animal mind was wrong. He insisted there was a great mental gap between the minds of men and even the highest of the brutes.[21] Obviously if Thorndike's conclusions were verified, and their implications grasped, the human instinct theory had a limited future in American psychology.

Yet so convinced were psychologists that man had instincts that none grasped the implications of Thorndike's work for the human instinct theory. The human instinct theory was a given— and therefore unquestioned—category of description, analysis, and interpretation of the new science of psychology. Even Thorndike did not perceive the difficulties he had introduced for the instinct theory, perhaps because he shifted from animal to human research.

Those who followed Thorndike in animal psychology saw no difficulty in embracing the human instinct theory *and* Thorndike's behavioristic methods and de-anthropomorphic assumptions. Perhaps they were so embroiled in their work, and in founding their specialty, that the pace of events permitted them little opportunity for reflection. In 1899, at Clark, E. C. Sanford established the rudiments of the first animal psychology laboratory, and offered the first college course. Yerkes launched a laboratory at Harvard while still a graduate student. In 1903, John B. Watson persuaded James R. Angell, his chairman at Chicago, to establish an animal psychology laboratory. Watson became director. By 1920, twelve other universities had established experimental laboratories and upper division or graduate courses devoted to comparative psychology.[22]

Yerkes and Watson soon became major figures in animal psychology, perhaps its leading researchers and theoreticians in America. And in many ways they were typical of the new post-1900 generation of experimentally-oriented psychologists, except that they both had positions in leading departments of psychology. They rose rapidly in animal psychology. But they had their own professional dissatisfactions. Their colleagues in human psychology were not always sympathetic toward their work. And increasingly Watson and Yerkes became disillusioned with the older generation's personal and professional behavior, especially their apparent lack of interest toward scientific research.

In a broad sense their experiences as young professional psychologists were remarkably similar, even though they had dissimilar personalities and different specific life experiences before becoming psychologists. Both were born and raised in an intensely pietistic, small town, middle class milieu in the 1870s— Yerkes in eastern Pennsylvania and Watson in a piney woods village of South Carolina. At adulthood both lost their childhood commitment to Protestantism and found satisfaction in a career in science. Yerkes lost his faith as the consequence of a genuine conversion to the panscientific and secular values of Cambridge, Massachusetts, and to the cosmopolitan values of the large city while a pre-medical student at Harvard in the late 1890s. Watson's family were evangelical Baptists, a most interesting background for the eventual founder of behavioristic psychology. In adolescence, Watson's faith apparently evaporated. He turned from theology and philosophy, which he now found utteringly meaningless, to science, particularly the new psychology. He entered the graduate program in psychology at the University of Chicago in 1900, just a year before Yerkes took his doctorate from Harvard. Each discovered a new world of intellectual challenges and professional opportunities, a world, as Yerkes put it later, of freedom, of liberality of view, broad tolerance, and world-mindedness, an atmosphere in which students were free to follow their own interests, to make their own discoveries, and to determine their own lives. The shy Yerkes shed his interest in medicine, and underwent a cultural renaissance in Cambridge; he attended lectures and concerts, and even went dancing with proper young ladies. The more rambunctious and extroverted Watson needed no such coaxing with the opposite sex; and he found Chicago a stimulating environment in which to develop his intellectual abilities. Both men did original experimental work in animal psychology; both published rapidly and extensively; both preferred research to teaching; and both became excited by the possibility that their work might have important implications for a new science of man that could predict and control human behavior.

Yet they differed in their interpretations of their work. Their interpretative differences mirrored larger differences within the profession at large. Yerkes had been greatly impressed by the new hereditarian biology—by the work he took in genetics at Harvard and by the contacts he made with biologists at the Marine Biological Laboratory in Woods Hole. Yerkes assumed the existence of inherited consciousness in animals. He approached the whole problem of animal intelligence from a structural perspective in which innate traits were most important. Watson, on the other hand, interpreted animal behavior as processes and functions, not as inherited structures, thanks in no small measure to the influence of such graduate teachers as John Dewey, George Herbert Mead, and Jacques Loeb. As Harvard stressed traits and structures, Chicago emphasized processes and environments. Eventually they came to understand that their differences were indeed fundamental.[23]

By 1905, Watson and Yerkes were good enough friends that they were in frequent postal contact.[24] They swapped tips about experimental equipment, they kept one another informed of their work, they collaborated to promote animal psychology, they schemed about the future, and they worked together on the *Journal of Animal Behavior* and a *Behavior Monograph* series. Even though they held good appointments and had many advantages many in their generation did not, they also had professional complaints. For example, Watson was bored teaching human psychology. And like many college professors he thought himself worthy of a far higher salary than he was receiving. He also thought the laboratory facilities at Chicago were very inadequate, and he often complained that his chairman was too tight-fisted in budgetary matters. Indeed, he believed he was simply not appreciated by his colleagues, which was probably true. In 1908 Watson resigned to take a better paying position at Johns Hopkins; he even toyed with the possibility, which Yerkes suggested, of going to Harvard the next year. But Watson remained at Hopkins, much happier than at Chicago. Yerkes was satisfied at Harvard so long as Charles W. Eliot was president; Yerkes had a full time research appointment during the regular academic year, and only taught in the summer. He directed the research of many graduate students in the psychological laboratory, and still had time for his own work. But circumstances changed with President Eliot's retirement in 1909. The new president, A. Lawrence Lowell, considered animal psychology a frill, and insisted that Yerkes teach full time. Yerkes became increasingly unhappy at Harvard.

But their discontents involved more than intramural institutional problems. Often they complained about the older generation of psychologists who seemed overinvolved in petty personal disputes. "My God what a condition of spirit for our leading men to be in," exclaimed Watson to Yerkes in June 1910, referring to a squabble over the organization of the International Congress of Psychology, scheduled to meet in Boston in 1913. "I am too busy for such rows—so are you. Why aren't they all that busy. I trust that when we younger men grow into the saddle we can set a better example to those who are to follow. I believe that we will," declared Watson.[25] Both Watson and Yerkes did work hard at teaching, research, and promotion of scientific psychology. They enjoyed their work and derived much satisfaction from it; but work had its minor frustrations. Initially Watson had to keep his monkeys at Johns Hopkins in a room without lights to save money. "They did not like the dark room and often defecated and urinated all over me," he complained to Yerkes. "I had always to make a complete change of clothing and to wear a rubber bathing cap to protect myself."[26]

Both Watson and Yerkes agreed that psychology was an experimental natural science whose practitioners should ignore consciousness and introspection and use only objective experimental methods. Psychology was, in their eyes, a consistently empirical natural science, to be purged of all philosophical concepts.[27] Yet each approached animal behavior differently. Yerkes continued to believe in innate mental traits, in inherited structures, and these beliefs eventually led him to head the Army's mental testing program during the World War and later to distinguished work on the mental life of primates. Over the years, Watson pursued a line of interpretation that led him in time to found the behaviorist movement, attack the instinct theory, and popularize behavioristic learning theory among educated Americans. By 1907, Watson had broken with the concept of consciousness. "To my mind," he wrote Yerkes then, "it is not up to the behavior man to say anything about consciousness."[28] Over the next several years Watson and Yerkes remained good friends, but their work pointed to different conclusions. Watson in particular emphasized habit-formation in the context of environmental stimulation. In 1912 Watson began to work with human subjects, using his methods. After considerable effort, he concluded that humans, like animals, learned by forming habits, through association, and without consciousness. Or, more precisely, he found it unnecessary to assume the existence of consciousness in order to explain learning to his satisfaction.[29] Then Watson disseminated his conclusions in a well attended series of public lectures he delivered at Columbia at Cattell's invitation and subsequently published in the *Psychological Review*.

Watson launched the behaviorist movement in American psychology in 1913. At this point, and for several years thereafter, Watson's behaviorism was essentially a methodological critique of the concept of consciousness and the introspective tradition, together with his redefinition of psychology as an experimental natural science recognizing no dividing line between man and brute and whose practitioners would study stimulus and response objectively. It had not yet occurred to him that by abandoning consciousness and stressing adjustment, formation of habit, conditioning, and the like, he had come very close to undercutting the human instinct theory. He was so preoccupied with his research and its immediate implications—especially the dazzling promise of a science of social control his work held out—that he did not immediately transfer his critique to the instinct theory.[30] Nor did any other psychologist who reacted to his article in the *Psychological Review* grasp its implications for the human instinct theory either. His call for a behavioristically-oriented psychology found much approval from the younger generation of psychologists, some of whom had arrived at that point independently of Watson.[31] Some of the older generation of psychologists, who blurred the distinction between philosophy and psychology, vehemently criticized Watson, but only for attacking consciousness and introspection. They too perceived no peril to the instinct theory.[32]

Perhaps one reason these psychologists did not anticipate the implications of Watsonian behaviorism for the human instinct theory was that to a considerable extent they paid it rhetorical homage but did not attempt to apply it directly to their research. The professional psychological literature of the era was barren of any reports of experimental investigations of human instincts. Psychologists disseminated and perpetuated the instinct theory chiefly in their college level textbooks, and probably in their lectures, but not in the pages of their research journals. And their textbook treatments of human instincts leave the unquantifiable impression that their authors included it for reasons that bordered on the ritualistic. At least some psychologists felt a vague disquietude about the human instinct theory's scientific validity even before Watson issued his original manifesto. In 1911, the American Psychological Association's Council planned a symposium on instincts for the upcoming December meeting. Those invited to participate initially accepted without hesitation. By the fall, however, several psychologists involved had second thoughts. Yerkes, for example, had doubts, even though he had recently written a textbook including a chapter on human instincts. As he reread the literature on human instincts, it struck him as too impressionistic and not adequately grounded in experimental verification. In November, Charles J. Herrick, an eminent neurologist and participant, told Yerkes that the symposium papers, with one exception, were not publishable because they said nothing new and were not rigorous or based on experimentation. "If it were not so late," Herrick continued, "I should be in favor . . . of calling the whole thing [the symposium] off."[33] Titchener wrote Yerkes in an amused tone that he was withdrawing from the symposium because the papers were based on personal opinion, not the methods of modern experimental psychology.[34] But it was too late to cancel the symposium. The symposium took place, and "provoked a lively discussion, which made evident . . . the need and importance of much patient, detailed observation and investigation of instinctive behavior."[35]

Watson needed no such prodding; investigation of human instincts was an integral and logical part of his larger behaviorist program. Initially he wanted merely to work out a rigorous experimental approach to human instincts congruent with his behavioristic outlook. In 1914 he published a textbook—in advance of much of his experimental work on the instinct problem—in which he severely modified the James-McDougall formulation; he argued, for example, that instincts did not serve teleological purposes or ends, that they were not transmitted as Mendelian unit-characters, that man's neurological structure was plastic rather than fixed, and that both heredity and environment interacted to produce the so-called "instinctive response."[36] In 1914 Watson began studying the behavior of infants in the Johns Hopkins Hospital. "Yes, I suppose I am monkeying a bit with human behaviorism," he wrote Yerkes in October, 1915. "I have been working now for nearly a year on the conditioned reflex, and it works so beautifully in place of introspection that I think it deserves to be driven home; we can work on the human being as we can on animals, and from the same point of view."[37] Watson made the conditioned reflex the subject of his presidential address before the American Psychological Association that December.[38] Watson and his research assistants continued their work throughout 1916, from initial infant responses to more complex behavior. "I am undertaking. . . a rather comprehensive study of the order of appearance and development of reflexes and instincts in the human child," he wrote Yerkes in the late fall. "We are also including. . . the genesis of the early habits of infants and the extent to which they can be taught."[39] Yerkes was delighted.[40] "I am glad you approve of the baby campaign," Watson replied. "It is a far more difficult field that [sic] I had anticipated. It is no wonder good work has never been done in it before."[41]

To his amazement Watson concluded that the James-McDougall human instinct theory had no demonstable exper-imental basis. He found the instinct theorists had greatly overestimated the number of original emotional reactions in infants. For all practical purposes, he realized that there were no human instincts determining the behavior of adults or even of children. Building upon his experiments with conditioning infants to make and break habits, his observations of infant native tendencies, and the work of Harvard physiologist Walter B. Cannon, Watson argued that there were only three general innate human emotions: fear, rage, and joy or love. Virtually all adult human behavior patterns were the result of environment and training rather than biopsychological inheritance. Behavior patterns were built up as conditioned reflexes, especially in in-fancy and childhood. Watson did not deny the existence of innate behavior patterns in man, but he insisted that those patterns were quickly and significantly modified by environmental conditioning. Heredity provided little more than the ability or the potential to learn, at least in man.[42]

## IV

Most psychologists and sociologists quickly accepted Watson's 1917 critique of the human instinct theory. What was remark-able about the instinct controversy was its brevity and its one sided character. Several score articles appeared in psychology, philos-ophy, and sociology journals, most highly critical of the human instinct theory. By the early 1920s the controversy was over. Subsequently, psychologists and sociologists joined hands to work out a new interdisciplinary model of the sources of human conduct and emotion stressing the interaction of heredity and environment, of innate and acquired characters—in short, the balance of man's nature and his culture.

Why the instinct controversy was so brief and one-sided cannot be adequately explained by either the scientific merits or the superior logic of Watson's critique, as estimable as they undoubtedly were. It is important to remember that Watson's investigation of human instincts was the first such inquiry ever, which is a telling commentary on the psychology profession before he began his work. And it was the only such experimental program on human instincts conducted before the controversy ended, which was equally revealing. Many critics of the instinct theory attacked it for not being supported by experimental work, even though they had done no experimental work themselves on the problem. This suggests that many psychologists and sociologists were prepared to accept a critique such as Watson's for reasons beyond its strictly scientific merits. Those reasons have to do with changes within the professions of psychology and sociology. Psychology changed even more after World War I than before it. By the time Watson published his attack, attrition had removed many of the pre-1900 pioneers of psychology; and the influence of the rest had been seriously diluted by the two generations that had emerged since 1900. The profession was simply too large after the war and in the 1920s for a small group to dominate; the number of young Ph.D.s entering the profession jumped from an average of 23 percent between 1908 and 1918 to almost 46 percent in the 1920s, and new graduate departments, such as those of the University of Iowa, the Ohio State University, the University of Pennsylvania, Stanford, Yale, and Minnesota, arose in the 1920s to challenge the leadership of the pioneering eastern departments of the 1890s and early 1900s. These changes became especially pronounced in the 1920s, the years when the instinct controversy flared up.[43]

The replacement of one generation by another may be better understood by considering the transformation of psychology at one of the new eminent departments of the twenties, at the University of Iowa. In 1887 the philosopher G. T. W. Patrick became professor of moral and mental science at the University. Patrick had taken his work with Hall at Johns Hopkins in philosophy. Patrick understood the new psychology well enough to inaugurate it at Iowa City and to teach undergraduate courses in psychology. He founded the psychological laboratory and organized a specialized library. But like many pioneer psychologists, he preferred philosophy. In 1895 he hired Carl Seashore, a young psychology Ph.D. from Yale committed to experimental psychology, to run the laboratory and teach the psychology courses so he could concentrate on philosophy. Seashore methodically built up a strong experimental program at Iowa, especially in such specialties as the psychology of music, the psychology of speech, educational psychology, and child psychology. To a considerable extent Seashore institutionalized the distinction between philosophy and psychology.[44] This happened elsewhere. At Columbia there was never any question that Cattell believed in psychology as an experimental natural science, and the men he appointed to carry on the work, such as E. L. Thorndike at Teachers' College and Robert S. Woodworth in the psychology department, pursued and promoted work in many fields of experimental psychology, including mental measurement and educational psychology.[45] The University of Chicago founded a psychology department in 1902, which eventually became an important center, despite Watson's resignation in 1908. The Chicago philosophers, notably John Dewey and George Herbert Mead, fully supported the new experimental psychology.[46] Although the details varied from one institution to another, by the late 1910s most of the major departments, and many of those devoted to undergraduate teaching as well, were dominated by the younger psychologists or at least now defined psychology as an experimental natural science. In many respects, then, Watson published his critique at a moment when many of his professional colleagues would be most receptive to it.

And the late 1910s was also a time when sociology was changing as both a profession and a discipline. A new generation of sociologists trained in sociology graduate programs and committed to original social investigation was beginning to call for the divorce of biological and social theory, which implied, at least for those of the new breed of sociologists interested in social psychology, a direct attack on the human instinct theory. In the 1920s these perspectives and commitments would lead sociologists to study, for example, immigrants in urban ghettoes and to conclude that the causes of their social disorganization were social, not biological or psychological. For present purposes it is enough to explain how and why three rising sociologists—Herbert A. Miller, Luther Lee Bernard, and Ellsworth Faris—attacked the human instinct theory in the late 1910s.

Miller took his doctorate in psychology in 1905 at Harvard. He wrote his thesis on the psychology of American blacks under the direction of Professors Münsterberg and Yerkes. At that point he believed fully in the power of race, instinct, and other natural determinants on human behavior. After graduation he taught psychology for six years at Olivet College, a small, denominational institution in Michigan—a setting that was incongruent with his ambitions for a career as a university researcher. He continued his interest in "folk psychology," perhaps perceiving in that field more room than in physiological psychology. In 1911 he took leave of absence to study "folk psychology" with William I. Thomas at Chicago. The rough-and-tumble atmosphere of the university and the city changed his outlook in several mutually reinforcing ways. Thomas suggested he study social disorganization among Chicago's Bohemian immigrants, a companion study to Thomas' work on the Polish immigrants. Gradually Thomas, John Dewey, and George Herbert Mead made Miller understand that natural science determinants such as race and instinct were not particularly useful for social scientific questions and inquiry; attributing a complex series of behavior patterns merely to instinct, for example, seemed an obstacle in the way of further analysis useful to the social scientist. And Miller was no cloistered student oblivious to the larger cultural and social milieu in which he lived and worked. Indeed, he became a socialist and conducted an unsuccessful campaign for the House of Representatives in 1912. As a socialist he found it more difficult to embrace hereditarian interpretations of man and society. Furthermore, Miller came to believe that sociology held out, for him, a brighter professional future than did psychology.[47] "The thing that is puzzling to me is how to break into a university," he complained to Yerkes in 1913. "I need the stimulus of a larger place but I do not seem to be able to get invitations."[48] Yerkes helped Miller have the opportunity to speak at the 1914 Race Betterment Congress, in Battle Creek, Michigan—a eugenics conference. Miller attacked eugenics as impractical because social problems had social, not biological, causes. "I started. . . with a criticism of [Charles] Davenport's statements," he reported to Yerkes, "and as I followed him on the program it made a good deal of stir."[49] Miller continued his work on immigrant groups, attacking eugenics, instinct, and other natural science determinants. He published an important study of immigrant groups with Robert E. Park in 1921, and eventually won a professorship of sociology at Ohio State University.

Charles Ellwood introduced Luther Lee Bernard to sociology at the University of Missouri, where Bernard was an honor student. In the early 1900s, Ellwood was much taken with the human instinct theory. So was Bernard. In 1906 Ellwood helped Bernard win a prestigious fellowship for graduate work in sociology at Chicago. There Bernard rapidly distinguished himself as a bright and energetic student. Professor Charles Henderson, who taught criminology and corrections in the department, had been much impressed by the new genetics and the new psychology. Off-handedly he suggested one day that Bernard take as his disserta-tion topic the hereditary determinants of criminal behavior. Bernard sensed the topic's relevance, and, in no small measure thanks to Ellwood's influence, at first he believed criminal be-havior had innate causes. All his professors told him that the first step to a doctoral dissertation was a systematic review of all exist-ing scientific literature on the topic. As a conscientious graduate student, Bernard began to read that literature, especially on heredity and instinct. To his considerable surprise, he discovered that natural scientists had none of the systematic definitions or logical concepts of instincts that he expected. He found instead a bewildering variety of definitions and lists of instincts that were contradictory, inconsistent, and for which there was not a scrap of experimental evidence. This sobering experience transformed him into a major critic of the instinct theory after he took his Ph.D. in 1910.[50] Throughout the 1910s, Bernard published a series of attacks on the instinct theory culminating in the classic study, *Instinct: A Study in Social Psychology* (1923). And Bernard himself was very much a young turk in the sociology profession; he held a series of positions in humble departments and felt considerable resentment toward the pioneer generation of sociologists and the American Sociological Society. "I have been seriously debating whether there is any advantage to the 'small fry' from belonging to this society. . . . I have felt that the organization itself existed primarily, if not wholly, for the benefit of a score or so of gentlemen who hold good positions."[51] It would be tempting to make too much of Bernard's ire, which could not have provided him, at least not directly, with the content of his attacks on the instinct theory. But we cannot dismiss, either, the contradiction he obviously perceived of an older generation of psychologists and sociologists espousing both the instinct theory and the exalted university ideal of original investigation.

Like Miller, Ellsworth Faris became a convert to sociology after having taken his doctorate in psychology. Born in the south in 1874, he spent his early adulthood as a missionary to Africa, and then as a professor of philosophy at Texas Christian University. In 1911 he began graduate work in philosophy and psychology at Chicago, where he was influenced by John Dewey, George Herbert Mead, and William I. Thomas. His training pushed him away from hereditary models of human behavior. He took his doctorate in 1914 and taught psychology at Chicago and the University of Iowa. After Thomas was dismissed by the University of Chicago's trustees, Faris succeeded him in the sociology department. In the early 1920s, Faris published several influential attacks on the human instinct theory in which he argued that culture operated apart from nature, and that instincts were merely hypotheses without substantiation.[52]

Two camps of anti-instinct writers appeared: behavioristically oriented psychologists who accepted Watson's definition of psychology as units of behavior or went even further to neurophysiological reductionism, and social scientists who insisted that culture could not be reduced to nature. Despite these differences in disciplinary outlook, the critics of the instinct theory found many areas of agreement among themselves. They faulted the instinct theory for its lack of experimental basis and its internal logical inconsistencies. They stressed the primacy of the cultural and social environment, the importance of habit over instinct, and, above all, the interaction of nature and culture, the notion that heredity and environment were not independent variables that could be studied separately but were interdependent variables interacting to produce adult behavior patterns.

Aside from Watson, at least four other behavioristic psychologists rapidly became the most visible critics of the instinct theory in the psychology profession. All were relatively young psychologists just embarking on their careers. They placed major emphasis on the theory's methodological difficulties. Knight Dunlap, a colleague of Watson's at Johns Hopkins, for example, attacked James' and McDougall's teleological definitions of instinct as unscientific. Purpose and teleology could not be read into natural processes, he insisted; McDougall's social psychology of instincts was "a teleology masquerading as psychology."[53] Hulsey Cason and Floyd H. Allport, who were younger than either Dunlap or Watson, argued, too, in their early professional papers that the instinct theory was teleology, not science, and not supported by any experimental evidence. Allport insisted, for example, that the fallacy of the instinct theory was the injection of social experience into the germ plasm. Zing Yang Kuo, as graduate student in psychology at the University of California, made the most radical critiques of the instinct theory in the 1920s; he continued even after his return to China. In a succession of papers, he argued that such categories of analysis and explication as instinct and heredity might be valid for biology, which studies biological structure, but not for psychology, which studies mental processes. Heredity and instinct were obstacles blocking the development of a truly scientific psychology, for once applied to a phenomenon they prevented further inquiry.[54] Few went to the extreme that Kuo did, however. The psychologists also discussed the instinct theory's logical weaknesses. Dunlap, for example, insisted that the instinct theorists offered lists of instincts serving the teleological functions their social psychology assumed to exist; given another list of instincts, a different social psychology would result. Not teleological schemes but observed social facts should be the basis of a scientific social psychology. And the psychologists also agreed that the instinct theorists undermined their argument by conceding that instincts could be extensively modified by social experience and conditioning.[55] And they minced no words about the lack of experimental proof of human instincts.[56]

The social scientists differed from the psychologists chiefly in their refusal to use the behaviorists' mechanistic and reductionistic stimulus-response formula and neurophysiological terminology. Most of the social scientists who took time to attack the instinct theory had close ties with the Chicago sociology, philosophy, and psychology departments. They wanted to construct a scientific social psychology. The most important members of this group were John Dewey, J. R. Kantor, Bernard, Park, Clarence M. Case, and Faris. Quite obviously building upon

Boasian anthropology, they argued that social psychology could not be reduced to the level of the physiological individual.[57]

Thus Dewey called for a distinction between the physiological and the cultural levels of human existence in an important article published in the *Psychological Review* in 1917, with which Kantor agreed. He criticized both the instinctivist and the behaviorists for not recognizing that the "logic of science dictates a humanistic description of human behavior."[58] Like the Boasians, Kantor argued man was an animal who lived in a unique cultural environment which social psychologists must study on its own terms.[59] And these champions of a scientific social psychology quite naturally stressed the primacy of habit over instinct. In *Human Nature and Conduct* (1922) Dewey took as his central thesis the interaction between the sentient human organism and its cultural environment. He did not deny the existence of original human tendencies. But he insisted these interacted with the culture to develop into the specific patterns of behavior that were the proper object of the new social psychology. The bewildering diversity of human customs throughout the world, Dewey continued, could hardly be attributed to McDougall's seven instincts, to James' three dozen instincts, or any other such list.[60] Bernard made essentially the same points in a very extensive analysis of the instinct problem and the instinct literature. After treating the instinct literature to devastating analysis, Bernard suggested that the new social psychology should be based on the conception of habit, not instinct, and on the recognition that culture and nature, while separable factors, interacted continually. The others made the same points.[61]

After 1918 a few philosophers and psychologists defended the instinct theory. The psychologists were, for various reasons, out of touch with the new experimental ethos of their science. Most often the defenders of the instinct theory stressed its importance to evolutionary teleology, or insisted it was an integral part of evolutionary science—to reject the instinct theory was to reject evolution. As Wesley R. Wells, a psychologist teaching at a small college, put it, critics of the instinct theory were guilty of "losing sight of the biological, evolutionary background of present human behavior."[62] J. R. Geiger, a philosophy professor at the College of William and Mary, insisted the very utility of instincts in social analysis proved they existed in man.[63] Edgar

James Swift, a philosopher, asserted in 1923 that until controlled experiments proved otherwise, "instincts, by whatever name they are called, remain unrefuted."[64] William McDougall, the British psychologist who had popularized the instinct theory, played the most active role in defending instinct after coming to Harvard in 1920. Yet McDougall had done no experimental work since 1911, none ever on human instincts, and he bluntly pegged his defense on his belief in a teleological universe. He complained about the inadequacies of "muscle and twitch" behaviorist psychology and appealed to the animal analogy—the existence of instincts in animals proved their existence in man. But McDougall was unaware of current ideas in American natural and social science, of the experimental ethos of natural scientists and of the interest of social scientists in cultural and social phenomena rather than the physiology of instinct or design in nature.[65] McDougall became something of a public personality in the 1920s; he proclaimed the superiority of the Nordic or Aryan races in the controversy over the Army tests, he often debated Watson over behaviorism, and he consistently defined psychology in a teleological way in an age in which most American psychologists had abandoned precisely that perspective. He left Harvard for Duke University in 1927, and there he revived the Neo-Lamarckian theory of heredity, believing it would substantiate the instinct theory; unfortunately, by the late 1920s, biologists had finally decided against the Neo-Lamarckian theory. Thereafter he became increasingly isolated from the American scientific community.[66]

Manifestly the instinct controversy was not a protracted debate among determined opponents over a sustained period of time. It demonstrated clearly the impact that the social and cultural dynamics of modern science often have on scientists' interpretations of natural phenomena. Watson's critique of 1917 was essentially a spark; the *coup de grace* had been applied to the instinct theory over the past two or three decades, not by the conscious decisions of scientists who drew a bead upon the theory, but by their separate and individual decisions to pursue certain intellectual lines of inquiry in their disciplines in modern university culture. The post-1900 psychologists, committed as they were to experimental inquiry, busily went about their work in the laboratory, work that had little to do with instincts or other teleological conceptions. When they thought of the instinct theory they paid it ritualistic homage. Some pioneers of sociology took the instinct theory more seriously and attempted to incorporate it into their "systems" of sociology. But the pioneers of American sociology were succeeded by the younger generation of sociologists committed to specialized social inquiry; and this new group discovered that determinants of natural science such as the instinct theory were impediments to further research, at most mere adjectives incapable of explaining the specific dynamics of social circumstances, and most probably phenomena that did not belong to the social sciences.

## V

After the mid-1920s, psychologists and social scientists created a new paradigm for the explanation of human behavior. In the new theory, they reconciled the perspectives of their disciplines. They integrated nature and culture. They assumed the interaction of heredity and environment as interdependent yet autonomous variables. In this way they pointed to a more general reorientation of evolutionary thought in American culture, for they helped provide the rationale for a unified, interdisciplinary, and coherent science of man based upon the precepts of evolutionary natural science and modern social science.

Behavioristic psychologists found the culture idea of the social sciences quite compatible with their point of view. Most behaviorists could easily believe that language habits and social conditioning influenced human behavior on the individual and the group levels because they assumed the human mind was plastic. At the same time, behaviorists did not reject evolutionary natural science in its own domain. "The moment the child forms the first language habit, he is forever differentiated from the beast and henceforth dwells apart in another world," wrote Watson. "The search for reasoning, imagery, etc., in animals must forever remain futile, since such processes are dependent upon language or upon a set of similarly functioning bodily habits put on after language habits."[67] Here Watson accepted the culture theory's distinction between animals and man. Yet he did not reject evolutionary science. Most behaviorists made similar accommodations with the social sciences in the post-1925 era. Karl S. Lashley, for example, one of Watson's few graduate students, broke with his teacher's dismissal of consciousness, and argued throughout the 1920s that culture could not be reduced to nature.[68] Charles J. Herrick, an eminent Chicago neurologist, agreed; he dismissed the notion that human behavior could be reduced to mechanistic models of human neurology. "It is a travesty of scientific method," he insisted, "to leave out of consideration in a total view of human nature just those characteristics which differentiate man from the brutes and upon which the future progress of civilization must depend."[69] Behaviorist psychologist Hulsey Cason also drew a sharp distinction between the minds of men and animals, thus affirming the difference between man's unique culture and the lesser psychic behavior of the brutes.[70] Another young behaviorist, Albert Paul Weiss of Ohio State University, dissociated himself from neurophysiological reductionism in explaining human behavior. He stressed man's cultural environment, which was, he insisted, unique in nature. It separated man from even the highest animals.[71] Even those behaviorists who refused to modify their reductionist models of individualistic social psychology conceded much importance to both the cultural environment and the interaction of culture and nature; and they recognized the distinction between group and individual behavior.[72] Most behaviorists followed Watson's cue. They embraced a formula in which they took for granted the interaction of environment and innate plasticity and in which they recognized the functioning of the human cultural group in man.[73] They did not yield on the dazzling prospect of an evolutionary science of human social control; they now said it was more complex—and more interdisciplinary—than they did a decade or so earlier.

There were other indications of the growing ascendancy of the new interactionist paradigm among psychologists after the early 1920s. In 1927, Harry Levi Hollingworth published the first comprehensive restatement of genetic psychology since Stanley Hall's massive *Adolescence* (1904). Hollingworth had taken his doctorate with Cattell at Columbia in 1909, and had done much experimental work since focusing on all the variables of human conduct. He took for granted the new interactionist view that heredity and environment interacted dynamically in the development of human nature, thus shifting dramatically away from Hall's biological determinism.[74] Animal psychologists also worked toward an accommodation with the new nature *cum* nurture paradigm. Customarily they studied rats or apes to find general functions common to both the higher animals and man. By the 1920s, both Yerkes and the Gestalt psychologists had worked with anthropoids. They discovered that apes had a social existence, complete with highly developed systems of social relationships, but they did not have a cultural existence in the sense of a cumulative cultural tradition transmitted from generation to generation as did man. What the animal psychologists of the interwar years emphasized was the recognition that there was a qualitative difference between the behavior of man and even the highest brutes, even though man himself was undeniably a product of evolutionary processes.[75]

With the demise of the instinct theory, psychologists sought to discover new theories to explain the sources of human conduct in line with current methodological canons. So-called "drive theories" emerged in several subspecialties. In psychiatry, for example, a new school of "Neo-Freudians"—Harry Stack Sullivan, Karen Horney, and others—came to the fore in the 1930s; they reinterpreted Freudian drive theories by deemphasizing Freud's biological determinism and admitting the social environment's importance within the context of the interaction of culture and nature. Psychologist Clark Hull and his Yale collaborators worked out major psychological theories based on a dynamic conception of the interaction of nature and culture in the development of drives in man. They worked out objective classifications of human drives and studied them experimentally without the barest hint of teleological thinking.[76]

Indeed in the 1930s American psychologists were so convinced of the interaction of heredity and environment, and of the natural and social science, that a number of them founded the new specialty of personality theory, which was a manifestation of the new interactionist model. The study of personality permitted American psychologists to embrace individualistic psychology and modern social science without triggering a revival of the nature versus nurture question. From the start, the founders of the study of human personality took a decidedly eclectic approach. Thus in the pioneering work *Approaches to*

*Personality* (1932), the Columbia psychologist Gardner Murphy and the psychiatrist Friedrich Jensen outlined the various approaches to the study of human personality that they believed were justified, and defended their nondoctrinaire approach by arguing that the field was too new for rigid models. Six years later Harvard psychologist Henry A. Murray and his associates published *Explorations in Personality*. Murray and his colleagues did not always agree on their views or their terminology. But they did agree on the methods to be used in studying life histories. They also assented to the proposition that it was legitimate to examine the whole personality: the animal drives, the cognitive processes, the social behavior, indeed social structure. Personality theorists now recognized, in other words, the interrelatedness and the distinctiveness of nature and culture, and of personality and culture.[77] Thus when American psychologists took up the study of the individual after the nature-nurture controversy, it was with this new interactionist paradigm.

Social scientists also recognized the interrelatedness of nature and culture. Indeed critics of the instinct theory such as Luther Lee Bernard and John Dewey had never denied the existence of original nature in man; rather they had consistently argued that cultural behavior was the special province of the social sciences and could not be explained with the determinants of the natural sciences. "It is culture and habit, not instinct, which must be the main concern of the sociologist," declared Charles A. Ellwood in 1924, "for it is the development of culture which distinguishes the social life of man from the social life of the brutes."[78] Social scientists had never denied the truth of evolution, of man's animal ancestry, or the importance of original tendencies within their own realm. In the decade following the instinct controversy, it became far easier for social scientists to admit the legitimacy of evolutionary natural science, because the controversy's resolution ended the threat that natural scientists would preempt the territory of the social sciences. And social scientists were reassured that in the 1920s and 1930s natural scientists were paying them more respect as legitimate specialists in their own right. The formation of the Social Science Research Council in 1923 was symptomatic of the new esteem the social sciences were beginning to enjoy in American academic and scientific culture. Social scientists did not say that culture shaped all human behavior; following Kroeber and the other Boasians, they distinguished between the organic and the super-organic levels of human behavior. They argued that culture comprised, in Ruth Benedict's telling words, those *patterns* of behavior men acquired and followed as the consequence of being members of a particular culture.[79] Perhaps the most impressive evidence of the impact of the new interactionist model upon the social sciences came in the decade following the instinct controversy, when several of Franz Boas' later doctoral students worked on research topics that combined natural and social science levels of behavior with Boas' full approval. Thus Melville J. Herskovits wrote a major treatise on the physical anthropology and ethnology of Africans, Margaret Mead investigated the psychocultural and sexual life of Polynesians, and Otto Klineburg studied migration patterns and group intelligence levels among American blacks.[80] The working out of the jurisdictionary boundaries between the natural and social sciences thus permitted natural and social scientists to go about the larger task of creating an interdisciplinary science of man based upon up-to-date conceptions of evolutionary science and social science. This was a breathtaking and exhilarating accomplishment for those involved.

# 7 MENTAL TESTING

*I*

" W E cannot conceive of any worse form of chaos than a real democracy in a population of an average intelligence of a little over thirteen years," declared George B. Cutten in his inaugural address as president of Colgate University in the fall of 1922. Cutten drew from the results of the Army's testing program the ominous lesson American democracy would survive only if its leaders came from the nation's "intellectual aristocracy." He argued that mental tests should be used to identify this elite, and its members should be granted all appropriate privileges of leadership to save American democracy.[1] Cutten's argument about the relationship between high intelligence and political leadership was not a new one in American civilization. One historian has argued convincingly that the gradual acceptance of the democratic ideology in the nineteenth century occurred largely because of the implicit assumption public education would prepare the masses to participate in politics intelligently and enable them to make sound choices.[2] In the post-Appomattox years, conservative social and political critics—especially those who identified with the New England intellectual and social elites—often called for either expansion of public education or restriction of the franchise. Reformers also campaigned to extend public education in these years; they saw in the public school an institution of social adjust-

ment and control that would supplant many of the traditional
agencies of the village culture, such as the family and the church.
On the basis of such ideologically diverse perceptions as these the
progressive education movement flourished, and so did the mental
testing movement.[3]

In post–World War America, many educated citizens, including President Cutten, began to sing the praises of mental tests as
instruments capable of evaluating a heterogeneous population in
a scientific and systematic fashion. Cutten's address was notable,
not for the relationship he made between intelligence and
political competence, but for his advocacy of mental tests as the
basis for selecting leaders. He mirrored the anxieties of many
educated Americans in the immediate post-war years who
thought the nation's very social fabric was threatened by conflict
and unrest. These were difficult years: an era of strikes, of an-
archist bombings, of race riots, of economic strains, of serious
ethnocultural conflicts that pitted black against white and
immigrant against native born. For about a decade, professional
psychologists had been working with various mental tests. Many
administered tests to members of the very groups educated Americans such as Cutten feared: immigrants from Southern and
Eastern Europe, nonwhites, the feeble-minded, the poor, and the
criminal elements of the population. They concluded these
groups were less intelligent than native born, middle class Anglo
Saxon Protestants, and that lower intelligence offered a genuine
threat to America's social stability.

By the early 1920s, in fact, mental testers in the psychology
profession had identified three areas of inquiry that bore on the
fears of educated Americans such as Cutten. In the interwar years
three distinct controversies over the results of their testing
programs occurred. One group of mental testers worked on racial
psychology. They assumed that the various average scores they
obtained from different elements of the population—native born
Anglo Saxons, immigrants, and nonwhites—signified innate
racial differences in intelligence. Many of the racial testers took
a strong interest in the immigration restriction controversy. The
ensuing controversy over racial mental testing fed to a considerable extent on the restrictionist debate in the halls of Congress;
it diminished after the passage of the immigration restriction laws
of 1924. Another group of mental testers specialized in the rela-

tionship between low intelligence as measured by the tests and such manifestations of anti-social conduct as crime, pauperism, and delinquency. They assumed that when the criminals, prostitutes, paupers, and delinquents they tested earned subnormal scores, this proved that low intelligence caused their charges to behave in unlawful and degenerate ways. A result of their work was a controversy over the relationship of subnormal intellect to good conduct. This debate was largely over by the late 1920s. A third group of mental testers inquired into the more general theoretical problem of the relationship between an individual's innate intelligence and his or her status in society. These testers assumed that persons of high social status—as measured by occupation, for example—possessed above-average intellect. This line of work dealt with individuals, and not groups; therefore the ensuing controversy over innate intelligence and social status lasted well into the 1930s because it was easier to prove the influence of innate intelligence in an individual than in a larger cultural group.

In each of the three mental testing controversies scientists took the positions they did for complex and often mutually reinforcing reasons. They responded to their scientific commitments and training, to be sure. But they also were powerfully influenced by their social and cultural backgrounds, by their commitments to social and public policy. In each of the controversies, those who took the side of heredity obviously did so in part because they were middle class, native born Protestants of Northwestern European ancestry, as their preoccupation with racial differences, subnormal intelligence among deviants from middle class culture, and high intelligence and social status, indicated. Yet their commitments to science mattered too, both before and after the controversies had taken place. Similarly, their environmentalist critics quite often had a far more liberal vision of American society than their hereditarian opponents; in some instances, at least, they were recent immigrants deeply offended by Anglo-Saxon bromides. Yet they too spoke the language of modern science and rationalism, and used its instruments to buttress their positions. Emphasis on either scientific or social policy commitments overlooks the fact that in most cases men and women of science responded to both in complex ways; sometimes the scientific and social policy commitments reinforced one another, at other times they came into conflict. It is this complexity, this ambiguity, that should be understood.

## II

After World War I, many mental testers and immigration restrictionists quickly seized upon the Army testing program's results as definitive proof that nonwhites and recent immigrants from Southern and Eastern Europe were less intelligent, and, therefore, less worthy of citizenship, than the native born Anglo-Saxons. In these hectic years of social unrest, the restrictionist movement blossomed, and its advocates often turned to science, or what they understood was science, for support of their position that the less intelligent "stocks" should be prevented from polluting the nation's germ plasm. Not all scientists who accepted the notion of racial superiority and inferiority necessarily believed in immigration restriction, at least not the more radical postures that some publicists such as Madison Grant took. Thus the eugenist leader Charles B. Davenport told Grant on several occasions he doubted immigration restriction would solve social problems.[4] Publicists for restrictionism were often less reserved than the men of science they quoted in their books and articles; probably their training made the men of science more cautious. Yet if the men of science were not as dogmatic as the popularizers of race psychology, in the fervered times it was sometimes difficult to distinguish between them. The pro-Nordic publicist Lothrop Stoddard, for example, argued the Army tests were irrefutable scientific proof that civilization "is. . . fundamentally conditioned by race."[5] Robert M. Yerkes, a careful scientist, nevertheless insisted in the *Atlantic Monthly* in 1923 that the Army tests proved races differed in innate intelligence. "Nordic" and "Alpine" races of Northwestern Europe were undeniably superior to "Mediterraneans" and nonwhites. He urged immigration restriction because the less gifted races were not as able to be good citizens. "Crime, delinquency, and dependency, as well as educability are intimately related to intellectual ability," Yerkes concluded.[6]

Yerkes' younger friend and colleague in the Army testing program, Carl C. Brigham, wrote perhaps the most celebrated

"Nordic" interpretation of the program's results, *A Study of American Intelligence* (1923). Brigham came from an old New England family. He took his Ph.D. from Princeton in 1916 at the age of twenty-six. He and Yerkes became good friends during the war. They believed in innate racial differences; thus both belonged to the Galton Society and supported the eugenics movement. After the Armistice, they kept in touch; Brigham returned to Princeton, Yerkes went off to the National Research Council and a Yale professorship thereafter. Yerkes often gave Brigham advice in preparing his study of the Army test results, and Brigham informed Yerkes of his progress. Brigham wanted to prove scientifically a correlation between racial type and racial intelligence. He told Yerkes that he had asked Madison Grant to assist by making "a percentage estimate of the amount of Nordic, Mediterranean, and Alpine blood in each of the nationalities."[7] Yerkes fully approved. He even let other psychologists know of Brigham's work. Thus he wrote the able, young Edwin G. Boring at Harvard that Brigham's work would justify the Army work on nativity differences and "be a truly notable contribution."[8]

Brigham argued bluntly the mental differences among races were innate and profound. He insisted his interpretations supported Madison Grant's thesis that the Nordic races were superior to all others. In support of his thesis, Brigham made some fanciful correlations between William Z. Ripley's anthropometric classifications of Nordics, Alpines, and Mediterraneans in *The Races of Europe* (1899) and the nativity groups in the Army results. Brigham apparently accepted Grant's personal estimate of how many of each nativity group could be classified according to Ripley's physical types. Then Brigham argued that the causes of the Nordics' superior performance on the tests was their superior intelligence. He found it necessary, however, to reconcile two potentially embarrassing findings of the Army tests with his interpretation of racial superiority. The first showed a direct correlation between length of residence in America and achievement in the tests; foreign born persons resident for twenty years or more, for example, scored as high as the "Nordics" regardless of nativity. Brigham argued that the quality of immigrants had declined disastrously in the last twenty years—or since 1890—as compared with before. Half the pre-1890 immigrants were "Nordics," which explained why the average performance of immigrants in the country for twenty years or more was higher. Since 1890, only one-fifth of the immigrants were "Nordics," which explained the lower average score, and the reason immigrants here for two decades or more did better. Brigham also tried to reconcile his racial superiority theory with the other finding, that northern blacks achieved higher scores than southern blacks. He did concede that northern blacks benefitted from superior environmental and cultural advantages. He remained convinced that northern blacks represented a better type because they had more white "blood" than the "purer" southern blacks. Northern blacks were thus more intelligent, and were in any case more ambitious because they migrated to the North. Brigham solemnly concluded his book by calling for immigration restriction to preserve America's precious heritage of "Nordic" germ plasm.[9]

Yerkes was so delighted with Brigham's book he asked his friend Boring to review it. Even though Boring sympathized with restrictionism, he was most dubious—on technical grounds.[10] Because of his rigorous training with Titchener at Cornell, as well as his considerable scientific expertise and caution, Boring was less intoxicated by the social policy ramifications of genetic and racial psychology than Yerkes and Brigham.[11] He found Brigham's methods faulty. "I do not believe that the examination of the Army records can be very safely extended to the foreign born population at large," he told Brigham in March 1923, "or at all safely extended to such general concepts as the 'Nordic race', the 'Alpine race', and the 'Mediterranean race'." Boring did not question Brigham's larger notion of racial superiority, but rather his methods; it "is the problem of the validity of a sample when the conditions of sampling are not surely known."[12]

Brigham's book rapidly became a focal point of the racial testing controversy because it was one of the few interpretations written by a professional scientist that was not tucked away in psychology journals. In the debate over Brigham's book, scientists responded to his arguments for complex reasons, both scientific and cultural, which were often mutually reinforcing. And it was easier for specialists in fields other than psychology to attack Brigham. Thus in the twenties, William C. Bagley, a professor of educational psychology at Columbia's Teachers' College, published several attacks on Brigham's work. In some respects Bagley was typical of many native-born American psychologists. Born in Detroit, Michigan, in 1874, he took his Ph.D. with Titchener in 1909. Before coming to Teachers' College in 1917, he held several academic posts, including one at the University of Illinois. He shifted on the nature-nurture issue at least twice. In the early 1900s he believed a good education could overcome any inherited deficiencies. He submitted an article containing that thesis to *Popular Science Monthly* in 1907. But in the fall of 1908 he asked the editor, James McKeen Cattell, to not publish his article because he had changed his mind; "the studies of [Frederick] Woods and the conclusions of such men as [E. L.] Thorndike and [Karl] Pearson have led me very radically to modify the views that I expressed in the paper."[13] Woods, Thorndike, and Pearson had argued with seemingly impressive statistical proof that heredity prevailed over environment in intelligence.[14] In the 1910s, however, Bagley became more and more convinced that the functional psychologists, who argued that innate nature adjusted to cultural environment, were closer to the truth than the hereditarians. At Columbia after 1917, Bagley came into contact with John Dewey, who was highly critical of racial testing, and Franz Boas, who was outraged by it. Bagley swung back again on the nature-nurture issue, largely on grounds of scientific argument. He now believed mind was a function that developed in a social medium and could not be measured apart from it. The social achievements of peoples depended, not so much on the innate capacities of brilliant individuals, but on the abilities of the mass or average man to share the thoughts of the most gifted through culture. In other words, Bagley came very close to embracing all the tenets of the culture idea.[15] Bagley used the culture idea to attack Brigham's study in 1924. Among its "glaring inconsistencies," he noted, was that the Southern states, which presumably had the highest proportion of "Nordic" blood, had also the lowest native born, white scores, whereas Massachusetts and Connecticut, which had been innundated by a "Mediterranean" tide since 1890, stood in the first rank in the Army tests. Brigham's study was "a most questionable and biased interpretation of certain facts that can be far more reasonably interpreted in quite another way."[16] Obviously for Bagley, scientific commitments were critical on the nature-nurture issue.

For other critics of racial mental testing, commitments to science *and* social policy commitments mattered. Several Jewish professionals also attacked Brigham and other proponents of the Nordic hypothesis. Both science and social policy commitments were intertwined in their critiques; and they were not psychologists, which made it easier for them to attack Brigham. In 1924, for example, Maurice B. Hexter, of the Federated Jewish Charities of Boston, and Abraham Myerson, a professor of neurology at Tufts Medical College, published an essay highly critical of Brigham's study. Doubtless their backgrounds and their obviously liberal political attitudes contributed much to their indignation over Brigham's Nordic hypothesis. But the specific arguments they used came from contemporary science, and they employed a scientific and technical level of discourse. Thus they made several rather damaging technical criticisms of Brigham's sampling methods, akin to Boring's. They also argued with Brigham's fundamental scientific and social policy assumptions. They insisted the tests were culture-bound. Man was both a biological and a cultural animal; therefore, the tests measured both cultural and biological factors. A narrowly racial interpretation could not explain such complex results as those of the Army tests. Hexter and Myerson declared further that all the data Brigham used merely demonstrated that recent immigrants were slower in their performance than either native-born or long-resident foreign born and had a lower average score. The data said nothing about the causes of these differences. Why resort to such tortured arguments as a decline in the "quality" of immigrants since 1890, they asked.[17] The Army tests, they concluded, merely measured "the inherent disaster attendant upon the transplantation of an adult group from one environment to another."[18] Obviously Hexter and Myerson responded to Brigham as men of science and as recent immigrants.

Gustave A. Feingold also criticized Brigham's work in ways suggesting the reinforcing impact of scientific and ethnocultural commitments. Born in a Jewish community in Russia in 1883, Feingold grew up in the United States. He graduated from Trinity College, in Hartford, Connecticut, in 1911; in 1914 he finished his Ph.D. in psychology at Harvard. Perhaps his Jewish identity prevented him from landing an appointment in a college.[19] As a teacher in the Hartford, Connecticut, public schools,

Feingold became an active psychological researcher. He had published ten articles and his dissertation and become a full member of the American Psychological Association by 1924. Like Hexter and Myerson, Feingold insisted that Brigham simply ignored the impact of the cultural milieu—home environment, language difficulties, and cultural traditions and values—of the various nationality groups in the Army sample. And Brigham also disregarded the influence of individual temperament on achievement, which was, he noted dryly, amazing, considering how much many native-born Americans stressed the Puritan ethic. Feingold also attacked Brigham's thesis of deterioration in the quality of post-1890 immigrants. This was too hasty a judgment; in any case, it was "absurd" to assume, as Brigham did, that a sudden mental decline had taken place in selected European countries at five-year intervals. Nor could the varied cultural conditions of the different European nations be so airily dismissed. Feingold then pointed to his own test results with several thousand Hartford pupils. He did find a rank order of nativity groups roughly similar to the Army tests. He insisted that the difference in average mental age between the native born and the lowest ranking foreign born was a mere nine months, not the two years the Army testers said they found. Was "it not more than probable that most of the original gap discovered in the Army tests. . . was due to environment, i.e., differences in educational opportunity, rather than to heredity?" Feingold asked.[20]

The brilliant political journalist Walter Lippmann proffered the most celebrated public attack on Brigham and other advocates of the Nordic hypothesis. Scion of a wealthy, educated German Jewish family, and something of an intellectual prodigy even in college, Lippmann took the Nordic hypothesis to task in the liberal magazine *The New Republic*. A major architect of modern liberal political theory, three generations removed from immigration, and a man almost totally secular, Lippmann approached the race superiority interpretation with different emphases than Feingold, Hexter, and Myerson, or, for that matter, than Bagley. His tone was less impassioned; he was more skeptical of human potentialities. More than they, he stressed the scientific and technical vulnerabilities of the Nordic hypothesis.

Lippmann took the pro-Nordic testers to task on both conceptual and methodological grounds. He based his theoretical arguments on the culture idea. The racial testers had not proven the Army tests measured innate ability apart from cultural influences, he insisted, or that each "race" had an innate mental endowment. And the testers blatantly ignored the formative preschool years, in which the child learned to speak and underwent much important intellectual and emotional development. To dismiss the preschool years, Lippmann insisted, was to obey the will to believe, not to respect the methods of modern science. Lippmann also supplied technical criticism. He declared that the racial testers did not understand how the tests measured the individual's ability to learn in a limited period of time, not the individual's learning ability in the real world, which was, he wryly presumed, what the testers were most interested in knowing. The testers, furthermore, defined intelligence narrowly, as the capacity to acquire "scholastic" information rapidly. Surely there was more to an individual's social competence and worth than mere scholastic ability, especially if the individual suffered from language handicaps. Lippmann directly conceded that innate ability probably did explain many of the differences among individuals, but not, he insisted, among whole groups or "races." In whole social groups cultural patterns and experiences determined their history and achievements. Finally, Lippmann argued, the testers err in assuming heredity and environment may be treated as independent variables. Nature and nurture were in fact interdependent yet separate variables that interacted in the making of mind and intelligence.[21]

"Lippmann's articles have been pretty generally read in this part of the country and have added . . . to popular misinformation," Yerkes told Lewis M. Terman, the Stanford University educational psychologist and inventor of the Stanford-Binet test, in 1923. "I hope to have opportunity to discuss several issues with him for I think he is educable."[22] While Yerkes attempted, without success, to educate Lippmann at a series of lunch meetings in New York restaurants, Terman took up the cudgels against Lippmann in the *New Republic*'s pages. He adopted a peevish tone. He belittled Lippmann for not having proper scientific credentials, thus implying Lippmann had no right to criticize either the tests or the testers. He also accused Lippmann of a naive belief in egalitarian democracy. And Terman reaffirmed that the tests, in his scientific judgment, measured innate intelligence apart from cultural influences. Finally, he attacked several of Lippmann's more technical arguments, notably that the test time factor influenced performance.

Terman soon discovered that Lippmann was not an opponent to be taken lightly. Lippmann reiterated most of his conceptual arguments, and correctly insisted he had never believed in egalitarian democracy. But his most devastating argument came over technical issues; he showed how Terman did not understand that the time factor in the tests, for example, did not influence the relative standing of the Army recruits who took the tests, but their absolute performance. Double the time, he insisted, and the whole group's performance increased accordingly. Lippmann declared, with much relish, that "Mr. Terman ought at the very least to interpret correctly work done by his own pupils under his own direction."[23] Terman sensed discretion was the better part of valor and dropped the issue.

The Boasians also jumped into the racial mental testing controversy. How influential professional versus social policy considerations were to each Boasian is difficult to estimate. Presumably both were mutually reinforcing. Characteristically, Boas became involved in the restrictionist controversy in the Congress by supplying technical information to the able New York Congressman battling the restrictionists, Emanuel Celler.[24] Yet leftist politics did not dominate the Boasians' stances in the testing controversy; the canons of modern science mattered powerfully to them too, apart from their feelings toward the restrictionist movement, and the treatment they had suffered over the years from physical and natural scientists in the National Research Council and in other contexts. They never took a dogmatic environmentalist stance. They never dismissed heredity, either in print or in private correspondence. Consistently they argued that mental traits were probably inherited through family lines, and that innate intellectual differences existed from one individual to the next. Their critique of racism centered on the misapplication of individual psychology to the cultural group, on, in other words, the assumptions of the culture idea. Both culture and nature were important, but on different levels of human behavior. In 1922, for example, Lowie wrote a scathing review of the most recent edition of Madison Grant's *The Passing of the*

*Great Race,* reiterating many of the arguments Bagley, Lipp-mann, and other critics had made. The Columbia paleontologist and eugenist Henry Fairfield Osborn chided Lowie for his review. Lowie's response to his former employer was that Grant was unscientific. "I do not hold the equality of all races as a *dogmatic* proposition and I do believe strongly in innate racial differences," Lowie told Osborn. "I even believe that a fairly plausible case might be made out for racial inequality." Grant, however, was a charlatan in Lowie's eyes.[25] Roland B. Dixon, a Harvard anthropologist and sometime ally of Boas in American anthropology's politics, wrote Lowie an approving letter about his review of Grant's book. He was glad Lowie handled the book "without gloves, for the book is a hopeless mess."[26] Dixon was even less willing to dismiss the racial superiority argument than Lowie; like Lowie, he thought it a scientific issue to be resolved by systematic, professional inquiry, and to be reconciled with the culture idea. Throughout the early and middle 1920s, Boas, together with several students—Melville J. Herskovits, Alfred L. Kroeber, Wilson D. Wallis, and Margaret Mead— attacked the racial interpretation of the Army tests, invoking the culture theory, a number of technical criticisms (such as the problem of linguistic handicaps), and declaring that the question of racial inferiority and superiority was entirely open.[27] Boas' friend and occasional ally, the Harvard anthropologist Alfred M. Tozzer, made perhaps the most systematic criticism of racial mental testing of any anthropologist in 1925:

Culture is not mainly made up of congenital factors. . . . Society is thus not a biological organism. Man's intellectual equipment, made possible by a large brain, coupled with his possession of articulate speech, places human society in an entirely different category from that found among non-human animals. Individual differences, race, and environment should be considered as factors in the diversities of cultures. All play a part, but not one occupies the entire role in the drama of peoples.[28]

By the mid-1920s, the popular nativist hysteria in American society was ebbing. When President Coolidge signed the Johnson-Reed National Origins Act on 26 May 1924, much of the immigration restrictionist movement's *raison d'etre* evaporated, even though the law would not take effect for several years. The passage of the National Origins Act, together with the decline of the Ku Klux Klan, of the Red Scare, and the growing middle class confidence in American institutions generated by the return of prosperity after 1923, worked a subtle but influential change on the scientific controversy over racial testing. These events in the public arena did not alter the inner content of the controversy's scientific ideas so much as they lowered the controversy's emotional temperature, so to speak. The sense of urgency with which men and women of science debated racial testing began to dissipate. The controversy's immediate political and cultural ramifications had been resolved, however crudely, by the restrictionist legislation. Men and women of science often remained bitter, especially some of the critics of race psychology. But generally speaking scientists could now discuss the issue without literally screaming at one another; they could now participate in the controversy with little of the public policy urgency that marked its earlier phase.[29]

It would be erroneous to assume that events in American civilization solely determined the race testing controversy in either of its phases. Even at the height of the restrictionist and testing controversies in the early 1920s a few psychologists began to report difficulties and technical problems in using the tests and in assuming they could validly measure innate intelligence apart from cultural milieu. Typically these doubters were most often young psychologists just embarking on their careers as testers, not individuals who had promoted testing for a decade or more. They were doubters because they found the tests had technical inadequacies they could not reconcile with what they thought were scientific standards of logic and evidence. Yet even a few pioneers of testing had sober second thoughts. James McKeen Cattell, for example, promoted mental measurement all his career. At the same time, he also believed in empirical, experimental investigation; and he had never taken a dogmatic stand on the nature-nurture issue. In 1923 he confessed to Robert Lowie that the Army tests were sadly inadequate as scientific measurements of racial intelligence, and much further work was needed before science could offer a definitive judgment on racial superiority and inferiority.[30]

Other psychologists shared their reservations with their colleagues in print. They noted in particular the linguistic, educational, and social handicaps from which minority group members suffered when they took tests standardized on middle class, Anglo-Saxon children. Many of these doubters had not joined the mental testing movement in the 1910s, but had turned to testing from curiosity after it became such a burning issue in the profession. Hence most had no obvious axes to grind and no career achievements to defend, and they were at least aware of the arguments of both sides. And unlike the testers who embraced the Nordic hypothesis, they assumed from the beginning that they should take into account such factors as language handicaps and other environmental influences. They approached the whole problem in a more even-handed, temperate fashion. Stephen S. Colvin, an established psychologist teaching at Brown University, for example reported that the results of his tests of Italian-American schoolchildren in Providence forced him to conclude that his subjects' linguistic, educational, and social handicaps unfairly penalized them when compared with native-born children.[31] Professor Martha MacLear of Howard University insisted that her experiences testing black school children in Washington demonstrated how the tests were culture-bound.[32] In 1922 Joseph Peterson voiced his disquietude. Peterson had become a mental tester after having taken his doctorate from Chicago in 1907. For some years he believed in innate racial differences. He served with Yerkes in the Army's testing program, and he even belonged to a eugenics society for some time. Yet he was committed to scientific methods in psychology as well as to a particular bundle of interpretations and assumptions. Initially the tests struck him as most impressive, because they seemed to be rigorous experiments. By the early 1920s, however, he believed, on reexamination, that the racial testers were using tenuous methods and making unwarranted assumptions. Thanks to his own testing program and to his careful scrutiny of the controversy's literature, by 1922 he had strong doubts concerning whether the tests really measured innate intelligence apart from environment. He was willing to concede that the tests were at least partly culture-bound—enough to account for the relative standings of nativity groups in the Army tests, and to explain some of the tests' more ambiguous findings, such as the universal correlation between length of residence in America and standing in the tests.[33]

By the mid-1920s, more psychologists criticized the tests as valid measures of innate racial intelligence, or, in other instances, argued that there were severe methodological difficulties with the racial interpretation. Their most persistent comments centered on such factors as language handicaps. They believed testing must be scientifically sound. For this generation, social policy commitments mattered, but commitments to science weighed more heavily, as the examples of David Wechsler and Stuart A. Courtis show. Both belonged to this generation; and both criticized the Nordic hypothesis. They had little else in common. Wechsler was born in Romania, immigrated to America as a child, and graduated from City College of New York in 1916 at the age of twenty. He took his master's degree in psychology at Columbia in 1917, studied psychology in London and Paris for five years, entered Columbia's doctoral program in 1922, and published four experimental articles on physiological psychology before defending his dissertation in 1924. He was quickly elected to full membership in the American Psychological Association on the basis of his research productivity, and in 1926 he became a psychologist for the Brooklyn Jewish Social Service Bureau. In that year he also published an important critique of racial mental testing, arguing that cultural factors such as educational and linguistic handicaps most probably explained the so-called "racial" differences on the tests.[34] Yet Wechsler was no opponent of mental testing, but rather, of sloppy methods; he is perhaps remembered today by psychologists for an important series of intelligence tests. Stuart A. Courtis' background differed sharply from Wechsler's. Born in Wyandotte, Michigan, in 1874, he grew up in a comfortable middle class home; after attending Massachusetts Institute of Technology in the 1890s, he taught mathematics and science in the Detroit public schools for years. Then he decided to study psychology and education. He took work in psychology at several universities, and received a B.S. in 1919; he joined the mental testing movement by publishing a series of articles on tests and designing a famous mathematical ability test. In the early twenties he became increasingly uneasy with racial mental testing (on methodological grounds) and in 1926 he criticized the Nordic interpretation of the Army test results on precisely the same technical grounds as had Wechsler.[35]

Thus many of the younger generation of psychologists were

becoming disenchanted with racial interpretations of testing for scientific as well as social policy reasons. As one young psychologist, Bertha M. Boody, implied in 1926, the first generation of racial mental testers simply had not done their work carefully. In summarizing her study of immigrant children at Ellis Island, she reported that although there were individual mental differences that were innate, "the curve of the scores seems not to differ in any marked degree from race to race, nor does it differ markedly, with possibly a slight allowance for differences in the strain of examination conditions, from the curves shown in the studies of unselected groups of American children."[36]

In the next several years most psychologists shifted on the issue of the racial interpretation of the tests, or, more precisely, a newer group of testers came to the fore who published such devastating technical criticisms of the earlier group's work that all psychologists felt obligated to abandon the assumption that races differed innately in their mental abilities. As late as 1925, most race psychology studies assumed innate racial differences; by 1930 none did. The change was that rapid and total.[37] When psychologists were presented with a conflict between their commitments to the larger Anglo-Saxon culture on the one hand and their commitments to professional science on the other, they resolved the tension by accepting the dictates of their professional subculture. Clearly they believed they had to change, regardless of personal wishes or social policy commitments, if they wished to maintain their standing in their profession.

A sense of how the process worked can be gleaned from the examples of Carl C. Brigham and Thomas R. Garth, two psychologists who made their careers as advocates of the racial superiority interpretation and then recanted before their profession. Brigham repudiated his controversial study in 1930. This was not an easy decision for him. The book constituted his major publication in psychology, and he remained personally convinced that racial differences probably did exist; he belonged, for example, to such organizations as the Galton Society and the Eugenics Research Association. Yet his book had become the object of so much methodological, technical, and conceptual attack he felt obligated to declare in a major scientific journal, "That study, with its entire hypothetical superstructure of racial differences, collapses completely."[38] Thomas R. Garth administered mental tests to the minority groups of the American Southwest for more than a decade after taking his doctorate from Columbia in 1917. He began his career, he later confessed, assuming he "would find clear-cut racial differences in mental processes."[39] At the same time, however, Garth was a careful scientist who never claimed too much for the results he found and who always interpreted them as tentative hypotheses. He often conceded there might be nonbiological factors influencing the results he found; he said this even when he found correlations between the percentage of "white blood" in nonwhite minority groups and their standing in the tests as compared to full blood whites. By 1925 Garth was satisfied experimental science had established that races differed in their innate endowments. He hastened to add that this did not damn nonwhites to permanent inferiority; they could rapidly approach the white standard by means of careful biological selection and controlled breeding.[40] But as his work continued after 1925 he began to perceive difficulties in his conceptual and methodological assumptions. As he became aware of the criticisms of the opponents of racial mental testing, he began to include in his research consideration of the cultural influences they mentioned. In a 1928 study, started to ease his mind about the critics' points, he found that the educational and other environmental influences of one thousand full blood Indians affected their test scores so positively that "these educational factors leave little room" for other influences.[41] For the next several years Garth reported the results of his continuing testing program without offering definitive interpretations. In 1931 he repudiated the doctrine of innate racial differences. He noted that the work done in the profession on the problem since 1925 was far more careful, from a scientific standpoint, than that done previously, and he said geneticists had shown that heredity was more complex and contingent than he or other psychologists had understood as late as the early 1920s. He declared that he now understood, on the basis of his recent work, and that of others, how much the tests were culture-bound. He concluded that heredity and environment together were measured in the tests.[42]

By 1930, then, most American psychologists had repudiated racial interpretations of mental testing. A few die-hards continued to publish such interpretations in the later 1920s, mainly the doctoral students of several psychologists who themselves were prominent advocates of the racial differences theory.[43] But in general psychologists had rejected this idea.[44] As a broader group of psychologists than the original champions of racial mental testing got involved in this line of inquiry, and as the public tensions of post-intervention American society dissipated, the "flaws" of racial mental testing, which may seem so obvious to contemporary Americans, came into view for the first time. Those flaws could not be reconciled with what that larger group of psychologists defined as scientific procedures and methods. Hence the abrupt decline of the race differences theory. Some psychologists—Robert Yerkes, for example—who continued to believe in innate differences withdrew from the field entirely because the whole field seemed too explosive a public issue and too dangerous a professional matter to pursue.[45] The racial mental controversy thus ended for complex reasons within and without the scientific community.

## III

On several Sunday afternoons in the early spring of 1915, Dr. Clinton P. McCord, Health Director of the Albany, New York, Board of Education, and several young associates, visited a total of eleven houses of prostitution. These good, sober burghers, with their neat business suits, celluloid collars, and freshly shined shoes, did not sin. Instead, armed with clipboards, pencils, and stacks of intelligence tests, they examined the ladies of the night to determine whether they had subnormal intelligence. Dr. McCord reported that the madams showed a spirit of friendly cooperation, and that "the tests were made under controlled conditions." The girls were relaxed and rested from their week's labor; "in all but two cases [they] were free from nervousness." After testing 100 women ranging in chronological age from twenty-two to forty-one years, Dr. McCord and his associates found the women's average mental age was ten years (definitely subnormal for adults), less than half had normal intelligence, and none had above average intelligence. Dr. McCord concluded that the test results demonstrated scientifically a causal relationship between low intelligence, which he assumed was inherited, and a life of crime. He added, almost parenthetically, that the difference between the ladies in the expensive and inexpensive houses was not intelligence but manners.[46]

Dr. McCord thus accepted the major assumptions of those psychologists, social workers, and corrections officials who insisted that innate low intelligence caused an individual to lead a life of dissipation and immorality. Dr. McCord, and men and women like him, were influenced by leading eugenists such as Charles B. Davenport, whose Eugenics Record Office sponsored several investigations of criminal families, and Henry H. Goddard of the Vineland Training School. Goddard in particular did much to promote interest in the menace of the feeble-minded to society, especially through his work with mental tests. Goddard was generous with his findings. He was delighted to encourage others to work in this field. "Here is an immense field, a magnificent opportunity," he wrote Edward B. Titchener in 1908. "I want to open it up to science. I am here on the ground and can do practically anything that I want. . . . I have no desire to exercise any proprietary rights."[47] Goddard was as able and enthusiastic a promoter as Davenport; and this was a key to his influence. He corresponded regularly with Davenport, with rising young psychologists such as Lewis M. Terman and Robert M. Yerkes, and he even recruited young psychologists—notably Arnold Gesell— to come to Vineland to do research.[48] The theme of the Vineland work was that innate subnormal intelligence was responsible for anti-social conduct.[49] Men and women of science such as Goddard, impressed by the new theories in biology and psychology, made their mark in the profession in the early days of mental testing by promoting the tests' utility for solving social problems. Perhaps understandably they were led to assume that they had discovered an important instrument of a new science of man. And quite apart from the assumptions they drew from science— which had profoundly hereditarian implications—they imbibed from the middle class WASP culture in which they were reared the reinforcing beliefs that the focus of criminology was the individual, not the social group or the cultural milieu, and that good brains and breeding always resulted in good conduct.

Men and women of Goddard's—and McCord's—outlook dominated the field before the 1920s. But a few dissented. In 1910, Lightner Witmer, a professor of psychology at the University of Pennsylvania and psychologist for the Pennsylvania

Training School for Feeble-Minded Children, published a report of his case studies of youthful offenders in which he declared that the individual case study approach he used had convinced him of the lack of any causal relationship whatsoever between heredity and crime, or, for that matter, between low intelligence and anti-social conduct. Witmer had studied with both Wilhelm Wundt at the University of Leipzig and with Cattell at Pennsylvania before going to Germany. Neither Wundt nor Cattell had much use for the nature versus nurture dichotomy; and this influenced Witmer. And Witmer's individual case study method provided him with a far different perspective with much more biological, psychological, cultural, and sociological information than group mental tests.[50]

Before the 1920s, it was not Witmer who studied the problem of deviant behavior with the individual case approach the most intensively, but Drs. William Healy and Augusta Bronner. Like Witmer they approached the problem from a different angle than Goddard or other proponents of the hereditarian view. Eventually their work became very influential. Healy came from a middle class, English family; he reacted against parental religious piety by seeking a life in science. After attending Harvard, he graduated from the Chicago Medical School in 1900 at the age of thirty-one. At Harvard, Healy learned from William James a life-long belief in indeterminacy. For several years Healy practiced medicine in Chicago; then he decided to specialize in nervous and mental diseases. Consequently he studied in Europe, and discovered Freud's work, which reinforced his individual, clinical approach. Upon his return in 1909, officials of the Chicago Juvenile Court made him head of the Court's new Psychopathic Institute, where he worked until he took an appointment at the Judge Baker Foundation in Boston in 1917. In 1913 he hired as his assistant Augusta Bronner. She finished her doctorate under E. L. Thorndike at Columbia the next year; her dissertation was a study of the intelligence of delinquent girls.[51] In the 1910s they worked together, at first embracing the individual or clinical method, in which they would study a few offenders intensively from as many perspectives as possible, and using whatever explanations seemed to fit the facts, including Goddard-like notions of inherited mental defect. Increasingly, however, they abandoned the notion their subjects were of subnormal intelligence, if for no other reason than many offenders they studied struck them as merely uneducated and intelligent, not innately stupid. Healy and Bronner came to assume in the later 1910s that heredity was far less important in causing social deviance than were such influences as mental repressions and conflicts—especially childhood sexual experiences—and social and familial relationships.[52] "It appears hazardous to offer any conclusions concerning the possible relationship of heredity to delinquency," they wrote in 1926. "Among the difficulties of interpretation is the fact that there are so often, surrounding youth, bad social situations created by socially unfit parents, the effects of which are not those of biological inheritance."[53] To a considerable extent, their approach gradually made them aware of a multiplicity of factors influencing social behavior and conduct besides heredity.

Their work began to influence other workers in the field after the mid-1910s, chiefly on methodological and conceptual grounds. In particular it began to count with some mental testers, as, for example, Edgar A. Doll, for many years a proponent of both mental testing and the hereditarian theory of anti-social conduct. Born in Cleveland, Ohio, in 1889, Doll graduated from Cornell University in 1912; for the next five years he was Goddard's research assistant at Vineland, and then he worked under Yerkes in the Army's testing program during the World War. By 1920, when he received his doctorate from Princeton, he had published over a dozen studies of the feeble-minded in which he assumed intelligence was innate and the major determinant of the individual's social conduct. But by 1922 he shifted his position substantially for several reasons: his own further graduate training, the rather hectic circumstances under which the Army tests were given, and, in particular, the work of Healy and Bronner. He now recognized that temperament was at least as important as intelligence in determining conduct, and he disavowed most of the work of the previous decade on the intelligence of criminals as "worthless," because it was based on faulty notions of genetics. Healy and Bronner's work made him realize that standard mental tests penalized offenders who did not have the requisite educational and social background to be measured fairly. We "must revise our thoughts about the feeble-minded in relation to delinquency," he wrote in 1923. "Feeble-mindedness alone is seldom an all-important cause of delinquency."[54] Yet Doll still believed that low intelligence contributed to anti-social conduct.

John E. W. Wallin was another advocate of mental testing and the hereditarian theory of anti-social conduct who changed his position in the early 1920s. Born in rural Iowa in 1876, Wallin took his doctorate in psychology at Yale in 1901. He then spent a year at Clark University as a research assistant, where he worked closely with Goddard's mentor, Stanley Hall. Hall's deterministic genetic psychology influenced Wallin no less than Goddard. Wallin specialized in child and deviant psychology, and held a series of positions in normal schools and various public institutions, especially municipal boards of education. He published extensively. Throughout the 1910s he firmly believed that innate intelligence "caused" good or bad conduct. Like Goddard and Doll, he gave mental tests to offenders; he found them substandard. Feeble-mindedness was, he argued, "a prolific source of poverty, destitution, all kinds of crime against property and persons, alcoholism, social immorality, illegitimacy, and of prolific and degenerate progeny."[55] Like Doll, he found it necessary to reverse himself in the early 1920s. In 1920 he became professor of psychology at Miami University and director of the University's Bureau of Special Education and Psycho-Educational Clinic. He began using the individual clinical method after running across Healy and Bronner's work, and this experience, combined with his knowledge that geneticists were drastically modifying the old nature over nurture theory of the early 1900s, forced him to shift. He was even doubtful by the early 1920s that intelligence was innate.[56]

Criticism of the hereditarian theory of anti-social conduct came too from a younger generation of psychologists who were taking their doctorates in the 1920s, and who had not made careers advocating application of mental tests in the 1910s. The contexts in which they worked and their experiences varied. Franklin S. Fearing, for example, took his doctorate with Lewis M. Terman at Stanford, but fought his professor on the hereditarian theory of anti-social conduct in no small measure because of his experiences working with social welfare agencies in Louisville, Kentucky, and with the National Committee for Mental Hygiene.[57] So did other students of Terman's, such as Ellen B.

Sullivan, who taught part-time at the University of California, Los Angeles, but who was mainly a full time social worker in southern California, not an academician, which made it easier for her to criticize Terman.[58]

Carl Murchison provided perhaps the most systematic critique of the hereditarian theory of crime of any psychologist of the 1920s. Murchison too had served under Yerkes in the Army testing program. "I am glad to see that you have been able to make army psychological data as well as army methods contribute somewhat to our scanty knowledge," Yerkes wrote Murchison in 1925, in commenting on Murchison's work on criminal intelligence.[59] Murchison indeed achieved much. He graduated from college in 1909, at the age of twenty-one. Then he took graduate work at the Harvard psychology department, where he met Yerkes, three years in theological seminary, and two more years of graduate work in psychology at Yale. In 1916 he became an assistant professor of psychology at Miami University. The next year Yerkes recruited him as a psychological examiner at Camp Sherman in Ohio. After the Armistice he returned to Miami, where he became a colleague of Wallin's just as Wallin was shifting on the hereditarian theory of conduct. In 1922 Murchison went to Johns Hopkins; he finished his doctorate under the behaviorist Knight Dunlap in 1923, and then landed a position at Clark University. "Criminal intelligence" was his specialty. Distinguishing him from the testers of the 1910s were his exposure to persons such as Wallin and Dunlap with different perspectives, conceptual and methodological, his far more intensive work with his "criminal" subjects, and his commitment to evaluating alternate hypotheses. In over a score of articles and a major monograph published in the 1920s, Murchison argued that there was no such thing as "criminal intelligence" apart from "civilian intelligence," no such thing as a criminal mental type. After administering Army Alpha, a major weapon of the hereditarians, to native born and foreign born criminals in four state prisons, he insisted, "It is just as invalid to speak of the foreign born criminal as representing a type as it would be to speak of the native born criminal as representing a type, or even to speak of the criminal class as a type."[60] Using a variety of research strategies, he attacked the whole concept of a "criminal intelligence," and thus the hereditarian notion of innate substandard intellect leading to a life of crime and dissolution, on both technical and conceptual grounds.[61]

The work of such students of the problem such as Healy and Bronner, Doll, Wallin, and Murchison, influenced other psychologists in the 1920s, although this is not meant to discount the impact of their own standards of research or the circumstances in which they functioned. In some instances, local institutional influences mattered considerably, as at the University of Chicago, where, by the early 1920s, the sociologists were studying social deviance from a wholly different and more specialized perspective and where they had considerable influence upon younger graduate students in psychology, such as Margret Wooster Curti and Curt Rosenow, who attacked the hereditarian theory of anti-social conduct as both psychologically and sociologically inadequate.[62] Quite independently of local situations, commitments to professional and methodological standards influenced some psychologists of the older generation to shift their postures in the 1920s. Lewis M. Terman, for example, found it necessary to admit the force of the criticisms of such persons as Healy and Bronner, and of his own students. Low intelligence alone did not cause anti-social conduct, he conceded.[63] Since Terman consistently believed throughout these years in the power of inheritance, obviously he shifted because of the now apparent methodological difficulties of the hereditarian theory of conduct. By the later 1920s, Terman finally succeeded in securing funding for a massive study of geniuses, a splendid subject for a man convinced of the power of psychobiological inheritance.[64]

Social workers also criticized the hereditarian theory of anti-social conduct. Significantly they had, as a group, a more varied constellation of experiences and training than did proponents of the hereditarian interpretation. Some, such as Frances Doughtery and E. J. Asher, had no training in psychology whatsoever, but considerable experience in social work.[65] Others were like Dr. W. B. Wolfe, a New York psychiatrist and sometime student of Alfred Adler, and approached deviance from a psychiatric and clinical case perspective. Wolfe, who had a medical, not a psychological, training, argued that delinquency had only social, not personal causes.[66] Others, such as Walter C. Reckless and Albert A. Owens, took their doctorates in sociology at Chicago in the 1920s and approached the problem of conduct from a statistical and a cultural perspective. Their work effectively pre-empted the biopsychological theory of anti-social conduct by focusing on the cultural and social context of whole subgroups of offenders.[67]

But the debate over the hereditarian theory of anti-social conduct also owed something to ethnocultural influences and commitments. Underneath much hereditarian literature of the 1900s and 1910s was the assumption that at least some immigrants constituted a "criminal type" with low-grade intellects. Three young Jewish social workers, all with training in psychology, vehemently attacked this notion in the 1920s. Sheldon Glueck, for example, became a leading expert in the twenties in the psychological and legal aspects of criminal behavior, and a famous opponent of the hereditarian theory of anti-social conduct. In 1925 he published an article in *The New Republic* that caused something of a sensation; he insisted that cultural, not biopsychological, explanations were appropriate. Born in Warsaw, Poland, in 1896, he grew up in Boston; he took advanced work in psychology and law at Harvard in the early 1920s. For a time he supported himself as a social worker in Boston. But he published over ten books, including two books in the 1920s on anti-social conduct, won an appointment at Harvard, and became shortly a nationally recognized expert.[68] Doubtless his personal experiences influenced his outlook. John Slawson, author of *The Delinquent Boy* (1926), took the hereditarian theory to task in that work. Born in Russia in 1896, he grew up in New York. He earned his doctorate in psychology at Columbia in 1924, supporting himself working as a research assistant for the city's Department of Education and as a psychologist for the New York State Board of Charities. Then he was research director for the Jewish Welfare Federation in Cleveland for four years. By the time he became executive director of the Jewish Welfare Federation in Detroit in 1928, he had published five articles on delinquency, in addition to his book. He made a number of arguments critical of the hereditarian theory of crime, the most common of which was that in his experience testing delinquent and nondelinquent boys, both did equally well on nonverbal tests, but the delinquent boys suffered on the verbal tests. The only fair conclusion, he insisted, was that their wretched social environment penalized them on nonverbal tests.[69] Obviously Slawson took the position he did for many reasons, not the least of which was the impact of his work and that of such specialists as Healy and Bronner upon his own conclusions. Yet the radically different social experiences he had as an immigrant, as compared with many native born Americans, prior to becoming a psychologist influenced the questions he asked, if not the answers he provided.

Perhaps the example of Samuel C. Kohs best illustrates such complexities. Kohs was born in New York in 1890. He graduated from City College in 1912. He then pursued both graduate work in psychology and social work. For a year he was Goddard's research assistant at Vineland. Then he took a master's degree in psychology at Clark, where he came into contact with Hall and some of Hall's disciples, especially Frederick Kuhlmann, who was advancing the hereditarian theory of crime. At Clark, Kohs worked at the Children's Institute; he even published an article on the "moral defective," which signalled his acceptance of the hereditarian theory.[70] From 1914 to 1916 he worked at the Chicago House of Correction as a social worker. His graduate training in psychology at Chicago and Stanford exposed him to both Chicago's cultural environmentalism and Terman's stress on innate talents. He finished his Ph.D. in 1919 while teaching at Reed College and working with the Portland, Oregon, Court of Domestic Relations as a psychologist. In 1923 he began working with a succession of Jewish social welfare agencies. By 1928 he was executive director of the Brooklyn Federation of Jewish Charities. At some point in the later 1910s and early 1920s, as he moved away from the perspectives of men such as Goddard, Hall, Kuhlmann, and Terman, he began to have gnawing doubts about the hereditarian theory of crime and the whole thrust of early twentieth-century biology and psychology as he had learned it in graduate school. In a major review of the literature of delinquency and crime in 1927, Kohs declared that the "biological sciences really give us no clue to the solution of the crime problem . . . . What happens, ultimately, in the direction of social behavior, depends not so much on the nature of the physical handicap, as on the individual's adjustment thereto, and the extent to which society, through its various agents, has aided in that adjustment."[71] To support his new position, he pointed, first, to the work of geneticists such as Thomas Hunt Morgan, which demonstrated biological inheritance was far more contingent and complex than the hereditarian theory of crime allowed; and, second, to the studies of crime and delinquency of Healy and Bronner and others, which strongly indicated high intelligence was not the only, or even a particularly important, prerequisite for proper moral and social conduct.

In the years following Kohs' review, a new generation of young psychologists published a number of specialized monographs focussing on urban crime: Nels Anderson on the hobo, Frederick Thrasher on adolescent gangs, Clifford Shaw and his associates on juvenile delinquency, Edwin H. Sutherland and John Landesco on adult criminals, and Walter C. Reckless on organized prostitution. Much of this work was sponsored by the Chicago sociology department and the Illinois Institute for Juvenile Research from the early 1920s on. For present purposes this new work replaced biological and individualistic perspectives on anti-social conduct and substituted new principles of interpretation stressing the primacy of the social group, the impact of the urban social order, the interactions between a particular group of offenders and others groups in society, and, above all, the influence of social disorganization upon human collective life. Perhaps William I. Thomas and Dorothy Swain Thomas best summarized the new sociological perspectives on social deviance in their *The Child in America: Behavior Problems and Programs* (1928), a study financed by grants from the Laura Spelman Rockefeller Memorial Fund, and obviously inspired by Chicago sociology. The Thomases attacked the individualistic perspective of the hereditarian natural scientists, arguing that social phenomena had social causes. In particular they insisted that intelligence tests were inadequate and artificial measures of human intellect, poor substitutes for intensive social research into social phenomena. Not innate intelligence, or the larger individualistic perspective, but the specific patterns of social conditioning and habit formation, for individuals, sub-cultures, and cultures, they insisted, were the grist of the social investigator.[72]

In the decade and a half following World War I, then, scientists of man in America shifted scientific perspectives, methods, and approaches on the issue of anti-social conduct and its causes.

In the early days of mental testing, a relatively small and homogeneous group of boosters of mental testing championed the hereditarian theory of anti-social conduct; perhaps they expected too much from the new biology and psychology. In the 1920s, a larger and more heterogeneous group of students of social deviance emerged. They were less interested in promoting new techniques and more interested in concrete, specialized social research. They found the hereditarian interpretation sorely inadequate on methodological, theoretical, and social policy grounds. The controversy over the hereditarian theory of anti-social conduct had ended much as had other phases of the larger heredity-environment controversy. Hereditarians and environmentalists had recognized the complexity of human nature and behavior. They had abandoned the nature versus nurture dichotomy of earlier years as artificial and misleading. In its place they assumed the interaction of culture and nature in human nature and behavior; they reaffirmed man's animal ancestry, his descent from the brutes, and at the same time they explained his social behavior in cultural terms.

## IV

The third mental testing debate centered on the causal relationship between an individual's social status and intelligence as measured by the tests. This controversy was not as dramatic as those over race psychology and anti-social conduct; it did not kick up the public attention they did. And it seemed to have relatively little direct relationship to current events, at least in the sense that the race intelligence controversy or even the social deviance debate had. The controversy over innate intelligence and social status was not simply a technical scientific discussion. Many Americans of middle class, Anglo-Saxon Protestant ancestry assumed that high social and occupational status resulted from superior intelligence. According to this view, a man with superior inborn intelligence and energy could achieve much; he could even rise from rags to riches. Quite obviously, those with lesser gifts were left behind in the struggle for existence. These popular attitudes were not cut out of whole cloth. Granted certain assumptions, test results did seem to bear them out. Those test results that included data on social and occupational status did demonstrate a consistent correlation between high intelligence quotient and enviable social and occupational status. Most testers—and many other psychologists as well—were inclined to interpret these results in light of the new hereditarian doctrines of the early 1900s. They argued that innate intelligence was the cause of social status. Most of the proponents of this view in psychology were brought up in middle class, WASP families and nurtured professionally upon hereditarian natural science doctrines—attitudes that were mutually reinforcing. In their work they gave mental tests to selected occupational groups, discovered a relationship between test standing and social status, and announced the cause of the relationship to be innate intelligence. The most celebrated study of this genre came from the Army tests.

The critics of the hereditarian position were a far more heterogeneous group than the advocates. Almost none were promoters of mental measurement. Among them were psychologists, geneticists, psychiatrists, sociologists, educational psychologists, social workers, and biologists. Some grew up in homes similar to the proponents of the hereditarian theory; others did not. Some contented themselves with technical criticisms, and concentrated their attention on alleged statistical flaws in their opponents' work. Others took a more holistic perspective and looked for social, not psychobiological, determinants of social status. Some began their work, quite obviously, with an eye to attacking hereditarian explanations. Others apparently arrived at environmentalist conclusions after years of study. In some instances, technical considerations seemingly mattered most; in others, social policy attitudes, or career commitments and experiences were probably more important.

The controversy lasted until the early 1940s, largely because it presented far more complex theoretical and methodological problems than the other two debates. It was one thing to demonstrate how mental tests were culture-bound and put members of racial and ethnic minorities at a disadvantage, to show that there was no such thing as a "criminal type," or to find methodological errors in hastily done testing programs. The question of intelligence and social status was more ambiguous. For example, investigators of both persuasions consistently found that those individuals of high social and occupational status scored well on virtually any kind of intelligence test, and those of middling and low social and occupational status were bunched beneath them. There was another difficulty besides this persistent hierarchy of status and achievement on the tests: the confusion of individual and group psychology. No one doubted individuals differed in intelligence at least partly because of their varying innate endowments. But a scientist could study an occupational group, defining it as a collection of individuals with their own differing inherited intellects, and announce with no sense of intellectual difficulty that the cause of the relationship between status and test score was inherited intelligence. Or a scientist could study an occupational group as a cultural group, or, more precisely, a subculture whose members had certain common social experiences, such as formal education, family background, and the like, and declare the cause of the relationship to be cultural rather than psychobiological. It was not until scientists abandoned the nature versus nurture dichotomy and accepted the idea that heredity and environment, nature and culture, man's biological nature and his cultural behavior, were at the same time intertwined and distinct factors in his development, that a resolution of the controversy came into sight.

In the 1910s, most psychologists and other scientists of man who took up the question of intellect and status assumed the hereditarian explanation. But a few argued the contrary posture, most probably because their professional and career experiences gave them nagging doubts about the hereditarian interpretation. Erville B. Woods of the University of Minnesota was one of the first academic sociologists to take up the question. Woods took his doctorate from Chicago in 1906. His work was typical of Chicago social science. In 1914 he reported the results of his statistical investigation of the occupational backgrounds and ambitions of one thousand local boys. Almost all boys in each occupational and class category possessed their parents' educational and occupational ambitions. He declared ambition to be at least as important as intelligence in shaping an individual's career and his or her social status. Mobility and achievement depended upon shared group attitudes and values, together with that group's position in the context of the larger society.[73] What Woods did, then, was to diminish the importance of innate intelligence. A Cincinnati psychologist, Helen Thompson Woolley, declared in 1915 that her experiences administering mental tests to local working class white children convinced her that native intelligence was but one factor in personal success; other factors, such as ambition, general appearance, manner, dress, social ease, and personal persistence, among others, were also important.[74] Woods and Woolley had roughly similar backgrounds. Both took their doctorates at Chicago, Woods in sociology, Woolley in psychology. Neither made the promotion of mental tests the central objective of their career. Both studied their subjects intensively, and came into continual contact with them. Woods taught social work courses at Minnesota. He had his students interview the socially deprived; often he participated in these interviews. Probably his continual face-to-face contact with his subjects helped him recognize many different influences. Woolley served as director of Cincinnati's Bureau for Investigation of Conditions of Working Children and director of the public schools' Vocational Bureau for several years. These were far different circumstances in which to investigate the relationship between intelligence and social status than those many mental testers experienced, who often limited their contacts with their subjects to the hour or so it took to give the tests, and who in any case were chiefly interested in measuring intelligence, which they assumed was innate. E. K. Strong made a discovery that was even more devastating for the hereditarian argument. Strong took his Ph.D. in psychology from Columbia in 1911 and then accepted an appointment at George Peabody College. He conducted a major research program, funded by the Rockefeller Foundation, on the impact of hookworm disease on the mental development of poor Southern children. Using the Stanford-Binet, Strong discovered a close correlation between the incidence of hookworm disease in these children and their standing on the test. He concluded that intelligence did not develop in a vacuum; it was influenced by a variety of post-natal and environmental influences, including, of course, hookworm disease.[75] Interestingly enough, Strong was no enemy of mental testing; he participated in the Army testing program during the War, and subsequently became professor of vocational education at Carnegie Institute of Technology, where he developed further his interests in advertising, educational psychology, the psychology of advertising, and other branches of applied psychology.

After 1915 or so, more psychologists turned to the problem of social status and intelligence. Frequently they became skeptical of the hereditarians' apparent problems with methodological rigor. Thus two of Robert Yerkes' graduate students published articles in which they voiced their discontents. In 1917, James W. Bridges argued that a correlation between test standing and social status had been established, but no one could tell whether its causes were due to original nature or acquired culture.[76] Bridges carried his doubts into the Army testing program. S. L. Pressey went even further than Bridges. Pressey gave intelligence tests to rural Indiana school children—270 from poor homes, 268 from prosperous families—and compared their averages with those of urban children in comparable classes. The city children stood higher than the ruralites. Pressey decided the obvious conclusion, that the country children were less intelligent was "decidedly not wholly adequate." The tests penalized the country children; they were far less comfortable than the city children in the examination situation, and they had vastly inferior educational backgrounds. Pressey concluded that new tests should be devised to measure the rural children's intelligence in accordance with their educational and cultural backgrounds.[77] Other psychologists not committed to the promotion of mental testing criticized the hereditarian interpretation, or at least argued that the matter was more complex than the hereditarians would admit. Some, like M. E. Cobb and Gladys G. Ide, were career social workers, not academicians; presumably they had more varied experiences than professors in universities and perhaps less vested interest in advocating a particular doctrine of the profession. Some professors joined the attack too; Melvin E. Rigg, for example, did some research on social status and intelligence after finishing his doctorate in comparative psychology with Robert Yerkes at Harvard. The behavioristic emphasis of comparative psychology helped push him away from individualistic and psychobiological explanations.[78]

In the early twenties, some hereditarian testers were sufficiently aware of these criticisms to try a new approach, that of testing related individuals in similar and dissimilar environments, to counter their critics' arguments that they had not taken environment into account. Arnold Gesell, a psychologist at Yale, argued in 1922 that the consistent mental and physical similarity in twins reared apart was based upon a common heredity, which was also the general conclusion of University of Chicago's H. H. Newman in a similar study.[79] Curtis Merriam, one of Lewis Terman's recent doctoral students, published a monograph on twins, using the Stanford-Binet, the Army Alpha, the National Intelligence Test, and teacher estimates of twin intelligence; he concluded that environment made no difference, for older twin pairs were no more alike than young twin pairs.[80] Generally speaking, however, these and other twin studies of the 1920s done by psychologists committed to hereditarian points of view did not probe very deeply into environmental circumstances, at least not as much as a social scientist might.[81]

Some scientists did criticize the twin literature. Hermann J. Muller, a brilliant member of Thomas H. Morgan's famous *Drosophila* group (and a man with left-liberal political views) tackled the problem. He gave various mental tests to a pair of identical twins reared apart. He found the twins had similar, but not identical, intelligence test scores—and remarkably dissimilar personalities and temperaments. He concluded that environment did influence these facets of human nature and could not be dismissed.[82] Eugene Shen, a native of China in the Stanford graduate program in education, criticized Merriam's study. As a specialist in the statistics of psychological measurement, Shen concentrated on what he believed were Merriam's methodological errors, notably his use of such statistical tools as the coefficient of correlation (which Merriam employed to demonstrate the similarities of different traits measured). Shen also insisted that Merriam had not studied the impact of the environment systematically. He further declared that Merriam's own data showed that the older pairs of twins were more dissimilar than the younger pairs of twins; wryly he concluded that this could hardly be attributed to heredity.[83] A. H. Winfield and Peter Sandiford, two psychologists committed to hereditarian explanations, criticized Shen, although the data they discussed in substantiating their position did show the differences of twins Shen claimed existed between the older and younger pairs of twins. The argument seemed to boil down to what the differences meant for the nature-nurture issue.[84]

By the later 1920s there were signs of change. For one thing, the drastic revisions of genetic and evolutionary theory were becoming diffused in the psychology profession, however dimly and imperfectly. It is highly doubtful that most (or even a few psychologists) kept up with the latest work in such technical journals as *Genetics*; but at least some psychologists were reading the more popular accounts of the new theories that leading geneticists wrote in the 1920s. Furthermore, most of the psychologists who entered the profession in the 1920s and took up mental measurement seemed a far less doctrinaire lot on the heredity-environment issue than the previous decade's pioneers of mental testing. Generally speaking, the tone of their publications seemed less confident, less assured, more tentative; they gave the impression that they believed they were testing various hypotheses rather than verifying known laws and truths of nature. Undeniably, to a certain extent this may have been a convenient tactic for younger researchers to present their work (and themselves) in the most "scientific" light before the larger profession; but quite obviously if they were deeply attached to the principle that intelligence was solely a matter of original endowment, it was inappropriate for them to attack the hereditarian views of senior people in the field. In any case, what mattered was not so much the motives but the consequences of their behavior, which were to shift the focus of the professional discussion from proving the power of inheritance to investigating all the variables in the making of intelligence. This in turn facilitated the crystallization of a new view of nature and nurture over the next several years: the conception of heredity and environment as interdependent yet distinct variables.

An even more tangible manifestation of this new model of the causes of intelligence and social status was the publication in 1928 of the mammoth study by the National Society for the Study of Education on nature and nurture. Various authors contributed their own studies to the yearbook, ranging from studies of the resemblance of siblings to those of Siamese twins, and various reports of investigations of foster children. Perhaps the most widely discussed study was that of 401 foster children conducted by University of Chicago educational psychologists Frank N. Freeman and his associates Karl J. Holzinger and Blythe C. Mitchell. Doubtless the intellectual milieu at the

University of Chicago reinforced whatever proclivities they had to be sensitive to the importance of environment, and to be even-handed on the nature-nurture issue. But there were other reasons, probably far more important ones, for their general conclusion that heredity and environment could not be artificially separated, that both functioned together in the development of intelligence. They tested the foster children before the children were placed in "superior" and "inferior" foster homes; after several years Freeman and his coworkers retested all the children. They discovered that the children in the superior foster homes had improved markedly when compared with those in the inferior foster homes. And the earlier the child was adopted by a family, the higher was his or her standing in the tests. They also tackled the problem of siblings' intellectual resemblances; they found that siblings reared apart resembled each other about half as much as siblings reared together, which suggested to Freeman and his colleagues that intelligence was influenced equally by heredity and by environment. Even though a larger percentage of the parents of the foster children were mentally substandard, furthermore, the mean intelligence quotient for the children was practically equal to the standard for children generally; Freeman, Holzinger, and Mitchell concluded that heredity was not the only factor in intelligence. They even took up the claim, which Henry H. Goddard and others had raised years ago, that feeble-mindedness was inherited according to Mendelian law. Twenty-six children had two feeble-minded parents; according to what Goddard had argued, this would mean all twenty-six were feeble-minded. Four of the children were slightly below 70 (the cut-off for feeble-mindedness) and the average for the twenty-six was 81, considerably higher than the expectation from the Mendelian argument, but of course considerably below that for the entire group of 401 children. Finally, Freeman and his associates argued that morals were learned, not inherited; many of the children had parents who were "morally defective," yet there were few cases of serious misbehavior among the children, which suggested environment did matter. Yet Freeman and his associates did not contend that environment alone explained intelligence or conduct or status; nature and nurture worked together as distinct, interdependent, mutually reinforcing variables.[85] This evenhandedness was quite conspicuous among the younger scholars who contributed to the NSSE yearbook. Thus Barbara Stoddard Burks of Stanford contributed an important report on foster parent and foster child resemblances as compared with true parent and true child resemblances. Burks' preference was to emphasize heredity; yet there was a world of difference between her work and that of similarly inclined psychologists in the past. She emphasized that she had selected a socially and ethnically homogeneous population; because of this environmental selection, environment would matter less in her results. Indeed, she insisted that mental development could not occur without heredity and environment. Her general conclusion was that about 17 percent of the variability in intelligence was due to differences in home environment; in other words, she argued, at most environment could make a difference of about 20 points on the tests. Of course 20 points (assuming 100 was average) was as a large spread for intelligence testers in the 1920s to contemplate as the gap, for example, between the average scores of "Nordics" and recent immigrants and blacks on the Army tests.[86]

What was even more interesting about the NSSE's yearbook was the reaction of a number of prominent testers to its results. Guy M. Whipple of the University of Michigan, the NSSE's secretary and a long-time advocate of the hereditarian interpretation of mental testing, declared that the studies in the yearbook proved that the problem of nature and nurture was one of degree; there was no sense in talking about nature versus nurture. "No one who reads the *Yearbook*—this is my impression at least —can put it away with the conviction that general intelligence is an absolutely fixed, immutable, innate capacity, but neither can one put it away with the conviction that general intelligence is readily susceptible to environmental influence. The truth lies between these extremes," concluded Whipple.[87] Lewis Terman, on the other hand, still seemed convinced that the nature-nurture dichotomy had meaning, but far more for social and educational policy than for laws of nature. Terman was in fact a liberal Democrat who believed that public education should provide the best possible training for all; if heredity mattered more, the educational system should be reformed accordingly to provide differential training, whereas if environment was chiefly responsible for intelligence, then appropriate changes in education should be made to reflect that. And Terman quite openly said that the

modern educational theorist was interested in the balance of
nature and nurture.[88] Frank N. Freeman, who preferred environmental interpretations (as Burks, Terman, and Whipple
looked to hereditarian explanations first) discussed the yearbook
too. After questioning the conclusions of some studies that
argued for nature rather than nurture, he insisted that modern
genetics had pretty well destroyed the conception of a unit-
character theory of intelligence, and, beyond that, the idea of
nature versus (or over) nurture.[89] William C. Bagley of Columbia Teacher's College attacked the yearbook as too hereditarian;
he insisted that the studies of Freeman and Burks showed that
nurture had some impact upon intelligence quotient, and that
most of the yearbook's authors made the erroneous assumption
that mental age was the undiluted product of heredity. He
argued that while individual differences were important, far
more important were the contributions that a sound educational
system could bring to the masses and thus uplift them.[90] It was
in particular this function of education that Bagley believed the
yearbook ignored; sound education would lead to a stable,
harmonious society.

But the larger point was that the old nature versus nurture
dichotomy was coming into disfavor in the psychology profession.
The new view of nature *cum* nurture, together with the recognition that intellectual development was highly complex, and not
simply the consequence of the transmission of a few (or one)
unit-characters, was gaining new ground in the later 1920s; it
seemed as though the more psychologists wanted to look for
complexity, the more they found it. Blanche C. Weill, a clinical
psychologist who worked with the Bakersfield, California, public
schools for many years, made a pioneering study of sibling relationships in families. She concluded that the family environment
was not the same for all children in any family, for each child
had a different relationship with the parents, and the social
chemistry of sibling interaction had an impact upon development
too.[91] D. Van Alystyne, a recent Ph.D. from Teacher's College,
studied three-year-old children intensively in many aspects, and
found that their intelligence was influenced by many post-natal
and cultural factors.[92] Several scientists who had taken emphatic
stands on nature versus nurture before the mid-1920s modified
their postures in the late 1920s and early 1930s. Thus H. H.

Newman, the University of Chicago zoologist who had been working on twin studies for some years, began to make somewhat different conclusions as his work progressed (and as he studied human rather than animal subjects intensively). In 1929, when he reported on the studies he had made of the mental and physical resemblances of three pairs of identical twins reared apart, he declared that the twin having the better education and home environment obtained a higher intelligence quotient by twelve points. By 1932, he had studied fifty pairs of identical twins reared apart. He said that the environment modified some physical traits, such as weight, general health, and so forth, but not those such as eye color, teeth, and the like, and it also modified intelligence and personality; in some instances the intelligence of a pair of separated twins was three times as different as the average intelligence of the fifty pairs of twins reared apart. Newman concluded that when full credit was given to environment, the fact stood out that "hereditary resemblances remain most strikingly close."[93] In 1930, Arnold Gesell of Yale argued in a major statement of his maturation thesis of individual development that "there is a profound interdependence between 'heredity' and 'environment' in the control of development."[94] Quite obviously Gesell did not find an emphatic either/or posture on nature-nurture useful for the purposes of scientific explanation. Other studies by psychologists in the early 1930s underlined the complexities of the problem of intelligence and social status; some explored the impact of rural and urban culture upon intelligence (and reenforced the conclusions of the Freeman study), others criticized the statistical methods of the hereditarian studies, and still others probed such newly discovered phenomena as the interaction of intelligence quotient and such factors as vitamin intake and glandular development, which put the nature-nurture problem in a rather different light.[95] In this context it became increasingly difficult for men and women of science to declare flatly that nature prevailed over nurture; even those who were personally convinced that this was the case had to make at least ritualistic overtures at the importance of environment, as H. H. Newman in some sense did in 1932.

In the middle and late 1930s several psychologists at the University of Iowa's Child Welfare Research Station conducted the most celebrated and publicized studies of foster children. The

Child Welfare Research Station had been founded in 1917 at the University, largely because of the lobbying of a number of women's groups in the state, for the purpose of studying the whole of the normal child so that children would grow up "right." Mrs. Cora Hillis, the Des Moines matron who led the campaign for the proposed station, had first approached Iowa State College, in nearby Ames, but when she was told by College officials that the State College did research on the normal pig, not the normal child, she repaired to Iowa City where she found University administrators, including psychologist Carl E. Seashore, dean of the graduate school and an eminent specialist in the psychology of music, most receptive to the idea of a $100,000 endowment and the expansion of the psychology department, which was thus implied. Although the Child Welfare Research Station sponsored studies on nutrition, as well as intelligence, it became famous in the 1930s—indeed, infamous in some circles— because a number of its staff, notably Beth L. Wellman, Ruth Updegraff, George D. Stoddard, and others, were able to study large groups of foster children in a variety of foster homes and come to conclusions that would enrage hereditarians. And within the context of a research institute given to the university in the spirit of prewar progressive reform, with the charge to work out those principles of science that would provide for a better generation in the future, it was not surprising that the members of the Iowa School, as they came to be known, argued not that environment determined all of intelligence, but that the environment made the intelligence quotient inconstant. For a group of scholars and scientists with heavy official commitments to public service and to liberal reconstruction of school and society, this was an understandable posture.[96] Yet it should be emphasized that the members of the Iowa School were not simon-pure environmentalists; they accepted the culture-nature theory of 1930s natural and social science, and they consistently argued that heredity and environment were intricately combined in development. They tested and retested selected control groups of foster children, carefully noting all factors, original and acquired. Harold M. Skeels, for example, published two studies in the late 1930s in which he declared that superior and inferior foster homes had a pronounced impact upon the intelligence quotients of foster children.[97] In 1938, Skeels, Updegraff, Wellman, and Harold M.

Williams published an important monograph, *A Study of Environmental Stimulation.* They studied two control groups of lower-class orphans, one that had attended a model preschool, another which had not. They found that after three years the foster children who attended the model preschool had made a substantial gain in intelligence test scores, whereas those who went to the inferior school had not improved at all. Since all orphans in both groups had such unpromising backgrounds before the study commenced, Skeels and his associates reasoned, heredity and environment obviously functioned together. They stressed the plasticity of the mind, but, in this and in other publications, they did not dismiss innate intelligence. The novel twist they gave to the IQ problem was their argument that the intelligence quotient would change, depending upon the quality of the foster home; that IQ itself was variable, if not exactly unstable. Aside from that, they hewed quite closely to the new orthodoxy of nature and nurture.[98]

In 1937 a new study of twins appeared in print. This was an unusual study, if for no other reason that it was the distillation of a ten-year research effort conducted by three University of Chicago scientists: Frank N. Freeman, the educational psychology, statistician Karl J. Holzinger, and zoologist H. H. Newman. This study of identical and fraternal twins was exhaustive and probably definitive. And it was symptomatic of the rise of the new nature *cum* nurture model in developmental and differential psychology, partly because Newman, a hereditarian, joined forces with Freeman, an environmentalist, but primarily because the complex conclusions the three scientists made suggested that the nature versus nurture dichotomy was pointless. That dichotomy, they declared, was meaningless, for every "character is an expression of the interplay of hereditary and environmental factors. . . . any statements as to the relative potency of heredity and environment are meaningless." They agreed that heredity and environment were indispensable, dependent variables in development; they could not be regarded as independent factors that could be added or subtracted from a given formula of intellectual development. Indeed, they despaired of any hope for a general resolution of the heredity-environment problem. They concluded that the nature-nurture problem was not a scientific question, that is, it could not be resolved by the concepts and methods of science. It was an abstraction that grossly oversimplified and distorted the real facts of intellectual development. The farther one penetrated into the complexities of genetic and environmental factors that together determined development, the more one was compelled to admit "that there is not one problem but a multiplicity of minor problems—that there is no general solution of the major problems or even of any one of the minor problems . . . . We feel in sympathy with Professor H. S. Jennings' dictum that what heredity can do environment can also do."[99]

## V

And, indeed, by the late 1930s there was much evidence that most psychologists were weary of the nature versus nurture dichotomy and regarded it as artificial, unproductive, and perhaps unscientific as well. There were, of course, always some who dissented, some who believed in heredity over environment, and other who affirmed the power of environment over heredity. But in the main there seemed little doubt that most psychologists wanted to take up other questions. Thus in December 1938, when psychologists Florence L. Goodenough and Lewis M. Terman wrote Arnold Gesell to argue for the necessity of a study for the NSSE yearbook of 1940, so that, as Terman put it, environmentalists such as Beth Wellman and George Stoddard would not "pack the Yearbook with evidence on one side," Gesell answered that he had no new work to write about at the moment; ". . . I am not in much of a mood to stir up anything additional," he told Terman.[100] He told Goodenough that although he was always interested in the problem of mental development, he and his associates "are trying to get away somewhat from a fixed dichotomy of Nature vs. Nurture."[101] That a scientist of Gesell's eminence and particular expertise wanted to put the nature versus nurture dichotomy aside suggested how much times had changed in American psychology and American science, for Gesell had devoted much of his career to studies of human development that allotted a large potency to original endowment and psychobiological factors.

Yet the exhaustion of the nature-nurture dichotomy was less

important, in retrospect, than might at first appear to be the case. No one proposed abandoning mental measurement; indeed, mental measurement now came into its own more than ever, with, for example, the rise of standardized achievement tests. Nor did anyone in the psychology profession or in any other profession whose main study was man seriously propose that the whole edifice of evolutionary natural and social science be jettisoned as useless baggage. What had really happened was that, at least for a generation or so, American scientists had discovered that an either/or question, which was as much inspired by social policy attitudes as by appropriate questions of nature, was too difficult and perhaps profitless to discuss further. Indeed, now that the nature-nurture controversy over mental testing had ended, and natural and social scientists had more or less resolved their jurisdictional disputes, a new era dawned in American cultural and scientific life in which the influence of the sciences of man upon the imagination and the institutions of many Americans could only increase. This was now the age of evolutionary science's most important impact upon American culture and society.

# 4 The Triumph of Evolution

# CONCLUSION

B
y the early 1940s the nature-nurture controversy had run its course. What had begun as a series of questions revolving around the perception that heredity and environment were independent, additive factors figuring into organic and psychological development ended as a series of nonquestions or nonsensical questions that no longer seemed to most American scientists to be answerable. As psychologist Anne Anastasi of Forham University put it in an address before the American Psychological Association in 1957, the once lively nature-nurture controversy had become a dead issue in recent decades in the eyes of most behavioral scientists because it hinged upon questions that were not answerable. If one asked whether nature or nurture was responsible for particular differences, one ran afoul of the apparent truth that nature and nurture were interwoven, in varying degrees, in virtually all behavioral patterns. Nor were the results much better, she noted, when scientists attempted to estimate the relative contributions of hereditary and environmental factors to particular patterns and traits; quite apart from the grave methodological difficulties encountered in experimental design for such investigations, there remained the serious conceptual flaw of regarding nature and nurture as independent factors that could be added to or subtracted from the phenomenon of development. This simply flew in the face of what

269

genetics, psychology, and social science had demonstrated about the continual interaction and interrelatedness of nature and nurture. She ended her address by calling for a whole new line of inquiry based on the simple question of how heredity and environment functioned in specific situations. Hopefully, she added, this might avoid the situation, common when the nature-nurture controversy was alive in the 1920s and 1930s, of scientists with different points of view offering conflicting interpretations of the same data.[1]

As Anastasi's comments clearly suggest, what American scientists have said and done about the scientific issue of nature and nurture since the collapse of the nature-nurture controversy in the early 1940s constitutes a distinct and different chapter in the history of evolutionary science in America. Its chronicling has no place in this book, for it represented a radically different way of looking at nature and the world than was manifest in the natural sciences before and during the nature-nurture controversy. Of course in recent years something approximating at least a minor revival of the old nature-nurture controversy seems to have arisen in American science and politics. It is certainly quite possible that this will lead to a full scale nature-nurture controversy in time, not simply because of the potential for a new model of nature that would permit a new debate, but also, as one historian has pointed out, because our own time, like the 1920s, has been a period of racial and ethnic polarization.[2] Obviously any further comment would be premature.

Yet perhaps we can begin to gain some glimmerings of understanding, not only of our own time but of the era of the nature-nurture controversy, if we can step back from the controversy and view its history from a distance. There seems little doubt that the nature-nurture controversy played an important role in the history of modern American culture and science, and in the continuing impact of evolutionary ideas and evolutionary science in our nation's history. Most historians have estimated the impact of evolution in American culture to be the Social Darwinist movement of the later nineteenth century. Yet the history of the sciences of man in the twentieth century—of which this book is a part—suggests that the influence of evolution became more pronounced in America with the demise of the Social Darwinist movement in the early twentieth century and its replacement by the new disciplines concerned with human nature and behavior. Virtually the only legacy of the Social Darwinist movement—aside from the contribution of a number of highly complex and ingenious schemes of social evolution—was the idea that it was possible to create a science of man, based on evolutionary science, that could facilitate the prediction of human behavior. This was a bold conception, and those natural scientists who took up experimentalism and academic professionalism in the early twentieth century simply assumed that it was true, that it was a given axiom of up-to-date science. Because the new, post-1900 professors of evolutionary science used experimental methods, not the loose analogies between nature and society of the Social Darwinists, it was easy for this new professoriate to supplant and displace the Social Darwinists and persuade far many more Americans, especially those who were educated and who were situated in positions of power and authority, that man was an animal whose behavior could be studied by the most precise methods known to modern science. Knowledge was power—the power to direct human progress.

At the heart of the nature-nurture controversy of the interwar years was not the validity of the general notion that a science of man was legitimate but merely the rather technical point that in such a science of man it was necessary to take account of the cultural as well as the biological levels of human behavior and conduct. To regard the environmentalist social scientists as either humanists who believed, as a group, in free will, or as professional experts who rejected the relevance of evolutionary science for the explanation and interpretation of human behavior is a serious misreading of the evidence. Social scientists were only too happy to join in the grand enterprise of erecting a science of man once their own particularistic professional and technical claims and modes of explanation were taken seriously by the natural scientists. Indeed, the emergence of the interactionist model of nature *cum* nurture, far from discouraging natural and social scientists about the possibilities of a science of man, instead stimulated them to new levels of thought, research, and action. Interdisciplinary cooperation was a rare phenomenon in the 1920s; by the 1940s and thereafter it was commonplace, as the evolving federal grant structure thereafter tacitly recognized. Certainly by the 1940s, scientists of man, whether they worked the natural or the social scientific sides of the street, could now present to educated Americans a unified, coherent, evolutionary science in the making, and thus popularize both their particular research findings and their larger message that social prediction and control was just around the corner, thanks to the resolution of the nature-nurture controversy. The processes of popularization were complex indeed, and would easily warrant the publication of several detailed monographic studies. Certainly the new ideas were disseminated in college lecture halls, in the newspapers, magazines, over the radio and television, and various special interest publications. In turn, members of the disciplines of man, if they were professors, had as a major component of their occupational role the educating of a whole new class of professional and semiprofessional human relations experts; both they and their students took up the work of advising the public on many aspects of human behavior, including the right rules of child-rearing, the management of labor and personnel relations in industry, the measurement of intelligence, the quality of marital relations, the authoritarian personality, and, in more recent times, the narcissism of contemporary self-help psychology.

Scientists of man also cheerfully took up the task of creating a demand for their services as consultants to foundations, business corporations, public agencies, even the military services. As professional experts they were able to lend a helping hand to the problems of social unrest at home, producing study after study of urban disintegration. And they served their nation's foreign policy patriotically, writing position papers and other documents of, for example, step-by-step scenarios of nuclear holocaust or of leftish revolutions in the third and fourth worlds.

How influential this science of man has been in our recent cultural history is still unclear. There is much we simply do not know. But there are good grounds for believing it has been of considerable importance. It is rather certain that the diffusion of evolutionary ideas in American culture, even before the celebrated trial of John Thomas Scopes in 1925, has been limited to less than the whole of the American population. Particular groups in the American population—nonwhites, Catholics, Protestant fundamentalists, and, generally speaking, those with scant formal education—have been either indifferent or in other cases actively hostile to evolutionary science, let alone the larger idea of a science of social control. In practical terms, exposure to evolutionary science in the wake of the Scopes trial has probably been limited to the college educated, for one of the consequences of the trial was that publishers of public school textbooks in biological science removed references to evolution in a compromise of science and education in American culture. Those Americans who were not going to enter the world of college—and thus, in a sense, those parts of our society most distinctively modern— were spared the necessity of confronting disturbing and possibly controversial ideas.[3] It has been only in recent years, as much of the literature of social mobility and social elites in America has shown, that persons who were not from middle-class, white, Anglo-Saxon Protestant homes and backgrounds have become much of a component of the technical and professional elites that comprise the modern occupational groups of our society. Thus, if we define our interest in the impact of evolutionary science in America in terms of the number of its adherents in the whole American population, we would probably say that it has been only a large minority of the population, those Americans who identify with our homogenized middle-class, WASP culture (regardless of their particular ancestry). And it would most probably be foolish to suggest that all such Americans have understood the magical promises of a science of social control, or, if they did, would accept it as a morally valid enterprise. Probably the advocates of this science of man comprise little more than the relatively small fraction of the population whose members either identify with the sciences of man professionally and occupationally or who obtain their services.

It may be no small irony that in a culture and a society in which the rhetoric and the reality of democracy and individualism are taken fairly seriously, at least by comparison with many other cultures of the day, the question of the influence of evolution and of the science of man probably resolves itself not into the number of American citizens who accept these ideas freely but to the extent to which these ideas are and have been embedded in the basic social institutions and social roles of our modern corporate social order, and to the extent these institutions and roles perpetuate the formulae and prescriptions of this science of man and influence the patterns of existence experienced by so many millions of Americans. Obviously there is much that is still unknown. But it may be suggested, however tentatively, that business corporations employ industrial psychologists, that schools and colleges use standardized achievement tests for admission, that governmental bureaucracies, the mass media, the advertising industry, the political parties, and other primary institutions operate on particular assumptions borrowed from modern human science. The long-range, historical function of the new evolutionary science was to resolve the basic questions about human nature in a secular and scientific way, and thus provide the possibilities for social order and control in an entirely new kind of society. Apparently this was a most successful and enduring campaign in American culture.

# Bibliographic Note, Notes, and Index

# BIBLIOGRAPHIC NOTE

It is my intention in this note to provide a general commentary on the most important materials for the preparation of this book. The notes to this book provide more citations, but in fact the literature is of such proportions that it would be a vain hope to attempt to provide a full guide to all relevant materials.

Scientific journals were an indispensable source of information for the history of the scientific disciplines and professions involved in the nature-nurture controversy, as well as for the controversy itself. They contained the technical papers of the disciplines, and, quite often, news of the profession. I used those journals starting in 1880, or at any point after then that they began publication; I stopped using them after 1941. The dates in parentheses indicate the first year of publication. Several general scientific journals were useful for understanding the development of the scientific community at large. *Science* (1883) was the most important such journal, especially after 1889, when it became a newsletter of the institutional history of science and a forum in which scientists published key papers on important developments in their own fields for the information of scientists in other fields. *Popular Science Monthly* (1872–1915), under the editorship of E. L. Youmans, often carried articles relevant to the nature-nurture issue, although its tone was popular; when James McKeen Cattell took over editing the journal in 1900, he converted it into a more rigorous journal and published a large number of articles relevant to the nature-nurture problem and the disciplines most involved in that problem. In 1915 Cattell sold the journal to a publisher who converted it into a popular magazine, and Cattell then founded *The Scientific Monthly* (1915) in which he continued the publishing program he had established in *Popular Science Monthly*. Several journals were helpful for the history of the biological sciences. The British journal *Nature* (1869) was useful for following the Neo-Lamarckian controversy, the rise of biometrics, Mendelism, and the mutation theory in an international context. Also helpful were *Popular Science Monthly, Scientific Monthly,* and *Science. The American Breeders' Magazine* (1910–12), which became *The Journal of Heredity* (1912), primarily a eugenics journal. *Biometrika* (1904) was the major journal of the British biometricians. *Eugenics Review* (1909) is useful for following the English eugenics movement, as was *Eugenics: A Journal of Race Development* (1928). Finally, several more specialized journals were useful: *The Journal of Comparative Neurology and Zoology* (1891), *The Journal of Experimental Zoology* (1904), *Biological Bulletin* (1899), *Proceedings of the Society for Experimental Biology and Medicine* (1902), *Genetics* (1916), and *The Quarterly Review of Biology* (1925).

Broadly speaking, journals in psychology and its related disciplines, such as psychiatry, philosophy, criminology, and education were more germane to the nature-nurture controversy than journals in the biological sciences. The most important journal in psychology was doubtless *The Psychological Review* (1893), which carried important articles on every conceivable topic of interest to psychologists. *The Psychological Bulletin* (1904) published invaluable bibliographical essays in every field of current interest in psychology, together with digests of important papers delivered at the annual meetings of the American Psychological Association, reviews of major monographs, and, occasionally, original articles. *The Psychological Index* (1891) was the other annual bibliographical guide to psychology. A number of journals already mentioned—*Popular Science Monthly, Science, Scientific Monthly,* and *The Journal of Comparative Neurology and Zoology*—occasionally published original articles of psychological interest or with implications for the new psychology. Important also was *The American Journal of Psychology* (1887), founded and edited by G. Stanley Hall, which carried important articles on mental testing, individual differences, abnormal psychology, animal behavior, child psychology, and instincts. Hall's other journal, *The Pedagogical Seminary and Journal of Genetic Psychology* (1891), was more specifically aimed at professional educators. Of distinctly lesser utility for this study were *The Journal of Religious Psychology* (1904) and *The Journal of Race Development* (1910). Occasionally *The British Journal of Psychology* (1904) published articles germane to this study, as on the instinct problem. *The American Journal of Physiology* (1898) published experimental papers on physiology sometimes relevant to instincts and reflexes. *Psychological Monographs* (1895) and *Archives of Psychology* (1906) frequently published important doctoral dissertations in many branches of experimental psychology. *The Journal of Animal Behavior* (1911–17) and *The Behavior Monographs* (1913–19), edited and founded by John B. Watson and Robert M. Yerkes, published important reports in comparative psychology with much relevance for the history of instinct theories. *The Journal of Experimental Psychology* (1916) carried reports of experiments in physiological psychology, animal and child behavior, and mental testing. *The Journal of Comparative Psychology* (1921) published work on animal behavior after Watson left science. Most philosophy journals were not, of course, useful, but *The Journal of Philosophy, Psychology, and Scientific Method* (1904) published articles on the mind-body problem, pragmatism, realism, and behaviorism. Occasionally useful were *The Philosophical Review* (1891) and two British journals,

*Monist* (1890) and *Mind* (1879). Of the several periodicals devoted to abnormal psychology, the most useful one was probably *The Journal of Abnormal Psychology* (1906–21), edited by the Boston physician Dr. Morton Prince and renamed *The Journal of Abnormal and Social Psychology* (1921) because of the large number of articles focusing on the sources of human conduct. Of the several other journals in abnormal psychology, the one I found more than modestly helpful was *Mental Hygiene* (1917).

There were a number of important journals in the field of education. Upon occasion, relevant articles appeared in such general, semipopular educational journals as *The Educational Review* (1890), *The Journal of Education* (1875), and *Education* (1880), although usually the germane articles were statements of opinion rather than reports of empirical investigations. *The Journal of Educational Psychology* (1910) was a major source for the history of mental testing. *The Journal of Experimental Pedagogy* (1911) often it published useful articles on mental testing. *School and Society* (1915), a weekly, performed the same general function for the educational community that *Science* did for the scientific community. It also carried some important articles on mental tests. Other journals that often carried relevant source materials on mental testing included *The Journal of Educational Research* (1920), *Progressive Education* (1924), *Childhood Education* (1925), *Child Development* (1930), *The Journal of Experimental Education* (1934), *Forum of Education* (1923), *The Elementary School Journal* (1900), *The Family* (1924), and *The Journal of Negro Education* (1932).

There were some more specialized journals in applied psychology that were good sources for the nature-nurture controversy. *The Journal of Social Psychology* (1930) carried some articles on human motivation, whereas such journals as *The Journal of Applied Psychology* (1917), *Industrial Psychology* (1926), *The Journal of Consulting Psychology* (1932), and *The Psychological Clinic* (1907) published articles on a wide variety of topics, including mental testing, instincts, and human motivations. Three journals were critical for studying "scientific" criminology: *The Training School Bulletin* (1904), edited by Henry H. Goddard, *The Journal of Delinquency* (1916–28), published by the Department of Research of California's Whittier State School.

For a variety of reasons, social scientists were not able to support and publish as many journals as natural scientists. The most important journal in anthropology was, of course, *The American Anthropologist* (1889), which carried key articles and news of the anthropological community and came to reflect the influence of the Boasians after 1910. *The Journal of American Folklore* (1888) was less germane to this study. Several journals already mentioned—*Popular Science Monthly, Science, Scientific Monthly,* for example—often carried articles written by anthropologists. *The American Journal of Physical Anthropology* (1919), edited by Aleš Hrdlička, restricted itself to physical anthropology and anthropometric studies. One can find some articles on physical anthropology in *Human Biology* (1929), the journal edited by the distinguished geneticist Raymond Pearl.

The most important journal in sociology was *The American Journal of Sociology* (1895); through reading it one can gain a clear picture of the rise of academic sociology. *The Publications of the American Sociological*

*Society* (1905) carried digests of important papers delivered at annual meetings of the American Sociological Society. *The Journal of Applied Sociology* (1915) occasionally carried articles of interest, and sometimes sociologists published germane articles in *The Annals of the American Academy of Political and Social Science* (1890). *The Journal of Social Forces* (1922–24), rechristened *Social Forces* (1924), published much of the new type of empirical, specialized social research of the 1920s.

Fortunately for historians of science in modern America there exist biographical directories with much useful information. *American Men of Science* (New York, 1906), edited by James McKeen Cattell and appearing in six editions before 1940, was most helpful. Each entry contained pertinent biographical information: birth place and year, formal education, disciplinary affiliation, professional appointments, memberships and offices in scientific societies, and research interests. Also, Cattell had the scientists in the disciplines rank their colleagues' scientific merit; while this "starring" system might not be a reliable guide always to the actual worth of an individual's scientific work, it was an objective indicator of his or her scientific reputation, especially those starred after the first edition. I have used the entries in the various editions of *American Men of Science* for much of the biographical information on individuals discussed in the text, as well as the "starring" system for identifying the most recognized or reputable scientists in a particular discipline at a given time. For American psychologists, I supplemented these entries with Carl Murchison, *ed., The Psychological Register* (Worcester, Mass., 1929), and later editions, which lists psychologists' publications, something *American Men of Science* does not. For information on scientists who were members of the National Academy of Sciences, I found the *Biographical Memoirs* of the National Academy of Sciences uneven but occasionally rewarding. George W. Stocking, Jr., "American Social Scientists and Race Theory: 1890–1915," (Ph.D. dissertation, University of Pennsylvania, 1960), has a most helpful biographical appendix of many social scientists and scientists interested in race in that era, for many of these individuals died before the first edition of *American Men of Science*. There is no full biographical directory of American sociologists; biographical information has to be pieced together from various sources, primarily obituaries in the sociological journals and in histories of sociology written by sociologists, the most comprehensive of which for these purposes is Howard W. Odum, *American Sociology: The Story of Sociology in the United States Through 1950* (New York, 1951).

Also indispensable were the personal and professional correspondence of important scientists and scientific administrators. Collections varied widely in utility, and upon occasion there was no relationship between the size of the collection and the help it could give me. Then, too, there were a number of collections instructive in a general sense, even though they did not contain specific materials that could be usefully cited. In the history of the biological sciences, and genetics in particular, there are a number of pertinent collections. Very useful are the collections gathered by the Library of the American Philosophical Society in Philadelphia in recent years. The Charles B. Davenport Papers have a mountain of material on genetics, biology, eugenics, and the institutions with which Davenport was associated, especially from the 1890s to the 1930s. The Leslie Clarence Dunn Papers, another large collection, are very useful after the 1920s for the history of genetics, the eugenics movement, and for understanding how American scientists reacted to the emigres and international events. The Herbert Spencer Jennings Papers are a small collection containing a wealth of information on the new biology, its relationship with psychology, scientific institutions, and the eugenics movement. The Milislav Demerc Papers contain valuable insights into the genetics profession after the late 1920s and on the Station for Experimental Evolution, which Demerec administered for many years. The Department of Genetics Archives, University of California, helped me understand some aspects of the rise of a particular local department and of the work its members did in the laboratory and the classroom. The Oscar Riddle Papers, another small collection, helped me understand the values and attitudes of biologists in the later nineteenth and early twentieth centuries. The Alfred F. Blakeslee Papers, and the George H. Shull Papers, gave a marvelously detailed picture of the development of agricultural hybridization in land-grant colleges in the early twentieth century. The Papers of the neurologist Henry H. Donaldson, contain a revealing autobiography. The Papers of geneticist Hubert H. Goodale were primarily records of genetics experiments. The Raymond Pearl Papers were especially rich for understanding the issues of biological science and democracy in the 1930s and 1940s. The small collections of the papers of William E. Castle and Thomas H. Morgan shed some light on personalities and professional standards in genetics. At the Manuscripts Division of the Library of Congress in Washington, D.C., the Jacques Loeb Papers have much incoming and outgoing correspondence connecting Loeb to many important biologists and geneticists, and are helpful for understanding Loeb's scientific and political attitudes. At Columbia University's Oral History Office in New York, I was able to use Theodosius Dobzhansky's oral history memoir, especially informative on the Morgan group in the 1920s and 1930s. In the Columbiana Collection, the Columbia University Department of Zoology Papers, constituting one box for the late 1920s and 1930s, contained information on the impact of the Depression on an important academic department. The Edwin G. Conklin Papers, Firestone Library, Princeton University, Princeton, comprise an extensive, major collection in the history of modern American biology, containing personal, professional, and organizational correspondence, and Conklin's unpublished books and speeches. Conklin corresponded with many important biologists; he was involved in many scientific institutions, such as the Marine Biological Laboratory, the American Society of Naturalists, and the National Research Council. Both the outgoing and the incoming correspondence are indispensable. On the other hand, the Theodore Cockerell Papers, Norlin Library, University of Colorado, Boulder, were of little use for me, even though Cockerell corresponded with many eminent biologists.

Historians of psychology in American must necessarily begin their research with the Robert M. Yerkes Papers, located in the Historical Division, Medical Library, Yale University, New Haven. Yerkes was an important scientist in his own right; he constantly corresponded with many major figures in psychology, eugenics, biology, mental measurement, animal psychology, and he was involved in many important scientific organizations. No study of the nature-nurture controversy could be written without extensive use of this massive, marvellously well-organized collection. At the Manuscripts Division, Library of Congress, there were other collections of the papers of American psychologists. The James McKeen Cattell Papers, which have been added to in recent years, cover Cattell's whole career as psychologist, editor, and organizer of scientific institutions to the early 1930s. The Archives of the American Psychological Association, spanning the years 1920–66, with special emphasis on the period since 1930, have materials dealing with the Association's inner history not germane to this study. The Arnold Gesell Papers, another large collection, were useful for understanding Gesell's ideas and his research plans. The Edward Lee Thorndike Papers, were a small collection, of little help; and the Papers of Columbia University psychologist Robert S. Woodworth, another small collection, were not available for scholarly use. There were several helpful collections in the history of American psychology at Harvard University, Cambridge, both at the Houghton Library and the University Archives. At the Houghton Library, I found the William James Papers, part of the James Family Papers, a small but immensely rewarding collection on American psychology's early days. John C. Burnham, "Oral History Interviews of William Healy and Augusta Bronner," typescript memoir, also at the Houghton, helped me understand two important students of delinquency. The Edwin G. Boring Papers, located in the Harvard University Archives, are a major collection for American psychology from the 1920s on, and contain much incoming as well as outgoing material. The Edward Bradford Titchener Papers, Cornell University Library, Ithaca, are not as large as the Boring Papers; nevertheless they contain a wealth of incoming and outgoing correspondence pertinent to American psychology from the 1890s to the 1930s. The Lewis M. Terman Papers, Stanford University Archives, Stanford, are a large collection especially helpful for the period 1920 to 1940. The Hugo Münsterberg Papers, located at the Boston Public Library, Boston, comprise a small collection helpful for understanding both American psychology and Münsterberg from the 1890s to his death in 1916. At Clark University in Worcester, Massachusetts, the fragmentary G. Stanley Hall Papers helped me understand Hall better than American psychology; the extensive University Archives contained much information on the history of Clark University. The Joseph Jastrow Papers and the William McDougall Papers at the Duke University Library in Durham were of little general use, although the McDougall Papers underlined McDougall's professional isolation after the mid-1920s.

Historians of American anthropology are fortunate to have access to the Franz Boas Papers at the Library of the American Philosophical Society, which constitute a massive collection of incoming and outgoing correspondence. The Boas Papers are arranged chronologically and provide a wealth of information about American anthropology, American science, the impact of war upon the scientific community, and the relationships of the social sciences to the natural sciences. Also very helpful for understanding Franz Boas as a promoter of scientific institutions were the four Franz Boas files in the Columbia University Central Files (the administrative papers of Columbia University) located in the Columbia University Libraries. The Elsie Clews Parsons Papers, at the Library of the American Philosophical Society, comprise a small collection of incoming correspondence not germane to this study. At the Manuscripts Division of the Library of Congress, the papers of William John McGee, provided important insights into the activities of an important evolutionary anthropologist and scientific administrator in the 1890s and 1900s—and on McGee's activities as chief of the Bureau of American Ethnology. At the Smithsonian Institution Archives, Smithsonian Institution, Washington, D.C., I consulted the collections of two important figures in Washington anthropology, William H. Holmes and John Wesley Powell, but found these collections small and not useful for my purposes. At the Harvard University Archives, I found the Frederick Ward Putnam Papers, a large collection, helpful for understanding American anthropology in the later nineteenth century. At the Bancroft Library, University of California, Berkeley, the Alfred L. Kroeber Papers and the Robert H. Lowie Papers were important supplements to the Boas Papers for understanding American anthropology and the activities and ideas of the Boasians. The Ruth F. Benedict Papers, Vassar College Library, Poughkeepsie, did not have much quotable material for my particular purposes, but they contained much information on the political activities and ideas of many American scientists in the 1930s and 1940s.

Generally speaking, the papers of American sociologists were disappointing for my purposes. The Franklin H. Giddings Papers at Columbia University consisted of one box of relatively unimportant material. The Harry Elmer Barnes Papers, Archives of Contemporary History, University of Wyoming, Laramie, are more extensive but emphasize the period since World War II. Similarly, the papers of Albion W. Small at the University of Chicago were scanty and of little use. Far more useful were the Edward A. Ross Papers, State Historical Society of Wisconsin, Madison, a major collection for the history of American sociology from the 1890s to the 1930s; the correspondence is arranged chronologically. Another important collection, the Luther Lee Bernard Papers, Pennsylvania State University Library, University Park, Pennsylvania, contains both interesting correspondence and Bernard's notes on the history of sociology departments, which I had microfilmed.

Finally, I consulted the papers of several important scientific or collegiate administrators. Very useful were the George Ellery Hale Papers; the originals are located at the Mount Wilson and Palomar Observatories Library, Pasadena, California. I used the microfilm edition on interlibrary loan from the Library of Congress. Now a number of libraries have the Hale Papers on microfilm; Daniel J. Kevles, "Guide to the Microfilm Edition of the George Ellery Hale Papers" (n.p., 1968), is very helpful. Hale was especially important as founder of the National Research Council, and he corresponded with many scientists who played a key role in the heredity-environment controversy. The Charles D. Walcott Papers in the Smithsonian Institution Archives contained some useful information on Walcott's role as Secretary of the Smithsonian Institution and on the Carnegie Institution of Washington. The Nicholas Murray Butler Papers, Columbia University Libraries, were not especially useful for my purposes, but the Charles W. Eliot Papers, Harvard University Archives, gave me important insights into the founding of several science departments at Harvard in the later nineteenth century. The John C. Merriam Papers, located at the Manuscripts Division, Library of Congress, are an immense collection; for my purposes they were especially helpful for understanding the National Research Council in the 1920s and 1930s.

I also found some autobiographical and reminiscenial literature useful, notably Carl E. Seashore, *Pioneering in Psychology* (Iowa City, 1942), a useful picture of the University of Iowa's department of psychology; Theodora Kroeber, *Alfred Kroeber: A Personal Configuration* (Berkeley and Los Angeles, 1970); Edward A. Ross, *Seventy Years of It* (New York, 1936); and Robert H. Lowie, *Robert H. Lowie, Ethnologist* (Berkeley and Los Angeles, 1959), are useful pictures of important social scientists, as is Melville J. Herskovits, *Franz Boas: The Science of Man in the Making* (New York, 1952). The autobiographies in Carl Murchison, ed., *A History of Psychology in Autobiography,* 3 vols., (Worcester, Mass., 1930–36), are uneven, and G. Stanley Hall, *Life and Confessions of a Psychologist* (New York, 1921), ought to be read with care.

Secondary works consulted and used varied widely in focus and quality. The reader is referred to the annual bibliography in *Isis,* the journal of the History of Science Society, for the most systematic international coverage of secondary materials. Among the general works on the functioning of the scientific community I found helpful were Thomas S. Kuhn, *The Structure of Scientific Revolutions* 2d ed., enlarged (Chicago, 1969), and especially Joseph Ben-David, *The Scientist's Role In Society* (Englewood Cliffs, 1971). For those interested in the sociology of science literature, a good starting point is Bernard Barber and Walter Hirsch, eds., *The Sociology of Science* (New York, 1962).

There is now a growing literature on the development of science in American culture. George H. Daniels, *American Science in the Age of Jackson* (New York, 1968); Daniels, "The Process of Professionalization in American Science: The Emergent Period, 1820–1860," *Isis* 58 (1967): 151–66; Sally G. Kohlstedt, *The Formation of The American Scientific Community: The American Association for the Advancement of Science, 1848–1860* (Urbana, 1976); and Dirk Struik, *Yankee Science in the Making* (New York, 1947), are all helpful introductions to the ante-bellum era, as are a number of the essays in George H. Daniels, ed., *Nineteenth Century American Science: A Reappraisal* (Evanston, 1972), and Alexandra Oleson and Sanborn C. Brown, eds., *The Pursuit of Knowledge in the Early Republic: American Scientific and Learned Societies from Colonial Times to the Civil War* (Baltimore, 1976). But see also Hamilton Cravens, "American Science Comes of Age: An Institutional Perspective, 1850–1930," *American Studies* 17 (Fall 1976): 49–70. Other works indispensable for general background include A. Hunter Dupree, *Science in the Federal Government: A History of Policies and Activities to 1940* (Cambridge, 1957); Dupree, *Asa Gray 1810–1888* (Cambridge, 1959); Edward Lurie, *Louis Agassiz: A Life in Science,* abr. ed. (Chicago, 1966 [1960]); and Howard S. Miller, *Dollars for Research: Science and Its Patrons in Nineteenth Century America* (Seattle, 1970). David D. Van Tassel and Michael G. Hall, eds., *Science and Society in the U.S.* (Homewood, Illinois, 1966), has some useful essays and reference information. On the twentieth century, see Ronald C. Tobey, *The American Ideology of National Science, 1919–1930* (Pittsburgh, 1971), an able, provocative account, as well as Barry D. Karl, "The Power of Intellect and the Politics of Ideas, *Daedalus* 86 (Summer 1968): 1002–35; Stanley Coben, "The Scientific Establishment and the Transmission of Quantum Mechanics to the United States, 1919–32," *American Historical Review* 76 (1970–71):

442–66, two articles which were of great assistance to me in working out the social dynamics of the modern scientific community. See also Edward H. Beardsley, *The Rise of the American Chemistry Profession 1850–1900* (Gainesville, 1964). Historians of science in American culture can also benefit from biographies of leading scientists. In addition to Dupree, *Asa Gray 1810–1888*, and Lurie, *Louis Agassiz: A Life in Science*, helpful background for the modern period may be found in Dean Conrad Allard, Jr., "Spencer Fullerton Baird and the U.S. Fish Commission: A Study in the History of American Science," (Ph.D. diss., George Washington University, 1967); a work with far broader implications is Edward H. Beardsley, *Harry L. Russell and Agricultural Science in Wisconsin*, (Madison, 1969). See also Hugh Hawkins, *Between Harvard and America: The Educational Leadership of Charles W. Eliot* (New York, 1972). Charles E. Rosenberg, *No Other Gods: On Science and American Social Thought* (Baltimore, 1976), has a number of insightful essays.

There are some useful books dealing with the impact of evolutionary thought in American culture. Especially good are Stow Persons, ed., *Evolutionary Thought in America* (New Haven, 1950); Persons, *American Minds: A History of Ideas* (New York, 1958); and Paul F. Boller, Jr., *American Thought in Transition: The Impact of Evolutionary Naturalism, 1865–1900* (Chicago, 1969); Edward A. Purcell, *The Crisis of Democratic Theory: Scientific Naturalism and the Problem of Value* (Lexington, 1973); Gilman Ostrander, *American Civilization in the First Machine Age: 1890–1940* (New York, 1970); and several of the sketches in Merle E. Curti, *The Social Ideas of American Educators*, new and rev. ed., (Paterson, N.J., 1965). On the other hand, I did not find Richard Hofstadter, *Social Darwinism in American Thought* (Philadelphia, 1944), especially useful; see Hamilton Cravens and John C. Burnham, "Psychology and Evolutionary Naturalism in American Thought, 1890–1940," *American Quarterly* 23 (1971): 635–57.

The secondary literature on the related topics of race relations, immigration restriction, and nativism is enormous. Among the works I found especially useful were John Higham, *Strangers in the Land: Patterns of American Nativism 1876–1925* (New Brunswick, 1955), a superb book; see also Barbara Solomon, *Ancestors and Immigrants: A Changing New England Tradition* (Cambridge, 1956). There are any number of excellent studies of race relations in the later nineteenth and early twentieth century; I found C. Vann Woodward's masterful and complex *Origins of the New South 1877–1913*, A History of the South, vol. 9 (Baton Rouge, 1951), indispensable for understanding the South, and Allan H. Spear's *Black Chicago: The Making of a Negro Ghetto 1890–1920* (Chicago, 1967), a good study of urban race relations in the North with broader implications beyond Chicago. Oscar Handlin, *Race and Nationality in American Life* (New York, 1957), focused attention on these phenomena in American life, whereas Thomas F. Gossett, *Race: The History of an Idea in America* (Dallas, 1963), is a reasonably systematic history of racial ideas. On the National Origins Act of 1924, see Robert A. Divine, *American Immigration Policy, 1924–1952* (New Haven, 1957).

I was also able to benefit from a number of works dealing with the history of the biological sciences. Of the several older general surveys,

Erik Nordenskiold, *The History of Biology: A Survey* (New York, 1936), was the most useful. Among the more recent books, I have found John C. Greene, *The Death of Adam* (Ames, Iowa, 1959); William Coleman, *Biology in the Nineteenth Century: Problems of Form, Function, and Transformation* (New York, 1971); Elizabeth Gasking, *The Rise of Experimental Biology* (New York, 1970); and Garland E. Allen, *Life Science in the Twentieth Century* (New York, 1975), authoritative guides. On the American Neo-Lamarckians, see Edward J. Pfeifer's "The Genesis of American Neo-Lamarckism," *Isis* 56 (1965): 156–67, an able account, but more such studies are needed; thus the major value of Alpheus S. Packard's *Lamarck: The Founder of Evolution* (New York, 1901), is its splendid bibliography of the neo-Lamarckians' writings. Conway Zirkle's "The Early History of the Idea of the Inheritance of Acquired Characters and of Pangenesis," *Transactions of the American Philosophical Society*, n.s. 35 (1946) provides helpful perspective, and Garland E. Allen, "T. H. Morgan and the Emergence of a New American Biology," *Quarterly Review of Biology* 44 (1969): 168–88, is a superb introduction to the rise of experimentalism in American biology at the end of the nineteenth century. See also Hamilton Cravens, "The Role of Universities in the Rise of Experimental Biology," *The Science Teacher* 44 (1977): 33–37. There are now several helpful histories of genetics, including Elof A. Carlson, *The Gene: A Critical History* (Philadelphia, 1966); Alfred H. Sturtevant, *A History of Genetics* (New York, 1965); and L. C. Dunn, *A Short History of Genetics* (New York, 1965); I found the last account especially useful for my purposes. Garland E. Allen has written a number of very able articles on genetics in America and on the mutation theory, including "Hugo de Vries and the Reception of the 'Mutation Theory'," *Journal of the History of Biology* 2 (1969): 55–87; "The Introduction of Drosophila into the Study of Heredity and Evolution: 1900–1910," *Isis* 66 (1976): 322–33; "Thomas Hunt Morgan and the Problem of Natural Selection," *Journal of the History of Biology* 1 (1968): 113–39; "Thomas Hunt Morgan and the Problem of Sex Determination, 1903–1910," *Proceedings of the American Philosophical Society* 110 (1966): 48–57, and his provocative essay, "Genetics, Eugenics, and Class Struggle," *Genetics* 79 (1975): 29–45. E. Carleton MacDowell, "Charles Benedict Davenport, 1866–1944, A Study in Conflicting Influences," *Bios* 17 (1946): 3–50 is revealing. See also Charles E. Rosenberg, "Charles Benedict Davenport and the Beginning of Human Genetics," *Bulletin of the History of Medicine* 22 (1967): 266–76; Alfred H. Sturtevant, "The Early Mendelians," *Proceedings of the American Philosophical Society* 109 (1965): 199–204; Elof A. Carlson, "The Drosophila Group: The Transition from the Mendelian Unit to the Individual Gene," *Journal of the History of Biology* 4 (1971): 149–70; Rosenberg, "Factors in the Development of Genetics in the United States: Some Suggestions," *Journal of the History of Medicine* 22 (1967): 27–46, helped me develop my ideas immeasurably. Of the several accounts of the American eugenics movement, I found Mark H. Haller, *Eugenics: Hereditarian Ideas in America* (New Brunswick, 1963), especially helpful on the intellectual and organizational aspects of the movement, and Kenneth Ludmerer, *Genetics and American Society: A Historical Appraisal,* (Baltimore, 1972), especially helpful on the history of the movement in American politics. William B. Provine, *The Origins of Population Genetics* (Chicago, 1971), is a clear discussion of its topic. There are two helpful historiographical articles; Garland E. Allen, "Genetics, Eugenics, and Society: Internalists and Externalists in Contemporary History of Science," *Social Studies of Science* 6 (1976): 105–22, and Ernst Mayr, "The Recent Historiography of Genetics," *Journal of the History of Biology* 6 (1973): 125–54. Also useful are L. C. Dunn, "Cross Currents in the History of Human Genetics," *American Journal of Human Genetics* 14 (1962): 1–13, and a number of the essays in Dunn, ed., *Genetics in the Twentieth Century* (New York, 1951).

There is also a growing literature on the history of psychology. Still the best general history is Edwin G. Boring, *A History of Experimental Psychology*, 2d ed., (New York, 1950). Frank M. Albrecht, Jr., "The Beginnings of Modern Psychology in America: 1880-1893," (Ph.D. diss., Johns Hopkins University, 1961) is a gold mine of information; Michael M. Sokal, "The Educational and Psychological Career of James McKeen Cattell, 1860-1904," (Ph.D. diss., Case Western Reserve University, 1972) is a thorough study, as is Dorothy Ross, *G. Stanley Hall. The Psychologist as Prophet* (Chicago, 1972); George Dykhuizen, *The Life and Mind of John Dewey* (Carbondale, Ill., 1973), and Geraldine Joncich, *The Sane Positivist: A Biography of Edward Lee Thorndike* (Middletown, 1968) are useful also. I have benefitted enormously from the work of John C. Burnham; see in particular his major essay "The New Psychology: From Narcissism to Social Control," in Robert Bremner, John Braeman, and David Brody, eds., *Change and Continuity in Twentieth-Century America: The 1920's* (Columbus, Ohio 1968), 351–98. See also Thomas Camfield, "The Professionalization of American Psychology, 1870-1917," *Journal of the History of the Behavioral Sciences* 9 (1973): 66–76, and David L. Krantz and David Allen, "The Rise and Fall of McDougall's Instinct Doctrine," *Journal of the History of the Behavioral Sciences* 3 (1967): 326–38. On the application of psychology, see, for example, Lawrence A. Cremin, *The Transformation of the School. Progressivism in American Education, 1876-1955* (New York, 1961), and Loren Baritz, *Servants of Power* (Middletown, 1960). Carol S. Gruber, *Mars and Minerva: World War I and the Uses of the Higher Learning in America* (Baton Rouge, 1975), has a detailed discussion of the dismissal of Cattell from Columbia University. John C. Burnham, "On the Origins of Behaviorism," *Journal of the History of the Behavioral Sciences* 4 (1968): 143–51, illuminates that subject.

With regard to anthropology, William Stanton, *The Leopard's Spots: Scientific Attitudes Toward Race in America, 1815-1869* (Chicago, 1960), and John S. Haller, Jr., *Outcasts from Evolution: Scientific Attitudes of Racial Inferiority 1859-1900* (Urbana, 1971), discuss pre-Boasian intellectual and institutional developments well, and George W. Stocking, Jr., *Race, Culture, and Evolution: Essays in the History of Anthropology* (New York, 1968), is a series of sophisticated, superbly executed essays; see also his helpful *The Shaping of the American Anthropology 1883-1911; A Franz Boas Reader* (New York, 1974). See also Regna Darnell, "The Professionalization of American Anthropology," *Social Science Information* 10 (1971): 83–103; useful departmental studies are Darnell, "The Emergence of Academic Anthropology at the

288 | *Bibliographic Note*

University of Pennsylvania," *Journal of the History of the Behavioral Sciences* 6 (1970): 80–92, and Timothy H. H. Thoresen, "Paying the Piper and Calling the Tune: The Beginnings of Academic Anthropology in California," *Journal of the History of the Behavioral Sciences* II (1975): 257–75.

Several of the essays on American sociology in Anthony Oberschall, ed., *The Establishment of Empirical Sociology: Studies in Continuity, Discontinuity, and Institutionalization* (New York, 1972) are useful. Helpful, too, are Roscoe C. and Gisela J. Hinkle, *The Development of Modern Sociology* (New York, 1954), on institutional matters; Harry E. Barnes, ed., *An Introduction to the History of Sociology* (Chicago, 1948), on personalities; John C. Burnham, *Lester Frank Ward in American Thought* (Washington, D.C., 1956), puts Ward's reputation in proper perspective; and Julius Weinberg, *Edward Alsworth Ross and the Sociology of Progressivism* (Madison, 1972), is an excellent biography. Finally, Mary O. Furner, *Advocacy and Objectivity: A Crisis in the Professionalization of American Social Science 1865–1905* (Lexington, Kentucky, 1975), is a stimulating account of the professionalization of economics, sociology, and political science in the later nineteenth century.

My research took me into the history of American higher education, where there is a vast amount of secondary material. Most of the books published before Laurence Veysey, *The Emergence of the American University* (Chicago, 1965), were histories of particular institutions, with little attempt to reconstruct a synthesis of broader developments. Nevertheless, I found many of these "biographies" of particular universities helpful as sources of specific information about departments, research programs, and other institutional matters, primarily because many universities and colleges have not kept their internal files on departmental affairs. And some were useful because they contained accounts written by contemporaries. Samuel Eliot Morison, ed., *The Development of Harvard University Since the Inauguration of President Eliot, 1869–1929* (Cambridge, 1930), contains valuable information on particular departments at Harvard through the 1920s. Thomas W. Goodspeed, *A History of the University of Chicago, the First Quarter Century* (Chicago, 1916); Earle D. Ross, *A History of Iowa State College of Agriculture and Mechanic Arts* (Ames, Iowa, 1942); and Waterman Thomas Hewett, *Cornell University. A History*, 2 vols. (New York, 1905), among others, were useful. But many of the post–World War II works were helpful, too, especially Merle E. Curti and Vernon Carstensen, *The University of Wisconsin. A History 1848–1924* 2 vols. (Madison, 1948); Hugh Hawkins, *Pioneer: A History of the Johns Hopkins University 1874–1889* (Ithaca, 1960); and Hawkins, *Between Harvard and America. The Educational Leadership of Charles W. Eliot* (New York, 1972). But the work of Veysey, *The Emergence of the American University*, mentioned above, is the only general synthesis, and it does not treat developments beyond 1910.

# NOTES

*Abbreviations*

SCIENTIFIC JOURNALS

| | |
|---|---|
| AA | *American Anthropologist* |
| AJP | *American Journal of Psychology* |
| AJS | *American Journal of Sociology* |
| AN | *American Naturalist* |
| AP | *Archives of Psychology* |
| BB | *Biological Bulletin* |
| JAbP | *Journal of Abnormal Psychology* |
| JASP | *Journal of Abnormal and Social Psychology* |
| JAB | *Journal of Animal Behavior* |
| JAF | *Journal of American Folklore* |
| J App Psych | *Journal of Applied Psychology* |
| JCP | *Journal of Comparative Psychology* |
| JCLC | *Journal of Criminal Law and Criminology* |
| JD | *Journal of Delinquency* |
| JEP | *Journal of Educational Psychology* |
| J Exp P | *Journal of Experimental Psychology* |
| JEZ | *Journal of Experimental Zoology* |
| JH | *Journal of Heredity* |
| JPPSM | *Journal of Philosophy, Psychology, and Scientific Method* |
| MH | *Mental Hygiene* |
| PSJGP | *Pedagogical Seminary and Journal of Genetic Psychology* |
| PSM | *Popular Science Monthly* |
| PSEBM | *Proceedings to the Society for Experimental Biology and Medicine* |

| | |
|---|---|
| PB | *Psychological Bulletin* |
| PC | *Psychological Clinic* |
| PR | *Psychological Review* |
| PASS | *Publications of the American Sociological Society* |
| SS | *School and Society* |
| SM | *Scientific Monthly* |
| SF | *Social Forces* |
| TSB | *Training School Bulletin* |

MANUSCRIPT COLLECTIONS

| | |
|---|---|
| LLB | Luther Lee Bernard Papers |
| FB | Franz Boas Papers |
| RFB | Ruth Benedict Papers |
| JMC | James McKeen Cattell Papers |
| CUA | Clark University Archives |
| CUCF | Columbia University Central Files |
| EGC | Edwin Grant Conklin Papers |
| CBD | Charles B. Davenport Papers |
| | Department of Genetics Archives, |
| DGAUC | University of California |
| LCD | L. C. Dunn Papers |
| CAE | Charles A. Ellwood Papers |
| AG | Arnold Gesell Papers |
| GEH | George Ellery Hale Papers |
| WJ | William James Papers |
| HSJ | Herbert Spencer Jennings Papers |
| ALK | Alfred L. Kroeber Papers |
| JL | Jacques Loeb Papers |
| RHL | Robert H. Lowie Papers |
| JCM | John C. Merriam Papers |
| HM | Hugo Münsterberg Papers |
| EAR | Edward A. Ross Papers |
| EBT | Edward B. Titchener Papers |
| CDW | Charles D. Walcott Papers |
| RMY | Robert M. Yerkes Papers |

*Introduction*

1. Richard Louis Dugdale, *The Jukes: A Study in Crime, Pauperism, Disease, and Heredity* (New York, 1910 [1877]), p. 12.
2. Ibid., p. 66.
3. See Franklin H. Giddings, "Introduction," in ibid., pp. iii–v, for an example of the deep impression Dugdale made upon a well-known American.
4. An excellent discussion of evolutionary naturalism is in Stow Persons, *American Minds: A History of Ideas* (New York, 1958), pp. 217–345. See also Edward A. Purcell, Jr., *The Crisis of Democratic Theory: Scientific Naturalism and the Problem of Value* (Lexington, 1973).

5. See, for example, Irwin G. Wylie, "Social Darwinism and the American Businessman," *Proceedings of the American Philosophical Society* 103 (1959), pp. 629–35.
6. John Higham, *Strangers in the Land: Patterns of American Nativism 1860–1925* (New Brunswick, 1955), has brilliant insights into the rise of nativism and ethnocultural tensions in America.
7. See in particular, C. Vann Woodward, *Origins of the New South 1877–1913* (Baton Rouge, 1951); Woodward, *The Strange Career of Jim Crow* (3rd ed.; New York, 1966). There is some debate over when and why segregation arose, but for present purposes this is not pertinent, for no one questions that segregation emerged in the South by the early twentieth century.
8. Sound accounts of the rise of black ghettoes in the North included Gilbert Osofsky, *Harlem: The Making of a Ghetto* (New York, 1966), and Allan H. Spear, *Black Chicago: The Making of a Negro Ghetto, 1890–1920* (Chicago, 1967); see also the masterful study of Chicago's black community by St. Clair Drake and Horace Cayton, *Black Metropolis* (New York, 1945).
9. Hamilton Cravens, "American Science Comes of Age: An Institutional Perspective, 1850–1930," *American Studies* 17 (Fall 1976): 49–70.
10. See Garland E. Allen, "T. H. Morgan and the Emergence of a New American Biology," *Quarterly Review of Biology* 44 (1969): 168–88.

## Chapter 1

1. Charles B. Davenport, "Animal Morphology in its Relation to Other Sciences," *Science*, n.s. 20 (1904): 698.
2. E. C. MacDowell, "Charles Benedict Davenport, 1866–1944. A Study in Conflicting Influences," *Bios* 17 (1946): 3–24.
3. Charles S. Rosenberg, "The Bitter Fruit: Heredity, Disease, and Social Thought in Nineteenth Century America," in *Perspectives in American History* 8 (1974): 189–235.
4. Edwin G. Conklin, "Recent Work in Biology," typed manuscript lecture, labelled Woman's Club of Evanston [Illinois], 18 October 1895, EGC.
5. My comments are based on much archival work cited below. Good discussions in print are: Laurence Veysey, *The Emergence of the American University* (Chicago, 1965), *passim*; Charles S. Rosenberg, "Science, Technology, and Economic Growth: The Case of the Agricultural Experiment Station Scientist, 1875–1914," in G. H. Daniels, *ed., Nineteenth Century American Science: A Reappraisal* (Evanston, 1972), pp. 181–209; Joseph Ben-David, *The Scientist's Role in Society: A Comparative Study* (Englewood Cliffs, 1971), pp. 139–68.
6. See the comments of Nathaniel Southgate Shaler, *The Autobiography of Nathaniel Southgate Shaler* (Boston and New York, 1909), pp. 95–100, about Agassiz's methods of training. The standard biography of Agassiz is Edward Lurie, *Louis Agassiz: A Life in Science* (Chicago, 1960).

7. On the antecedents of genetics research at Harvard, see William E. Castle to Charles B. Davenport, 8 September 1893, 24 August 1894, Sir Francis Galton to Charles B. Davenport, 5, 27 May, 27 July 1897, 20 October 1899, 16 October 1900, 20 September 1902, H. J. Muller, "An Episode in Science," typescript speech, 25 July 1921, *CBD*; William E. Castle, "Memorandum on the beginning of Genetic Studies at Harvard University (1901–1905)" *DGAUC*; G. H. Parker to Jacques Loeb, 30 September 1906, *JL*.
8. A general discussion of developments at Harvard in zoology is Samuel Eliot Morison, ed., *The Development of Harvard University Since the Inauguration of President Eliot 1869–1929* (Cambridge, 1930), pp. 378–93; Mary Alice and Howard Ensign Evans, *William Morton Wheeler, Biologist* (Cambridge, 1970), pp. 178–204, has a satisfactory chapter on the Bussey Institution, as does Morison, ed., *The Development of Harvard University Since the Inauguration of President Eliot 1869–1929*, pp. 508–17. A good recent biography of Eliot is Hugh Hawkins, *Between Harvard and America: The Educational Leadership of Charles W. Eliot* (New York, 1972).
9. Hugh Hawkins, *Pioneer: A History of the Johns Hopkins University, 1874–1889* (Ithaca, 1960), pp. 135–45; see also Dennis McCullough, "W. K. Brooks's Role in the History of American Biology," *Journal of the History of Biology* 2 (1969): 411–38, an able analysis based on sophisticated scientific judgments.
10. Henry E. Crampton, *The Department of Zoology of Columbia University, 1892–1942* (New York, 1942), pp. 10–52.
11. Thomas W. Goodspeed, *A History of the University of Chicago: The First Quarter Century* (Chicago, 1916), pp. 301–6, 322, 325–26; see also H. H. Donaldson, "Memories for My Boys," typescript autobiography, 1 December 1930, pp. 94–99, *HHD*.
12. Morris Bishop, *A History of Cornell University* (Ithaca, 1962), pp. 110–11, 172–73; Waterman Thomas Hewett, *Cornell University: A History*, 2 vols. (New York, 1905), 2: 203–24.
13. Dorothy Ross, *G. Stanley Hall: The Psychologist As Prophet* (Chicago and London, 1972): 186–230, on the early years of the University, pp. 181, 195, 197, 230, on biology there.
14. See the comments in L. C. Dunn, "Autobiographical Notes," p. 5, *LCD*; see also incoming and outgoing correspondence, 1907–14, *AFB*.
15. Benjamin Ide Wheeler to Jacques Loeb, 10 July 1902, *JL*; see also Benjamin Ide Wheeler to Jacques Loeb, 30 September 1902, *JL*.
16. Merle E. Curti and Vernon Carstensen, *The University of Wisconsin, A History 1848–1925*, 2 vols (Madison, 1949), 2: 355–59; Lowell E. Noland, "History of the Department of Zoology, University of Wisconsin," *Bios* 21 (1950): 83–109.
17. H. L. Fairchild, "The History of the American Association for the Advancement of Science," *Science*, n.s. 59 (1924): 414; F. R. Moulton, "The AAAS and Organized American Science," *Science*, n.s. 108 (1948): 573–77.
18. On these nineteenth-century scientific societies, see for example: Walter B. Hendrickson, "Science and Culture in the American Middle West," *Isis* 64 (1973): 326–40; Ralph S. Bates, *Scientific Societies in the United States* (2nd ed. New York, 1958 [1945]),

pp. 28-84; Alexandra Oleson and Sanborn C. Brown, eds., *The Pursuit of Knowledge in the Early American Republic: American Scientific and Learned Societies from Colonial Times to the Civil War* (Baltimore, 1976), has several valuable studies of particular societies J Kirkpatrick Flack, *Desideratum in Washington: The Intellectual Community in the Capital City, 1870–1900* (Cambridge, 1975), shows the persistence of this sort of scientific society in the nation's capital in the later nineteenth century.

19. C. S. Minot, C. B. Davenport, W J McGee, William Trelease, S. A. Forbes and James McKeen Cattell, "The Relation of the American Society of Naturalists to Other Societies," *Science*, n.s. 15 (1902): 242.

20. I calculated the growing number of persons with a Ph.D. admitted, using membership lists in *Records of the Society of Naturalists, E.U.S.* [title varies] (n.p., 1884–1920); on the difficulties natural historians sometimes had in the Society's affairs, see, for example, E. M. East to Jacques Loeb, 15 November 1913, *JL*; H. S. Jennings to Charles B. Davenport, 17 July 1911, H. S. Jennings to P. P. Boyd, 12 April 1918, *HSJ*.

21. "Report of the Society for Experimental Biology and Medicine," *Science*, n.s. 19 (1904): 829.

22. See the perceptive essay, Francis Haber, "Sidelights on American Science As Revealed in the Hyatt Autograph Collection," *Maryland Historical Magazine* 46 (December 1951): 233–56, and the comment on the *American Naturalist* is on p. 240.

23. H. J. Jennings to the President of the Carnegie Institution of Washington, 1 May 1902, *HSJ*.

24. Circular, *The Journal of Experimental Zoology*, 1 January 1904, *HSJ*; see also Memorandum, Meeting of Directors of the Journal of Experimental Zoology, Philadelphia, 27 December 1904, and the Ross G. Harrison files, *EGC*.

25. A. Hunter Dupree, *Science in the Federal Government: A History of Policies and Activities to 1940* (Cambridge, 1957), pp. 236–38; W. H. Dall, *Spencer Fullerton Baird: A Biography* (Philadelphia, 1915), pp. 420–29; Dean Conrad Allard, Jr., "Spencer Fullerton Baird and the U.S. Fish Commission: A Study in the History of American Science" (Ph.D. diss., George Washington University, 1967), pp. 87–179, 251–56, 317–36.

26. Edwin G. Conklin, "The Marine Biological Laboratory," *Science*, n.s. 11 (1900): 336.

27. See Frank R. Lillie, *The Woods Hole Marine Biological Laboratory* (Chicago, 1944). There is an avalanche of materials on the MBL— its administrative functioning and its role on developing a sense of professional self-consciousness in many archival collections—see, for example, the Frank R. Lillie and Charles O. Whitman folders, *EGC*.

28. MacDowell, "Charles Benedict Davenport, 1866–1944," pp. 15–24.

29. Carnegie Institution of Washington, *Year Book No. 6 1907* (Washington, 1908), p. 76. The Station's development can be followed in the *Year Book*.

30. Jacques Loeb, "The Recent Development of Biology," *Science*, n.s. 20 (1904): p. 781; Donald H. Fleming supplies biographical information on Loeb in "Introduction," *The Mechanistic Conception of*

*Life,* by Jacques Locb, ed. by Donald H. Fleming (Cambridge, 1964 [1912]), pp. vii–xli. Much of the correspondence in *JL* is most useful in understanding Loeb.

31. Edward J. Pfeffier, "The Genesis of American Neo-Lamarckism," *Isis* 41 (1965): 156–67; A. S. Packard, *Lamarck: The Founder of Evolution* (New York, 1901), is the indispensable bibliographical guide to the Neo-Lamarckians' writings.

32. See, for example: A. S. Packard, "A Half-Century of Evolution, with Special Reference to the Effects of Geological Changes on Animal Life," *AN* 32 (1898): 673–74.

33. See, for example: Edward Drinker Cope, *The Origin of the Fittest: Essays on Evolution* (New York, 1887), pp. 378–89. Nor was Cope alone in this attitude. Lester Frank Ward, for example, clearly repudiated Weismann's ideas for Neo-Lamarckism in the 1890's—and, in the same breath, embraced eugenics as the only means of permanently improving the lives of workers and the poor; see Lester Frank Ward, "Neo-Darwinism and Neo-Lamarckism," *Proceedings of the Biological Society of Washington, D.C.* 6 (1890–91): 1–71.

34. Readily available sources for Weismann's ideas are August Weismann, *Essays Upon Heredity and Kindred Biological Problems,* A. B. Poulton, Selmer Schonland, and Arthur E. Shipley, eds. and trans., 2 vols. (2d ed; Oxford, 1891–92); August Weismann, "Prof. Weismann's Theory of Heredity," *Nature* 41 (1889–90): 317–23; August Weismann, "Inheritance of Injuries," *Science* 14 (1889): 93–94. On Weismann, see Frederick B. Churchill, "August Weismann and a Break from Tradition," *Journal of the History of Biology* 1 (1968): 91–112, and Garland E. Allen, *Life Science in the Twentieth Century* (New York, 1975), pp. 7–8.

35. Weismann, *Essays Upon Heredity and Kindred Biological Problems,* 2: 31–70.

36. George Gaylord Simpson, *The Meaning of Evolution* (special rev. and abr. ed.; New York, 1951), pp. 128–35, summarizes the Neo-Lamarckian controversy. The controversy may be followed in *Nature, Science, Popular Science Monthly,* and *The American Naturalist* between 1887 and 1898; much of it took the form of letters to the editor or short notes.

37. Edmund B. Wilson, *An Atlas of the Fertilization and Karyokinesis of the Ovum* (New York, 1895), p. 4.

38. See, for example, the following: William E. Castle, "Mendel's Laws of Heredity," *Science,* n.s. 18 (1903): 405–6; Edmund B. Wilson, "The Problem of Development," *Science,* n.s. 21 (1905): 292; Castle, "Heredity," *PSM* 76 (1910): 417–28; Castle, *Heredity in Relation to Evolution and Animal Breeding,* pp. 27–32, 33–50, 72–86, 90–142; Herbert E. Walter, *Genetics: An Introduction to the Study of Heredity* (New York, 1913), p. 231.

39. Good monographs covering part or all these new discoveries in biology in considerably more depth are: L. C. Dunn, *A Short History of Genetics: The Development of Some of the Main Lines of Thought, 1864–1939* (New York, 1965), pp. 3–77; Elof Axel Carlson, *The Gene: A Critical History* (Philadelphia, 1966), pp. 1–32; Garland E. Allen, *Life Science in the Twentieth Century* (New York, 1975),

pp. 1–72. See also Allen, "Hugo de Vries and the Reception of the 'Mutation Theory'," *Journal of the History of Biology* 2 (1969): 55–87.

40. Charles B. Davenport, "Mendel's Law of Dichotomy in Hybrids," *BB* 2 (1900–1901): 307.

41. Dunn, *A Short History of Genetics*, pp. 81–87; G. H. Shull, "A New Mendelian Ratio and Several Types of Latency," *AN* 42 (1908): 443–51, is an example of such an "exception."

42. Edmund B. Wilson, "Some Recent Studies on Heredity," in *The Harvey Society Lectures: Delivered Under the Auspices of the Harvey Society of New York 1906–1907* (Philadelphia, 1908), p. 219.

43. For painstaking discussions of Morgan's attitudes toward Mendelism in these years, see: Garland E. Allen, "Thomas Hunt Morgan and the Problem of Sex Determination, 1903–1910," *Proceedings of the American Philosophical Society* 110 (1966): 48–57; Allen, "Thomas Hunt Morgan and the Problem of Natural Selection," *Journal of the History of Biology* 1 (1968): 113–39; Allen, "The Introduction of *Drosophila* into the Study of Heredity and Evolution: 1900–1910," *Isis* 56 (1975): 322–33; Allen, *Life Science in the Twentieth Century*, pp. 52–55. An example of Morgan's initial skepticism in these years is Thomas H. Morgan, *Experimental Zoology* (New York, 1907), pp. 167–68.

44. See, for example: Charles B. Davenport, *Inheritance in Poultry* (Washington, D.C., 1906), pp. 98–99; William E. Castle, "Recent Discoveries in Heredity and Their Bearing on Animal Breeding," *PSM* 67 (1905): 193–208.

45. See, for example: William Bateson, *Mendel's Principles of Heredity* (Cambridge, England, 1909), pp. 1–234; R. C. Punnett, *Mendelism* (3d ed.; Cambridge, England, 1911), pp. 1–43; J. Arthur Thompson, *Heredity* (London, 1908), pp. 2, 42–46, 67, 83–85, 100, 513, 525; A. D. Darbishire, *Breeding and the Mendelian Discovery* (2d ed.; London, 1912), p. 96ff.

46. See Hugo de Vries, *The Mutation Theory: Experiments and Observations on the Origin of Species in the Vegetable Kingdom*, J. B. Farmer and A. D. Darbishire, eds. and trans., 2 vols. (Chicago, 1909–10). See also Allen, *Life Science in the Twentieth Century*, pp. 10–18, and Allen, "Hugo de Vries and the Reception of the 'Mutation Theory'," *Journal of the History of Biology* 2, no. 1 (1969): 58–65.

47. "Recent Scientific Work in Holland," *Nature* 64 (1901): 209.

48. Castle, "Mendel's Laws of Heredity," pp. 405–6.

49. Allen, "Hugo de Vries and the Reception of the 'Mutation Theory'," pp. 57–58. Some examples of American biologists who criticized the mutation theory are: O. C. Glaser, "Autonomy, Regeneration, and Natural Selection," *Science*, n.s. 20 (1904): 149–53; J. A. Allen, "The Probable Origin of Birds," *Science*, n.s. 22 (1905): 431; Thomas L. Casey, "The Mutation Theory," *Science*, n.s. 22 (1905): 307–9.

50. Allen, "Hugo de Vries and the Reception of the 'Mutation Theory'," p. 68. A recent biography of Merriam is Keir B. Sterling, *Last of the Naturalists: The Career of C. Hart Merriam* (New York, 1974).

51. Vernon L. Kellogg, *Darwinism Today* (New York, 1907).
52. Allen, "Hugo de Vries and the Reception of the 'Mutation Theory'," pp. 59–87, is a profound analysis; in particular he attributes much of the cause for opposition to Darwinian selection and support for de Vriesian mutation then to a massive and general lack of understanding of the species problem.
53. D. T. MacDougall to Charles B. Davenport, 19 July 1904, *CBD*.
54. Hugo de Vries, "The Evidence of Evolution," *Science,* n.s. 20 (1904): 399–400.
55. "The Mutation Theory of Evolution," *Science,* n.s. 21 (1905): 525–26.
56. Allen, "Thomas Hunt Morgan and the Problem of Natural Selection," pp. 113–39.
57. C. F. Hodge to G. Stanley Hall, 14 October 1906, *CUA*.
58. In general, see Hugo de Vries file, *CBD*; in particular, see Hugo de Vries to Charles B. Davenport, 11 February, 25 February, 11 April 1906, *CBD*.
59. Allen, "Hugo de Vries and the 'Mutation Theory'," p. 68.
60. See, for example, the discussions of the mutation theory among these experimentalists: D. T. MacDougal to Charles B. Davenport, 14 and 19 May, 19 and 23 July, 1, 8, 19, and 27 August, 11 October 1904, 10 April, 9 and 31 July 1906, Thomas Hunt Morgan to Charles B. Davenport, 23 January 1906, George Harrison Shull to Charles B. Davenport, 18 December 1905, 3 January 1906, *CBD*; George Harrison Shull to Albert F. Blakeslee, 22 April 1911, *AFB*.
61. Charles B. Davenport to Hugo de Vries, 4 January 1908, *CBD*.
62. H. E. Jordan, "Heredity as a Factor in the Improvement of Social Conditions," *ABM* 2 (1911): 249.
63. David Starr Jordan, "Prenatal Influences," *JH* 5 (1914): 39.
64. A. Gartley, "A Study of Eugenic Genealogy," *ABM* 3 (1912): 243.
65. Mark H. Haller, *Eugenics: Hereditarian Attitudes in American Thought* (New Brunswick, 1963), pp. 91–92.
66. David Starr Jordan, *The Blood of the Nation* (Boston, 1910). On Jordan's role in the peace movement, see David S. Patterson, *Toward a Warless World: The Travail of the American Peace Movement, 1887–1914* (Bloomington, Ind., 1976), pp. 66, 69–71.
67. Henry H. Goddard, *The Kallikak Family* (New York, 1912). A contemporary discussion of this "cacogenic" literature is L. W. Crafts, "A Bibliography on the Relations of Crime and Feeble-Mindedness," *JCLC* 7 (1916): 544–54.
68. Madison Grant, *The Passing of the Great Race* (New York, 1916); Edward Alsworth Ross, *The Old World in the New* (New York, 1914).
69. MacDowell, "Charles Benedict Davenport 1866–1944," pp. 28–30; Haller, *Eugenics,* pp. 65–66.
70. Charles B. Davenport, *Heredity in Relation to Eugenics* (New York, 1911); Davenport, *The Trait Book* (Cold Spring Harbor, 1912); Davenport, "How Did Feeble-Mindedness Originate in the First Instance?" *TSB* 9 (1912–13): 87–90; Davenport, "The Feebly Inhibited. I. Violent Temper and Its Inheritance," *Journal of Nervous and Mental Disease* 42 (1915): 593–628; Davenport, *The*

*Feebly Inhibited: Nomadism . . . with Special Reference to Hered-ity* (Washington, 1915), pp. 7–26.
71. Castle, "Heredity," pp. 417–28; Castle, *Heredity in Relation to Evolution and Animal Breeding,* pp. 27–32, 33–50, 72–86, 90–142. See also William F. Castle, *Genetics and Eugenics* (Cambridge, 1916).
72. Edward M. East, "Hidden Feeble-Mindedness," *JH* 8 (1917): 215–17.
73. Edward M. East and Donald F. Jones, *Inbreeding and Outbreeding: Their Genetic and Sociological Significance* (Philadelphia, 1919), p. 253.
74. Herbert E. Walter, *Genetics: An Introduction to the Study of Heredity* (New York, 1913), p. 231.
75. Alfred F. Blakeslee, "Corn and Men," *JH* 5 (1914): 511–18; Samuel J. Holmes, *Studies in Evolution and Eugenics* (New York, 1923); Michael F. Guyer, *Being Well-Born: An Introduction to Eugenics* (Indianapolis, 1916); Leon F. Cole, "Biological Eugenics," *JH* 5 (1914): 305–12; Charles W. Hargitt, "A Problem in Educa-tional Eugenics," *PSM* 83 (1913): 355–67; Edwin G. Conklin, "Value of Negative Eugenics," *JH* 6 (1915): 538–41; William E. Kellicott, *The Social Direction of Human Evolution* (New York, 1916); Henry Fairfield Osborn, "The Fighting Abilities of Different Races," *JH* 10 (1915): 29–31; David Starr Jordan, *The Heredity of Richard Roe: A Discussion of the Principles of Eugenics* (Boston, 1913); Vernon Kellogg, *Mind and Heredity* (Princeton, 1923). As an objective check on these scientists' reputations as scientists, I looked in the first three editions of James McKeen Cattell, *American Men of Science,* to see if they were starred, i.e., voted distinguished by their co-professionals. All were starred in either the first or the second edition. Admittedly such a rating system has its flaws—and it was not well-refined in the first edition, in particular—but it is still valuable as a crude objective yardstick to be used with other indices of prestige and reputation.
76. Edwin G. Conklin, *Heredity and Environment in the Development of Men* (2d ed.; Princeton, 1916), p. 4.
77. Ibid., p. 213. See also ibid., pp, 297–456.
78. Guyer, *Being Well-Born: An Introduction to Eugenics,* pp. 97–120, 195, 208–27, 228–93; Kellicott, *The Social Direction of Human Evolution,* pp. 10–44, 76–77, 128, 147.
79. H. W. Conn, *Social Heredity and Social Evolution* (New York and Cincinnati, 1914).
80. Garland E. Allen, "Genetics, Eugenics, and Class Struggle," *Genetics* 74 (1975): 33.
81. Kenneth Ludmerer, *Genetics and American Society: A Historical Appraisal* (Baltimore, 1972), pp. 91–92.
82. See, for example, ibid., pp. 87–112.
83. See James McKeen Cattell, "Families of American Men of Science," in A. T. Poffenberger, ed., *James McKeen Cattell 1860–1944, Man of Science,* 2 vols., (Lancaster, Pa., 1947), 1: 478–519, on the social backgrounds of American scientists.

Chapter 2

1. Charles W. Eliot to Robert M. Yerkes, 24 December 1918, RMY.
2. Robert M. Yerkes to Charles W. Eliot, 30 December 1918, RMY.
3. James McKeen Cattell, "Families of American Men of Science," in A. T. Poffenberger, ed., *James McKeen Cattell 1860–1944, Man of Science*, 2 vols. (Lancaster, Pa., 1947), 1: 478–519.
4. William James to James Sully, 8 July 1890, *WJ*.
5. E. G. Boring, *A History of Experimental Psychology* (2d ed; New York, 1950), pp. 275–452.
6. Robert S. Harper, "Tables of American Doctorates in Psychology," *AJP* 62 (1949): 580–81.
7. C. R. Garvey, "List of American Psychological Laboratories," *PB* 26 (1929): 652–60.
8. Edwin G. Boring, "Statistics of the American Psychological Association in 1920," *PB* 17 (1920): 271–278; Samuel W. Fernberger, "The American Psychological Association: A Historical Summary, 1892–1930," *PB* 29 (1932): 7–11.
9. See, for example, William James, "Note on the Appointment of Professors Wundt and Hitzig," *The Nation* 20 (1875): 377–78.
10. William James to Hugo Münsterberg, 21 February, 23 March, 13 and 19 April 1892, *WJ*.
11. Hugo Münsterberg to William James, 5 November 1899, *WJ*; see also Charles W. Eliot to Hugo Münsterberg, 3 March 1896, *HM*.
12. Robert M. Yerkes, "The Scientific Way," typescript autobiography, pp. 83–109, *RMY*. See also Hugo Münsterberg, *Report of the Psychological Laboratory 1909, Report to the President–1904, HM*.
13. Hugo Münsterberg to James McKeen Cattell, 25 February 1898, *JMC*.
14. S. E. Morison, ed., *The Development of Harvard University Since the Inauguration of President Eliot 1869–1929* (Cambridge, Mass., 1930), pp. 216–22.
15. Harper, "Tables of American Doctorates in Psychology," pp. 580–81.
16. Dorothy Ross, *G. Stanley Hall: The Psychologist as Prophet*, (Chicago, 1972), pp. 3–230.
17. President's Report to the Trustees of Clark University for 1902–3, Department of Experimental and Comparative Psychology, *CUA*.
18. Nathan G. Hale, Jr., *Freud and the Americans: The Beginnings of Psychoanalysis in the United States, 1876–1917* (New York, 1971), pp. 3–23; Ross, *G. Stanley Hall*, pp. 386–94.
19. I have based my interpretation of Hall upon Ross' excellent *G. Stanley Hall, passim*.
20. Waterman Thomas Hewett, *Cornell University. A History*, 2 vols. (New York, 1905), 2: 66–85.
21. Edward Bradford Titchener to Robert M. Yerkes, 4 August 1914, *RMY*.
22. Boring, *A History of Experimental Psychology*, pp. 410–12.
23. Edward Bradford Titchener to James McKeen Cattell, 26 October 1898, *JMC*.
24. Based on: Boring, *A History of Experimental Psychology*, pp. 532–40, 548–49; Michael M. Sokal, "The Education and Psychological

Career of James McKeen Cattell, 1860–1904," (Ph.D. diss., Case Western Reserve University, 1972) pp. 39–593; and much of the correspondence in *JMC*.

25. Carl E. Seashore, *Pioneering in Psychology* (Iowa City, 1942), p. 12.

26. Compare the list of laboratories in Garvey "List of American Psychological Laboratories," pp. 652–60, and the compilation of Ph.D.-granting departments in Harper, "Tables of American Doctorates in Psychology," pp. 579–87. For interesting comments of new Ph.D.s about the difficulties of establishing laboratories, see for example: J. B. Miner to James McKeen Cattell, 6 November 1903, E. B. Huey to James McKeen Cattell, 15 September 1904, *JMC*; Thomas H. Haines to Hugo Münsterberg, 22 January 1908, 9 November 1909, *HM*; Robert M. Ogden to E. B. Titchener, 18 September, 17 October, 1903, *EBT*.

27. Boring, *A History of Experimental Psychology*, p. 414; E. C. Sanford to Edward Bradford Titchener, 8 January, 19 January 1904, Max Meyer to Edward Bradford Titchener, 19 January 1904, Lightner Witmer to Edward Bradford Titchener, 20 January, 25 January 1904, *EBT*.

28. Boring, "Statistics of the American Psychological Association in 1920," pp. 271–78; Fernberger, "The American Psychological Association: A Historical Summary, 1892–1930," pp. 7–11. The Philosophical and Psychological Associations continued to meet together for several years, but chiefly because of personal relationships, not commonality of interest; W. V. Bingham, "Proceedings of the Twenty-First Annual Meeting of the American Psychological Association, Cleveland, Ohio, December 30 and 31, and January 1, 1913," *PB* 10 (1913): 42, points out that at least three-fourths of the papers read were reports of experimental investigations. See also Hugo Münsterberg to Edward Bradford Titchener, 30 January 1904, *EBT*.

29. An excellent portrait of the quarrels among the pioneer generation in the 1890s is in Ross, *G. Stanley Hall: The Psychologist as Prophet*, pp. 231–50.

30. Walter B. Pillsbury, *The Essentials of Psychology* (New York, 1912), pp. 10–11.

31. Charles R. Darwin, *The Expression of Emotions in Man and the Animal* (New York, 1872); Darwin, *The Descent of Man and Selection in Relation to Sex* (New York, 1871).

32. George John Romanes, *Animal Intelligence* (London, 1882); Romanes, *Mental Evolution in Animals* (London, 1883); Romanes, *Mental Evolution in Man* (London, 1888).

33. William James, *The Principles of Psychology*, 2 vols. (New York, 1890), 2: chapt. 24 *et passim*.

34. James, *The Principles of Psychology*, 2: chapt. 24 *et passim*; see also James, "The Laws of Habit," *PSM* 30 (1886/87): 433–51; James, "Some Human Instincts," *PSM* 31 (1887): 160–70, 666–81.

35. Virtually the only American psychologist of the 1890s who understood the implications of Weismann for the instinct theory was James Mark Baldwin; on this point, see Hamilton Cravens and John C. Burnham, "Psychology and Evolutionary Naturalism in American Thought, 1890–1940," *American Quarterly* 23 (1971): 640–41, as

well as James Mark Baldwin, *Development and Evolution* (New York, 1902), pp. 34–118 *et passim*.

36. See Hamilton Cravens, "American Scientists and the Heredity-Environment Controversy, 1883–1940," (Ph.D. diss., University of Iowa, 1969), pp. 32–37; and Luther Lee Bernard, *Instinct: A Study in Social Psychology* (New York, 1924), for more bibliographical citations.
37. G. Stanley Hall, *Adolescence, Its Psychology and Its Relations to Physiology, Anthropology, Sociology, Sex, Crime, Religion, and Education* (New York, 1904).
38. William McDougall, *An Introduction to Social Psychology* (14th ed.; Boston, 1916 [1908]), pp. 10–124 *et passim*.
39. See Bernard, *Instinct: A Study in Social Psychology*, who argues, pp. 76–80ff., that McDougall really popularized the instinct theory outside psychological circles.
40. Charles H. Cooley, *Human Nature and the Social Order* (New York, 1902); Charles A. Ellwood, *Sociology in its Psychological Aspects* (New York, 1912).
41. Ordway Tead, *Instincts in Industry* (New York, 1918), p. 220. See Stow Persons, *American Minds: A History of Ideas* (New York, 1958), pp. 256–61, 298–315.
42. As cited in Lawrence Cremin, *The Transformation of the School: Progressivism in American Education 1876–1957* (New York, 1961), p. 187.
43. Boring, "Statistics of the American Psychological Association in 1920," p. 274.
44. Karl Pearson, "On the Inheritance of the Mental Characters in Man," *Nature* 65 (1901–2): 118.
45. See, for example, James McKeen Cattell and Livingston Farrand, "Physical and Mental Measurements of the Students of Columbia University," *PR* 3 (1896): 648ff.
46. See Cravens, "American Scientists and the Heredity-Environment Controversy," p. 48.
47. Henry H. Goddard, "A Measuring Scale of Intelligence," *TSB* 6 (1909–10): 146–55; Goddard, "Two Thousand Normal Children Measured by the Binet Measuring Scale of Intelligence," *PSJGP* 18 (1911): 232–59.
48. Henry H. Goddard, *The Kallikak Family* (New York, 1912); Goddard, *The Criminal Imbecile* (New York, 1915).
49. Lewis M. Terman, *The Measurement of Intelligence* (Boston, 1916); Terman, *The Stanford Revision and Extension of the Binet-Simon Measuring Scale of Intelligence* (Baltimore, 1917).
50. For insights into testers' professional motivations, see, for example: E. B. Huey to James McKeen Cattell, 18 March 1902, 15 September 1904, *JMC*; T. H. Haines to Hugo Münsterberg, 22 January 1908, *HM*; W. H. Burnham to Arnold Gesell, 23 October, 7 November 1906, 10, 24, 28, 29, and 31 May, 3 and 24 June 1907, 18 February 1908, 24 May 1909, 7 July 1914, *AG*; see also the comments in Ross, *G. Stanley Hall: The Psychologist as Prophet*, pp. 352–55. My general comments are based on career-line studies of all such testers who published between 1910–20 drawn from standard biographical directories.

51. See, for example: J. E. W. Wallin, "The Problem of the Feeble-Minded in Its Educational and Social Bearings," SS 2 (1915): 115–16; E. B. Huey, "The Present Status of the Binet Scale of Tests for the Measurement of Intelligence," PB 9 (1912): 160–68; F. Kuhlmann, "The Results of Grading Thirteen Hundred Feeble-Minded Children with the Binet-Simon Tests," JEP 4 (1913): 261–68; S. C. Kohs, "The Problem of the Moral Defective," TSB 11 (1914–15): 18–22.
52. See, for example: M. J. Mayo, *The Mental Capacity of the American Negro*, AP, no. 28 (New York, 1913); H. E. Jordan, "The Biological Status and Social Worth of the Mulatto," PSM 82 (1913): 573–582; J. Morse, "A Comparison of White and Colored Children Measured by the Binet Scale of Intelligence," PSM 84 (1914): 78. G. O. Ferguson, Jr., *The Psychology of the Negro: An Experimental Study*, AP, no. 36 (New York, 1916); S. Z. Pressey and G. F. Teter, "A Comparison of Colored and White Children," *J. App. Psych.* 3 (1919): 282ff; W. D. Partlow and T. H. Haines, "Mental Rating of Juvenile Delinquents and Dependents in Alabama," *J. App. Psych.* 3 (1919): 291–309; K. T. Waugh, "A Comparison of Oriental and American Student Intelligence," PB 18 (1921): 106.
53. Shepherd Ivory Franz to Robert M. Yerkes, 23 April 1917, RMY; Daniel J. Kevles, "Testing the Army's Intelligence: Psychologists and the Military in World War I," *Journal of American History* 55 (1968): 565–81.
54. Robert M. Yerkes to Raymond Dodge, 5 June 1917, RMY; Yerkes, "Psychology in Relation to the War," PR 25 (1918): 85–115.
55. Robert M. Yerkes, ed., *Psychological Examining In the United States Army*, Memoirs of the National Academy of Sciences, vol. 15 (Washington, D.C., 1921), pp. 202–11, 235–58, 421–22, 551–657, 699, 704–42, 764, 779, 820–37, 853–61, *et passim*.
56. Robert M. Yerkes to Carl C. Brigham, 26 October 1921, RMY.

### Chapter 3

1. Robert H. Lowie, *Culture and Ethnology* (New York, 1917), p. 66.
2. For European uses of the word and idea of culture see A. L. Kroeber and C. Kluckhohn, *Culture: A Critical Review of Concepts and Definitions* (Cambridge, Mass., 1952), pp. 3–73.
3. George W. Stocking, Jr., *Race, Culture, and Evolution: Essays in the History of Anthropology* (New York, 1968), pp. 110–32, 234–69.
4. Chiefly those Americans were James Mark Baldwin, whose ideas were summarized in *Development and Evolution* (New York, 1902) and William Graham Sumner, *Folkways* (Boston, 1906).
5. See, for example, Clark Wissler, *Man and Culture* (New York, 1923), pp. 284–312; Margaret Mead, *Coming of Age in Samoa* (New York, 1928), *passim*.
6. Kroeber and Kluckhohn, *Culture: A Critical Review of Concepts and Definitions*, pp. 81–154.
7. Franz Boas to Seth Low, 25 February 1899 CUCF.
8. See, for example, Daniel G. Brinton, *Races and Peoples*, (Philadelphia, 1890), pp. 51–52.

9. Stocking, *Race, Culture, and Evolution*, pp. 277–78.
10. Ibid., p. 278; N. M. Judd, *The Bureau of American Ethnology: A Partial History* (Norman, Okla., 1967); W J McGee, "Bureau of American Ethnology," and J. W. Fewkes, "Anthropology," in G. B. Goode, ed., *The Smithsonian Institution, 1846–1896: The History of Its First Half Century* (Washington, 1897); Curtis M. Hinsley, Jr., "The Development of a Profession: Anthropology in Washington, D.C., 1846–1903" (Ph.D. diss., University of Wisconsin, Madison, 1976), *passim*.
11. Stocking, *Race, Culture, and Evolution*, pp. 278–80; S. E. Morison, ed., *The Development of Harvard University Since the Inauguration of President Eliot* (Cambridge, 1930), pp. 202–15.
12. George Grant MacCurdy, "The Academic Teaching of Anthropology in Connection with Other Departments," *AA*, n.s. 21 (1919): 49–60.
13. Precise figures on the number of full-time positions are virtually impossible to find. However, it would appear that the number was quite small; between 1898 and 1915 American universities granted only 33 doctorates in anthropology, which indicates a very small academic profession; see "Doctorates Conferred by American Universities," *Science*, n.s. 42 (1915): 557. One estimate in 1907 was that nine universities had 27 professional positions in anthropology from lecturer to full professor; see Edward A. Ross to President Charles Van Hise, 27 March 1907, *EAR*.
14. "Doctorates Conferred by American Universities," p. 557. Harvard granted 15, Columbia 11.
15. Stocking, *Race, Culture, and Evolution*, pp. 133–59.
16. *Department of Anthropology [Clark University], 1889–1909* (n.p., 1909), pp. 4–7, *CUA*.
17. Dorothy Ross, *G. Stanley Hall, The Psychologist as Prophet* (Chicago, 1972), pp. 207–30, has the best account of the whole episode in print; see also Franz Boas to G. Stanley Hall, 4 June 1891, G. Stanley Hall Papers, Clark University Library.
18. My discussion of Boas is drawn from my research in *FB* and the following: Stocking, *Race, Culture, and Evolution*, pp. 133–233, 270–307; Melville Herskovits, *Franz Boas: The Science of Man in the Making* (New York, 1952), pp. 1–25 *et passim*.
19. Franz Boas to Mrs. Zelia Nuttall, 16 May 1901, *FB*.
20. See, for example: Franz Boas to Seth Low, 12 April 1900, 4 February, 8 October 1901, Franz Boas to Nicholas Murray Butler, 16 January, 26 January, 5 March, 18 April 1903, 16 February, 4 March, 17 March, 27 April, 24 May, 8 July 1904, 8 December, 12 December 1905, 5 January, 10 February, 28 February, 2 March, 5 March, 5 April, 22 May, 25 May, 12 October, 30 October, 30 November, 5 December, 10 December 1906, 27 February 1907, 20 January 1909, *CUCF*.
21. See, for example, Franz Boas to Seth Low, 12 April 1900, Franz Boas to Nicholas Murray Butler, 8 December 1905, *CUCF*.
22. Franz Boas to Starr Murphy, 23 November 1906, *FB*.
23. John Michael Kennedy, "Philanthropy and Science in New York City: The American Museum of Natural History 1868–1968," (Ph.D. diss., Yale University, 1968), p. 152.

24. Franz Boas to Nicholas Murray Butler, 15 June 1905, *CUCF*. See also Franz Boas to Nicholas Murray Butler, 25 May 1905, *CUCF*.
25. Franz Boas to Nicholas Murray Butler, 5 March, 15 April 1903, 9 February, 4 March (3 letters), 17 March, 27 April, 10 May, 12 May, 23 May 1904, 3 January, 25 February 1905, Nicholas Murray Butler to Franz Boas, 16 April, 24 October 1903, 7 January, 1 February, 8 February, 10 February, 16 February, 3 May, 11 May, 13 May 1904, 3 March 1905, *CUCF*. See also Franz Boas to Nicholas Murray Butler, 13 June, 15 June, 2 August, 12 August, 28 September 1905, 27 January 1906, Nicholas Murray Butler to Franz Boas, 19 June 1905, 2 October 1905, 31 January 1906, *CUCF*. See also Franz Boas to James McKeen Cattell, 30 June 1905, *JMC*.
26. Franz Boas to Nicholas Murray Butler, 23 March 1905, *CUCF*.
27. Franz Boas to Nicholas Murray Butler, 25 November 1916, *FB*.
28. Robert H. Lowie, *Robert H. Lowie Ethnologist: A Personal Record* (Berkeley and Los Angeles, 1959), p. 3.
29. See the extensive Boas-Sapir correspondence in *FB*, 1905–18.
30. My comments on Boas' relationships with his students are based on their correspondence, chiefly in *FB*, and upon Herskovits, *Franz Boas*, pp. 22–24, Lowie, *Robert H. Lowie, Ethnologist*, pp. 1–14, and Theodora Kroeber, *Alfred L. Kroeber: A Personal Configuration* (Berkeley and Los Angeles, 1970), pp. 45–52.
31. W J McGee to Franz Boas, 4 January 1902, *FB*.
32. Franz Boas to W J McGee, 25 January 1902, *FB*.
33. See, for example: W J McGee to Franz Boas, 21 January, 19 February, 27 February 1902, W J McGee to Livingston Farrand, 26 February 1902, Franz Boas to W J McGee, 26 February 1902, *FB*.
34. See, for example: Franz Boas to James McKeen Cattell, 5 July 1902, *JMC*.
35. George Grant MacCurdy, "Anthropology at the Providence Meeting," *AA*, n.s. 13 (1911): 99–120; "Anthropology at the Washington Meeting," *AA*, n.s. 14 (1912): 142–77; "Anthropology at the Cleveland Meeting," *AA*, n.s. 15 (1913): 87–108; Robert H. Lowie, "Procs. AAA for 1914," *AA*, n.s. 17 (1915): 357–63; George Grant MacCurdy, "Anthropology at the Washington Meeting," *AA*, n.s. 18 (1916): 131–40; "Anthropology at the Baltimore Meeting," *AA*, n.s. 21 (1919): 102–12; Alfred L. Kroeber to R. B. Dixon, 18 December 1907, Alfred L. Kroeber to S. A. Barrett, 5 February 1908, *ALK*; Franz Boas to Alfred L. Kroeber, 2 November 1903, 6 October 1908, R. B. Dixon to Franz Boas, 1 January 1912, *FB*.
36. Franz Boas to Alfred L. Kroeber, 27 April 1908, *FB*. See also Alfred L. Kroeber to Council, American Folklore Society, 7 December 1905, Alfred L. Kroeber to Franz Boas, 9 March 1906, 14 January 1908, Franz Boas to Alfred L. Kroeber, 6 October 1908, Franz Boas to Roland B. Dixon, 15 November 1907, 5 March 1908, Roland B. Dixon to Franz Boas, 18 November 1907, 7 March 1908, 18 January 1911, *FB*.
37. Franz Boas to Nicholas Murray Butler, 18 April 1903, *CUCF*.
38. See, for example, Franz Boas to Daniel Coit Gilman, 27 November

1903, Daniel Coit Gilman to Franz Boas, 25 November 1903, *FB*; Franz Boas to Charles B. Davenport, 10 October 1913, *CBD*.

39. A good discussion of Boas' work with the Commission is Stocking, *Race, Culture, and Evolution*, pp. 175–80; see also Franz Boas to J. W. Jenks, 19 March, 23 March, 23 April, 2 May 1908, J. W. Jenks to Franz Boas, 8 April, 14 April 1908, *FB*.

40. Franz Boas to Nicholas Murray Butler, 25 November 1916, *FB*.

41. Alfred L. Kroeber to Ellsworth Huntington, 6 December 1913, *ALK*. For proof Kroeber understood August Weismann's distinction between culture and nature, see A. L. Kroeber to Edward P. Fitch, 11 April 1914, *ALK*.

42. Kroeber and Kluckhohn, *Culture*, p. 85, say that Boas knew Tylor and was influenced by his definition of culture.

43. See, for example: *Department of Anthropology [Clark University] 1889–1909*, pp. 4–7, CUA; Stocking, *Race, Culture, and Evolution*, pp. 155–233.

44. Alfred L. Kroeber to John Wesley Powell, 4 October 1901; Alfred L. Kroeber to the Honorary Advisory Committee, Department of Anthropology, University of California, 8 November 1902, *ALK*; Robert H. Lowie to "Risa" [Lowie], 27 July 1906; Robert H. Lowie to "Kid" [Risa Lowie], 21 July 1907; Robert H. Lowie to "Uncle" [I. L. Lowie] 20 July 1907, *RHL*.

45. Alfred L. Kroeber to Franz Boas, 11 January 1919, *FB*. Thomas S. Kuhn, *The Structure of Scientific Revolutions* (2d ed; Chicago, 1969), pp. 10–11, 19, 22.

46. Robert S. Woodward to James McKeen Cattell, 24 January 1906, *JMC*. See also Robert S. Woodward to James McKeen Cattell, 2 February, 16 February 1906, *JMC*.

47. Alexander A. Goldenweiser, "Totemism, An Analytical Study," *JAF* 23 (1910): 179–293.

48. Alfred L. Kroeber, "The Morals of Uncivilized Peoples," *AA*, n.s. 12 (1910): 437–47.

49. Robert H. Lowie, "On the Principle of Convergence in Ethnology," *JAF* 25 (1912): 24–42.

50. Clark Wissler, "The Psychological Aspects of the Culture-Environment Relation," *AA*, n.s. 14 (1912): 217.

51. Edward Sapir, "Language and Environment," *AA*, n.s. 14 (1912): 226–36.

52. Alexander A. Goldenweiser, "Culture and Environment," *AJS* 21 (1915–16): 632; Clark Wissler, "The Relation of Culture to Environment from the Standpoint of Invention," *PSM* 83 (1913): 164–68; Wissler, "The North American Indian of the Plains," *PSM* 82 (1913): 436–44.

53. W. H. Babcock, "Proceedings of the Anthropological Society of Washington," *AA*, n.s. 15 (1913): 355–56.

54. Clark Wissler, "Psychological and Historical Interpretation for Culture," *Science*, n.s. 43 (1916): 200.

55. Clark Wissler, "Opportunities for Coordination in Anthropological and Psychological Research," *AA*, n.s. 22 (1920): 3.

56. Alfred L. Kroeber, "The Eighteen Professions," *AA*, n.s. 17 (1915): 288.

57. Alfred L. Kroeber, "The Superorganic," *AA*, n.s. 19 (1917): 169.
58. Alfred L. Kroeber to Alexander A. Goldenweiser, 23 October 1917, *ALK*. See also Alexander A. Goldenweiser to Alfred L. Kroeber, 17 October 1917, *ALK*. See also Alfred L. Kroeber to Mr. [Wilson D.?] Wallis, 11 June 1916, *ALK*.
59. See, for example, Edward Sapir, "Do We Need a 'Superorganic'?" *AA*, n.s. 19 (1917): 441, 446. See also Alfred L. Kroeber to Edward Sapir, 23 October 1917, *ALK*.
60. William H. Homes to George Ellery Hale [no date, roll 18, frame 741, probably 1915 or 1916], George Ellery Hale to William H. Holmes, 16 January, 27 September 1916, William H. Holmes to George Ellery Hale, 29 September 1916, *GEH*.
61. Edwin Grant Conklin to George Ellery Hale, 9 February 1917, *GEH*.
62. Franz Boas to A. Wisbrun, 22 March 1917, *FB*.
63. Alfred L. Kroeber to Franz Boas, 2 August 1917, *FB*.
64. Walter P. Metzger, *Academic Freedom in the Age of the University* (New York, 1961, paper ed. [1955]), pp. 224–27; James McKeen Cattell to Jacques Loeb, 23 May 1913, Thomas Hunt Morgan to Jacques Loeb, 25 May 1913, 4 October 1917, *JL*. See also Thomas Hunt Morgan to Edwin G. Conklin, 15 October 1917, *EGC*.
65. Franz Boas to Jacques Loeb, 18 October 1917, Jacques Loeb to Franz Boas, 20 October 1917; Jacques Loeb to James McKeen Cattell, 23 November 1917; James McKeen Cattell to Jacques Loeb, 17 November 1917, Jacques Loeb to Thomas Hunt Morgan, 23 November 1917, *JL*; Franz Boas to Edwin G. Conklin, 29 October, 8 November 1917, Edwin Grant Conklin to Franz Boas, 1 November, 9 November 1917; Edwin G. Conklin to G. H. Parker, 24 October 1917, G. H. Parker to Edwin G. Conklin, 26 October 1917, E. G. Conklin to Thomas Hunt Morgan, 12 October 1917, Thomas Hunt Morgan to Edwin G. Conklin, 15 October 1917, Edwin G. Conklin to Thomas Hunt Morgan, 22 October 1917, *EGC*.
66. Franz Boas to Robert H. Lowie, 3 December 1917, *FB*. On the Cattell case, see Carol S. Gruber, *Mars and Minerva: World War I and the Uses of the Higher Learning in America* (Baton Rouge, 1975), pp. 187–206.
67. Kroeber, *Alfred L. Kroeber*, pp. 46–50, discusses Kroeber's upbringing.
68. See, for example: Robert H. Lowie to "Kid" [Risa Lowie], 18 August 1908, Robert H. Lowie to "Kidlet" [Risa Lowie], 29 December 1910, Robert H. Lowie to "Kiddo" [Risa Lowie], 2 August 1912, Robert H. Lowie to James McKeen Cattell, 19 December 1914, Robert H. Lowie to "Kid" [Risa Lowie], 5 July, 7 September 1915, 10 April 1918, Robert H. Lowie to M. Yves-Guyot, 6 January 1919, Robert H. Lowie to Alfred L. Kroeber, 6 January 1919, Robert H. Lowie to Paul [Radin?], 2 October 1920, Robert H. Lowie to Clark Wissler, 8 November, 18 November 1920, Elsie Clews Parsons to Robert H. Lowie, 6 February 1914, *RHL*.
69. Based on examination of pertinent documents in *FB, ALK, RHL*, Wissler and Boas did not agree on political matters, and their relationship had been cool since Wissler succeeded Boas as curator of anthropology at the American Museum of Natural History (cf.

Franz Boas to Nicholas Murray Butler, 6 July 1905, Clark Wissler to Franz Boas, 20 June, 27 June, 1 July 1905, Franz Boas to Clark Wissler, 22 June, 28 June [2 letters] 1905, *CUCF*). Wissler had been trained as a psychologist and stood between the psychologists and the anthropologists in his theoretical commitments (cf. Clark Wissler to James McKeen Cattell, 18 August 1899, *JMC*).

70. Charles B. Davenport to James McKeen Cattell, 23 October, 4 November 1908, James McKeen Cattell to Charles B. Davenport, 24 October 1908, *CBD*.
71. See, for example, Charles S. Minot to James McKeen Cattell, 26 January 1906, *JMC*.
72. Franz Boas to Bernard Marcus, 14 July 1906, *CUCF*. Franz Boas, "Eugenics," *SM* 3 (1916): 471–78.
73. See, for example, Alfred L. Kroeber to Frank D. Watson, 11 April 1914, *ALK*.
74. Ronald C. Tobey, *The American Ideology of National Science 1919–1930* (Pittsburgh, 1971), pp. 20–61. The standard biography of Hale is Helen Wright, *Explorer of the Universe: A Biography of George Ellery Hale* (New York, 1966).
75. The Hale-Breasted relationship can be followed in roll 5, frames 64–687, *GEH*. Much material on the Save the Redwoods League is in both *GEH* and *JCM*. Professor James Penick first pointed out the League's significance to me.
76. William H. Holmes to George Ellery Hale, [n.d. roll 18, frame 741], George Ellery Hale to William H. Holmes, 27 September 1916, 16 January 1917, William H. Holmes to George Ellery Hale, 29 September 1916, *GEH*.
77. Charles B. Davenport to George Ellery Hale, 29 August 1916, *GEH*.
78. Madison Grant to Henry Fairfield Osborn, 9 March 1918, *CBD*. Grant and Davenport were good friends; see, for example, Madison Grant to Charles B. Davenport, 2 April, 20 November 1917, Charles B. Davenport to Madison Grant, 5 April, 22 November, 26 November 1917, *CBD*.
79. Franz Boas to Edwin G. Conklin, 18 March 1919, *FB* and *EGC*.
80. Edwin G. Conklin to Franz Boas, 20 March 1919, *FB*.
81. Franz Boas to Henry H. Donaldson, 9 April 1919, *FB*. See also Ales Hrdlicka to Franz Boas, 27 March, 12 April 1919, Franz Boas to Edwin G. Conklin, 9 April 1919, Franz Boas to Ales Hrdlicka, 11 April 1919, *FB*; Franz Boas to Edwin G. Conklin, 9 April, 14 April 1919, Edwin G. Conklin to Franz Boas, 10 April 1919, *EGC*.
82. See, for example, Franz Boas to Alfred L. Kroeber, 7 November 1907, 6 January 1908, *FB*.
83. Franz Boas to Edward Sapir, 6 August 1919, *FB*. See also Franz Boas to Edward Sapir, 28 September 1920, *FB*.
84. Stocking, *Race, Culture, and Evolution*, pp. 273–307, has a full account of the whole episode and its implications.
85. William H. Holmes to "To Whom It May Concern," 12 January 1920, attached to Charles D. Walcott to Nicholas Murray Butler, 10 January 1920, *CDW*.
86. Michael Pupin to Charles D. Walcott, 12 January 1920, *CDW*.
87. Charles D. Walcott to Nicholas Murray Butler, 10 January 1920, A. Mitchell Palmer to Charles D. Walcott, 20 January 1920, *CDW*.

See also Russell H. Chittenden to Charles D. Walcott, 5 January 1920, *CDW*. See also Charles D. Walcott to Edwin G. Conklin, 29 December 1919, *EGC*.

## Chapter 4

1. Luther Lee Bernard, "Neuro-Psychic Technique," *PR* 30 (1923): 437.
2. Unpublished and untitled manuscript attributed to William F. Ogburn, as cited in "Introduction," Otis Dudley Duncan, ed., *William F. Ogburn on Culture and Social Change Selected Papers* (Chicago, 1964), p. xviii.
3. Here I am borrowing Thomas S. Kuhn's definition of a discipline in *The Structure of Scientific Revolutions* (2nd ed., Chicago, 1969), pp. 1-22.
4. Edward A. Ross to Charles A. Ellwood, 18 February 1933, *CAE*.
5. Frank L. Tolman, "Study of Sociology in Institutions of Learning in the United States," *AJS* 7 (1901-2): 797-838; ibid. 8 (1902-3): 85-121, 251-72, 531-58; Luther Lee Bernard, "The Teaching of Sociology in the United States," *AJS* 15 (1909/10): 164-213.
6. "History of Sociology Departments," Box 4, File 2, "Cornell University" folder, "Harvard University" folder, *LLB*.
7. Thomas Haskell, "The American Social Science Association, 1865-1909" (Ph.D. diss., Stanford University, 1972), *passim*; Anthony Oberschall, "The Institutionalization of American Sociology," in Oberschall, ed., *The Establishment of Empirical Sociology: Studies in Continuity, Discontinuity, and Institutionalization* (New York, 1972), 187-215; Luther Lee and Jessie Bernard, *Origins of American Sociology. The Social Science Movement in the United States* (New York, 1943), pp. 527-669.
8. Bernard, "The Teaching of Sociology in the United States," p. 211.
9. Based in large part on my reading of the materials in "History of Sociology Departments," Box 4, Files 1-8, *LLB*, which contains histories of 97 collegiate departments of sociology written in the late 1920s.
10. Edward Cary Hayes, "Albion Woodbury Small," in Howard W. Odum, ed., *American Masters of Social Science* (New York, 1927), pp. 149-59; Thomas W. Goodspeed, "Albion Woodbury Small," *AJS* 32 (1926/27): 1-8; Oberschall, "The Institutionalization of American Sociology," p. 203. Unfortunately, the Albion W. Small papers at the University of Chicago have little useful information.
11. Goodspeed, "Albion W. Small," p. 9; Albion W. Small, "Fifty Years of Sociology in the United States, 1865-1915," in *American Journal of Sociology Index to Volumes 1-52* (1895-1947), pp. 218-19.
12. Goodspeed, "Albion W. Small," p. 11; Oberschall, "The Institutionalization of American Sociology," p. 221.
13. Oberschall, "The Institutionalization of America Sociology," pp. 232-40; Robert E. L. Faris, *Chicago Sociology 1920-1932* (San Francisco, 1967), *passim*.

14. Oberschall, "The Institutionalization of American Sociology," p. 211.
15. John L. Gillin, "Franklin Henry Giddings," in Odum, ed., *American Masters of Social Science*, pp. 191–99; Odum, *American Sociology: The Story of Sociology in the United States through 1950* (New York, 1951), pp. 86–94; unfortunately, the Franklin H. Giddings papers at Columbia University contain little useful information.
16. Oberschall, "The Institutionalization of American Sociology," pp. 225–32; Clarence H. Northcott, "The Sociological Theories of Franklin Henry Giddings: Consciousness of Kind, Pluralistic Behavior, and Statistical Method," in Harry Elmer Barnes, ed., *An Introduction to the History of Sociology* (Chicago, 1948), pp. 744–64. Also useful are official reports of the University; see, for example, *Eleventh Annual Report of President Low to the Trustees of Columbia University* (New York, 1900), p. 183; *Twelfth Annual Report of President Low to the Trustees of Columbia University* (New York, 1901), pp. 118–19; *Annual Reports of the President and Treasurer to the Trustees with Accompanying Documents* . . . (New York, 1903), pp. 127–33.
17. Bernard, "The Teaching of Sociology in the United States," pp. 164–213. See comments below on the composition of the American Sociological Society.
18. E. H. Sutherland, "Edward Cary Hayes, 1868–1928," *AJS* 35 (1929–30): 93–99; Harry Elmer Barnes, "The Sociological Theories of Edward Cary Hayes," in Barnes, ed., *An Introduction to the History of Sociology*, pp. 869–71; Odum, *American Sociology*, pp. 119–22.
19. Edward Cary Hayes, "The Development of Sociology at the University of Illinois," 48-page typed manuscript, pp. 1–20, "History of Sociology Departments," Box 4, File 3, *LLB*.
20. Julius Weinberg, *Edward Alsworth Ross and the Sociology of Progressivism* (Madison, Wisc., 1972), pp. 3–56.
21. Edward A. Ross to Mrs. [Mary D.] Beach, 8 March 1887, *EAR*.
22. Weinberg, *Edward Alsworth Ross and the Sociology of Progressivism*, pp. 29–35, 40.
23. Ibid., pp. 39–55. Walter P. Metzger, *Academic Freedom in the Age of the University* (New York, 1961 paperback ed. [1955]), pp. 162–66.
24. See, for example, Richard T. Ely to Edwin R. A. Seligman, 22 January 1901, Richard T. Ely to Edward A. Ross, 24 January 1901, Edwin R. A. Seligman to Edward A. Ross, 25 November 1900, Albion W. Small to Edward A. Ross, 19 April 1901, *EAR*.
25. Edward A. Ross to Richard T. Ely, 6 August 1900, *EAR*.
26. Richard T. Ely to Edward A. Ross, 24 March 1906, *EAR*.
27. Weinberg, *Edward A. Ross and the Sociology of Progressivism*, pp. 123–48 et passim; see also Ross' clippings in *EAR*.
28. Weinberg, *Edward A. Ross and the Sociology of Progressivism*, pp. 191–202. I have also depended upon materials in *EAR* extensively.
29. Harry Elmer Barnes, "Charles Abram Ellwood: Founder of Scientific Psychological Sociology," in Barnes, ed., *An Introduction to the History of Sociology* pp. 853–58. I have also gathered much

biographical information on Ellwood from the materials in *CAE*.
See also, Odum, *American Sociology*, pp. 128–31.

30. Charles A. Ellwood, "A History of the Department of Sociology in
the University of Missouri," 44-page typed manuscript, pp. 1–11,
"History of Sociology Departments," Box 1, File 1, *LLB*.

31. Ibid., pp. 16–25, 30–43. On Ellwood's conception of sociology, see,
for example, *Sociology and Modern Social Problems* (New York,
1910); *The Reconstruction of Religion* (New York, 1922); *Christianity and Social Science* (New York, 1923); and *Methods in Sociology—A Critical Study* (Durham, N.C., 1933).

32. Ulysses G. Weatherly, "Sociology at Indiana University," 24-page
typed manuscript, pp. 1–10, "History of Sociology Departments,"
Box 4, File 3, *LLB*; Odum, *American Sociology*, pp. 126–28.

33. Edwin B. Reuter, "Sociology in Iowa," 26-page typed manuscript,
pp. 1–24, "History of Sociology Departments," Box 4, File 3, *LLB*.
See also Odum, *American Sociology*, pp. 165–67.

34. Emory S. Bogardus, "Sociology at Southern California," 9-page
typed manuscript, pp. 1–9, "History of Sociology Departments,"
Box 4, File 7, *LLB*. See also Odum, *American Sociology*, pp.
158–61.

35. Bernard, "The Teaching of Sociology in the United States," pp.
164–213, lists 110 teachers in 1909. See the list of sociology textbooks in Odum, *American Sociology*, p. 250.

36. [C. W. A. Veditz], "Organization of the American Sociological
Society. Official Report," *AJS* 11 (1905–6): 558.

37. Ibid., p. 567.

38. See the quoted remarks of Small and Ross in ibid., pp. 555–56, and
of Ross, Small, F. G. Young, Henderson, Ellwood, and Frank W.
Blackmar, in ibid., pp. 558–61.

39. See, for example, the quoted remarks of Henderson and Small in
ibid., pp. 555–56, 558–59, 560, who clearly wanted an academically-
oriented group even though Small was a promoter of sociology and
Henderson's background before coming to Chicago was as a "practical" sociologist. They did not want to be "swamped" by the "practical" sociologists.

40. The standard biography of Ward is Samuel Chugerman, *Lester F.
Ward: The American Aristotle* (Durham, N.C., 1939), which is
badly out of date. For comments on the gap between Ward and the
academic sociologists, see, for example, Robert E. L. Faris, "American
Sociology," in G. Gurvitch and W. E. Moore, *eds., Twentieth
Century Sociology* (New York, 1945), p. 541; Albion W. Small,
*Origins of Sociology,* (Chicago, 1915), p. 342ff. See also John C.
Burnham, *Lester Frank Ward in American Thought* (Washington,
D.C., 1956), pp. 3–10, *et passim*, which, although quite hostile to
Ward, amply documents what the academic sociologists really
thought of Ward. Ross was the only full-time academic sociologist
who genuinely liked and respected Ward; most of the others considered him a prima donna. See, for example, Albion W. Small to
Edward A. Ross, 24 November 1903, Edward A. Ross to Albion W.
Small, 28 November 1903, Albion W. Small to Edward A. Ross,
7 December, 10 December 1903, *EAR*.

41. For evidence Ward was elected for symbolic and organizational purposes, see Burnham, *Lester Frank Ward in American Thought,* p. 8, and Albion W. Small to Edward A. Ross, 9 January 1906, C. W. A. Veditz to Edward A. Ross, 13 October 1906, 6 November 1908, *EAR.*
42. W. P. Meroney, "The Membership and Program of Twenty-Five Years of the American Sociological Society," *PASS* 25 (1930): 55–68. See also the membership lists and programs in *PASS* before the 1920s.
43. See, for example, Edward A. Ross to Charles H. Cooley, 7 March 1918, *EAR;* this is also evident from reading *PASS.*
44. Edward A. Ross, *Seventy Years of It* (New York, 1936), p. 180. I compiled membership statistics from *PASS.*
45. Edwin Sutherland, "Social Pathologies," *AJS* 50 (1944–45): 429, points out that 25 percent of *AJS* authors between 1895 and 1900 were members of the NCCC and/or the American Prison Congress.
46. Oberschall, "The Institutionalization of American Sociology," pp. 231–32.
47. The seventeen sociologists and their main appointments as teachers of sociology are: Frank W. Blackmar (Kansas University, 1889–1931); Charles H. Cooley (University of Michigan, 1894–1929); James Q. Dealey (Brown University, 1895–1928); Charles A. Ellwood (University of Missouri, 1900–1930); Franklin H. Giddings (Columbia University, 1891–1931); John M. Gilette (University of North Dakota, 1907–47); John L. Gillin (University of Wisconsin, 1912–43); Edward Cary Hayes (University of Illinois, 1907–29); George E. Howard (University of Nebraska, 1879–1928); Albert Galloway Keller (Yale University, 1900–42); James P. Lichtenberger (University of Pennsylvania, 1909–40); Edward A. Ross (University of Wisconsin, 1906–37); Albion W. Small (University of Chicago, 1893–1926); William Graham Sumner (Yale University, 1872–1910); William I. Thomas (University of Chicago, 1894–1919); George E. Vincent (University of Chicago, 1895–1911); Ulysses G. Weatherly (Indiana University, 1899–1940). I excluded Lester Frank Ward primarily because he was a federal employee for virtually his whole career; he won an academic appointment a few years before his death and did not build or promote sociology as a discipline except from his own highly particularistic point of view.
48. My calculations of the seventeen's articles and book reviews in *AJS.* I have biographical information on them in Hamilton Cravens, "American Scientists and the Heredity-Environment Controversy, 1883–1940," (Ph.D. diss., University of Iowa, 1969), pp. 378–83.
49. See the list of textbooks in Odum, *American Sociology,* pp. 250–51.
50. See ibid., pp. 75–171 for convenient biographical sketches of these men.
51. Albion W. Small and George E. Vincent, *An Introduction to the Study of Society,* (New York, 1894), p. 76, 305. Small and Vincent could not emancipate themselves from biological analogies; see ibid., p. 369.
52. Franklin H. Giddings, *The Principles of Society* (New York, 1896), p. 63.

53. Franklin H. Giddings, "The Psychology of Society," *Science,* n.s. 9 (1899): 16.
54. Charles A. Ellwood, *Prolegema to Social Psychology* (Chicago, 1901). Ellwood's remark about James is in Ellwood, "A History of the Department of Sociology at the University of Missouri," pp. 24–25, "History of Sociology Departments." Box 4, File 3, *LLB*.
55. Edward A. Ross, *Social Control* (New York, 1901); the quote is in Ross, *The Foundations of Sociology* (New York, 1905), p. 54.
56. Ross, *The Foundations of Sociology,* pp. 75–76, 159–71, 181; see also Ross, "The Present Problem of Sociology," *AJS* 10 (1904–5): 469.
57. Edward A. Ross to A. M. Hayes, 15 February 1906, *EAR*.
58. See, for example: Albion W. Small to Richard T. Ely, 3 August 1900, Albion W. Small to Edward A. Ross, 19 December 1901, 17 March 1904, 15 September 1906, 20 November 1907, 8 June 1908, Franklin H. Giddings to Edward A. Ross, 3 February, 22 November 1901, 25 December 1902, 10 December 1906, 18 June 1908, E. R. A. Seligman to Edward A. Ross, 24 June 1905, John M. Gilette to Edward A. Ross, 15 November 1907, Charles A. Ellwood to Edward A. Ross, 18 September 1907, Albert G. Keller to Edward A. Ross, 10 July 1908, S. N. Patten to Edward A. Ross, 24 June 1908, Edward Cary Hayes to Edward A. Ross, 19 June 1908, Charles H. Cooley to Edward A. Ross, 19 June 1909, *EAR*; S. N. Patten to Franklin H. Giddings, 4 May 1897, 29 October 1898, 31 January 1901, 6 November 1901, *FHG*; Franklin H. Giddings to Charles A. Ellwood, 15 September 1905, Charles H. Cooley to Charles A. Ellwood, 4 December 1910, 15 October 1912, Robert H. Gault to Charles A. Ellwood, 27 November 1910, *CAE*.
59. Franklin H. Giddings to Edward A. Ross, 22 November 1901, *EAR*.
60. Franklin H. Giddings to Edward A. Ross, 10 December 1906, *EAR*.
61. William I. Thomas, "The Gaming Instinct," *AJS* 6 (1900–1901): 750–63; Thomas, "The Province of Social Psychology," *AJS* 10 (1904–5): 445–55; F. W. Blackmar, *The Elements of Sociology* (New York, 1905), pp. 12, 17–21, 32–34, 44–46, 51–62, 69–70, 75–76, 134, 218–30; Albion W. Small, *General Sociology* (Chicago, 1905), pp. ix, 3, 7, 17, 21–22, 35, 39, 46, 49–82, 100–114, 131, 142–43, 147–49, 153, 167, 170, 184–88, 197–99, 203–9, 213–23, 339, 371, 442, 446–47, 472, 474.
62. Albion W. Small, "Points of Agreement Among Sociologists," *AJS* 12 (1906–7): 635, 643.
63. William Graham Sumner, *Folkways* (Boston, 1907): iii–iv, 2–4, 19–20, 34–38, 59–74, 75–118, 119–57, 172–260. Sumner's shift was political as well as "scientific"; see Bruce Curtis, "William Graham Sumner 'On the Concentration of Wealth'," *Journal of American History* 55 (1969): 823–32 and Curtis, "The Middle Class Progressivism of William Graham Sumner," (Ph.D. diss., University of Iowa, 1964), *passim,* which show that Sumner became converted to an anti-trust posture late in life.
64. Edward A. Ross, *Social Psychology* (New York, 1908), pp. 1–3, 7–8, 11–41, 43–337. Ross still believed that human nature was biosocial and that human society had evolved. See Ross, "The Nature and Scope of Social Psychology," *AJS* 13 (1907–8): 557–83.

65. Charles H. Cooley, *Social Organization* (New York, 1909), pp. 29–31, 36–37, 63, 68, 70, 80–81, 88, 113, 121–24, 208–15, 229, 237.
66. Edward Cary Hayes, "Sociology and Psychology; Sociology and Geography," *AJS* 14 (1908–9): 371–407.
67. George Herbert Mead, "Concerning Animal Perception," *PR* 14 (1907): 383–90; Mead, "Social Psychology as a Counterpart to Physiological Psychology," *PB* 6 (1909): 401–8; Mead, "The Psychology of Social Consciousness Implied In Instruction," *Science*, n.s. 31 (1910): 688–93; Mead, "What Social Objects Must Psychology Presuppose?" *JPPSM* 6 (1910): 174–80; Mead, "The Mechanism of Social Consciousness," *JPPSM* 9 (1912): 401–6; Mead, "The Social Self," *JPPSM* 10 (1913): 374–80.
68. William I. Thomas, ed., *Source Book For Social Origins* (Chicago, 1909), pp. 4–26, 130–33, 316–17, 436–39, 530–34, 635, 733–35, 856–58. As late as 1907 Thomas still accepted at least some naturalistic assumptions; see Thomas, *Sex and Society* (Chicago, 1907).
69. Charles A. Ellwood, *Sociology and Modern Problems* (New York, 1910); Ellwood, *Sociology in its Psychological Elements* (New York, 1912); Ellwood, "The Instinctive Element in Human Society," *PSM* 80 (1912): 263–72; Ellwood, "The Psychological View of Society," *AJS* 15 (1909–10): 394–404. On Ellwood and eugenics, see, for example, Mrs. Huntington Wilson to Charles A. Ellwood, 7 and 12 December 1912, *CAE*.
70. Charles A. Ellwood to Floyd N. House, 14 November 1935, *CAE*. H. R. Hayes, *From Ape to Angel: An Informal History of Social Anthropology* (New York, 1959), pp. 139–42, discusses Marett. English anthropologists maintained the evolutionary outlook much longer than did American anthropologists, which may explain why such a devout evolutionist as Ellwood was attracted to Marett. On Ellwood and instinct, see E. L. Thorndike to Charles A. Ellwood, 16 August 1913, G. Elliot Smith to Charles A. Ellwood, 8 August 1916, L. T. Hobhouse to Charles A. Ellwood, 4 August 1917, Luther Lee Bernard to Charles A. Ellwood, 29 April 1922, *CAE*.
71. Charles A. Ellwood, review of H. W. Conn, *Social Heredity and Social Evolution*, *PB* 12 (1915): 472–73; Ellwood, review of A. de Gobineau, *The Inequality of the Human Races*, *PB* 13 (1916): 481.
72. Charles A. Ellwood, *Introduction to Social Psychology* (New York, 1917); Ellwood, "The Instinct in Social Psychology," *PB* 16 (1919): 71.
73. Edward Cary Hayes, *Introduction to the Study of Sociology* (New York, 1915).
74. Charles H. Cooley, *Social Process* (New York, 1918).
75. William I. Thomas and Florian Znanecki, *The Polish Peasant in Europe and America* 5 vols. (n.p., 1918), 1: 1–86.
76. This is obvious from the sociologists' careers in the 1920s. See, for example: Odum, *American Sociology*, pp. 75–140. On Giddings, see Northcott, "The Sociological Theories of Franklin Henry Giddings: Consciousness of Kind, Pluralistic Behavior, and Statistical Method," in Barnes, ed., *An Introduction to the History of Sociology*, pp. 763–64; on Ross, see William L. Kolb, "The Sociological

Theories of Edward Alsworth Ross"; in Barnes, ed., *An Introduction to the History of Sociology*, pp. 819, 821, 830–31, and Weinberg, *Edward A. Ross and the Sociology of Progressivism*, pp. 191–238. On Keller, see Albert G. Keller to Edward A. Ross, 17 October 1916, *EAR*.

77. Charles A. Ellwood to Dr. Charles W. Flint, 15 April 1925, *CAE*. See also Odum, *American Sociology*, pp. 75–157, for discussions of the careers of the other sociologists.

78. Robert E. L. Faris, *Chicago Sociology 1920–1932* (San Francisco, 1967), pp. 20–87; Roscoe C. and Gisela J. Hinkle, *The Development of Modern Sociology: Its Nature and Growth in the United States* (New York, 1954), pp. 18–40; Robert E. Park and Ernest W. Burgess, *An Introduction to the Science of Sociology* (New York, 1921).

79. Odum, *American Sociology*, pp. 147–52, 172–76. See also William F. Ogburn, "The Historical Method in the Analysis of Social Phenomena," *PASS* 16 (1922): 70–83; Ogburn, "An Analysis of the Standard of Living in the District of Columbia in 1916," *Journal of the American Statistical Association* 16 (1918–19): 374–89. See William F. Ogburn to Robert H. Lowie, 7 March, 28 April, 25 July 1917, 26 March 1922, *RHL*, for Ogburn's interest in the culture theory. On Chapin, see also his "The Statistical Redefinition of a Societal Variable," *AJS* 30 (1924/1925): 154–71.

80. Luther Lee Bernard, "Some Historical and Recent Trends of Sociology in the United States," *Southwestern Political and Social Science Quarterly* 9 (1928–29): 289.

81. Albion W. Small, *Origins of Sociology* (Chicago, 1924), pp. 349–50.

82. Ibid., p. 347.

83. John L. Gillin, "The Development of Sociology in the United States," *PASS* 21 (1927): 25. W. P. Meroney, "The Membership and Program of Twenty-Five Years of the American Sociological Society," *PASS* 25 (1930): 63, noted that prior to 1922 American sociologists did not seem to be sure of their field, their subject matter, or their method, and they tended to argue and discuss "many questions which now seem of little moment."

## Chapter 5

1. Theodosius Dobzhansky, and M. F. Ashley-Montagu, "Natural Selection and the Mental Capacities of Mankind," *Science*, n.s. 105 (1947): 590.

2. Edwin G. Conklin to Thomas Hunt Morgan, 30 September 1915, *EGC*.

3. "Doctorates Conferred by American Universities," *Science*, n.s. 42 (1915): 557, for example, shows the more than double increases in both disciplines through 1916 alone!

4. Based on my examination of the institutional affiliations, research programs, and careers of several dozen geneticists who launched their careers in the 1910s and whose work found acclaim.

5. A. H. Sturtevant, "Thomas Hunt Morgan, September 25, 1866–

December 4, 1945," in National Academy of Sciences, *Biographical Memoirs,* vol. 33 (New York, 1959), pp. 283–85.

6. Theodosius Dobzhansky, "The Reminiscences of Theodosius Dobzhansky," typescript oral history memoir, Oral History Research Office, Columbia University, 1962, Vol. I, Part II, pp. 248–51, 255–56. I could find no evidence of precisely when Morgan became so anti-religious in any archival document or printed source. His early experimental interests and work, however, suggest he had formed his attitudes toward religion by then.

7. Thomas H. Morgan, "Recent Experiments on the Inheritance of Coat Colors in Mice," *AN* 43 (1909): 505–9. For an example of the critical attitude Morgan held toward Weismann's germ plasm theory, see Thomas H. Morgan, *The Physical Basis of Heredity* (Philadelphia and London, 1919), pp. 234–40.

8. See the comments on Davenport's slipshod work by Oscar Riddle, "Biographical Memoir of Charles Benedict Davenport 1866–1944" in National Academy of Sciences, *Biographical Memoirs, XXV* (Washington, D.C., 1949), pp. 85–90, which are devastating when it is considered that Riddle was a friend of Davenport's and that the purpose of these memoirs is memorial, not critical. See also Robert M. Yerkes to G. H. Parker, 13 May 1946, *RMY*.

9. Sturtevant, "Thomas Hunt Morgan," pp. 290–93; H. J. Muller, "An Episode in Science," typescript copy of speech, 29 pages, delivered 25 July 1921, Cold Spring Harbor, pp. 8–9, H. J. Muller folder, *CBD.* Sturtevant and Muller were among Morgan's earliest *Drosophila* students and their accounts agree. See also W. E. Castle, "The Beginnings of Mendelism in America," in L. C. Dunn, ed., *Genetics in the 20th Century* (New York, 1951), p. 73.

10. Thomas H. Morgan, "Chromosomes and Heredity," *AN* 64 (1910): 449–96.

11. Henry E. Crampton, *The Department of Zoology of Columbia University, 1892–1942* (New York, 1942), pp. 68–71, lists all Ph.D.s and dissertation titles. I have identified and followed those named in the text through successive generations of James McKeen Cattell, *American Men of Science.* The question of priority of discovery of genetics ideas by the Morgan school has cropped up in the recent historical literature. The ultimate effect of this revisionist work has been to scale down Morgan's reputation as a source of new ideas in genetics, but not his leadership of the group. See, in particular, Garland E. Allen, "The Introduction of *Drosophila* into the Study of Heredity and Evolution: 1900–1910," *Isis* 66 (1975): 322–33; Allen, "Thomas Hunt Morgan and the Problem of Natural Selection," *Journal of the History of Biology* 1 (1968): 113–39; Elof Axel Carlson, "An Unacknowledged Founding of Molecular Biology: H. J. Muller's Contributions to Gene Theory, 1910–1936," *Journal of the History of Biology* 4 (1971): 149–70; Carlson, "The Drosophila Group," *Genetics* 79 (1975): 15–27.

12. Thomas Hunt Morgan, "Eight Factors that Show Sex-linked Inheritance in *Drosophila*," *Science,* n.s. 35 (1912): 472–73; Morgan, "The Explanation of a New Sex-ratio in *Drosophila* and Complete Linkage in the Second Chromosome of the Male,"

*Science,* n.s. 36 (1912): 718–19; Morgan, "The Masking of a Mendelian Result by the Influence of the Environment," *PSEBM* 9 (1912): pp. 73–74; Morgan, "Heredity of Body Color in Drosophila," *JEZ* 13 (1912): pp. 27–43.

13. Edmund B. Wilson, "Some Aspects of Cytology in Relation to the Study of Genetics," *AN* 46 (1912): 60.
14. Thomas H. Morgan, "Factors and Unit Characters in Mendelian Heredity," *AN* 47 (1913): 5–16.
15. Hubert H. Goodale and Thomas H. Morgan, "Heredity of Tricolor in Guinea-Pigs," *AN* 47 (1913): 340.
16. A. H. Sturtevant, "A Third Group of Linked Genes in *Drosophila Ampelophila,*" *Science,* n.s. 37 (1913): 990–92.
17. Thomas H. Morgan, "Multiple Allelomorphs in Mice," *AN* 48 (1914): 449–58; A. H. Sturtevant, "The Reduplication Hypothesis as Applied to *Drosophila,*" *AN* 48 (1914): 535–49; Calvin B. Bridges, "The Chromosome Hypothesis of Linkage Applied to Cases in Sweet Peas and *Primula,*" *AN* 48 (1914): 524–34; Charles W. Metz, "An Apterous *Drosophila* and Its Genetic Behavior," *AN* 48 (1914): 690.
18. Thomas H. Morgan, "The Role of the Environment in the Realization of a Sex-Linked Mendelian Character in *Drosophila,*" *AN* 49 (1915): 385–429; M. A. Hoge, "The Influence of Temperature on the Development of a Mendelian Character," *JEZ* 18 (1915): 241–85.
19. Calvin B. Bridges, "Non-Disjunction as Proof of the Chromosome Theory of Heredity," *Genetics* 1 (1916): 1–52; 107–63.
20. Thomas H. Morgan, Alfred H. Sturtevant, H. J. Muller, and C. B. Bridges, *The Mechanism of Mendelian Heredity* (New York, 1915). For Morgan's views on the preformation issue, see, for example, Thomas H. Morgan to Charles B. Davenport, Mar. [?] 1910, *CBD.*
21. H. H. Plough, "The Effect of Temperature on Crossing-Over in *Drosophila,*" *JEZ* 24 (1917): 147–209; F. Payne and M. Denny, "The Heredity of Orange Eye Color in *Drosophila Melanogaster,*" *AN* 55 (1921): 377–81; S. R. Safir, "Buff, a New Allelomorph of White Eye Color in *Drosophila,*" *Genetics* 1 (1916): 584–90; H. J. Muller, "The Mechanism of Crossing-Over," *AN* 50 (1916): 193–221, 284–305, 350–66, 421–34; Muller, "Are the Factors of Heredity Arranged in a Line?" *AN* 54 (1920): 97–121.
22. L. C. Dunn, *A Short History of Genetics: The Development of Some of the Main Lines of Thought: 1864–1939* (New York, 1965), pp. 81–115, 131–66.
23. George H. Shull, "Color Inheritance in *Lychnis Dioica L.,*" *AN* 44 (1910): 90.
24. George H. Shull to John M. Coulter, 11 February 1911, *DGAUC.*
25. Two excellent recent studies suggest the political relationship between land-grant colleges and their political constituencies: Charles E. Rosenberg, "Science, Technology, and Economic Growth: The Case of the Agricultural Experiment Station Scientist, 1875–1914," in G. H. Daniels, ed., *Nineteenth Century American Science: A Reappraisal* (Evanston, 1972), pp. 181–209; Edward

H. Beardsley, *Harry L. Russell and Agricultural Science in Wisconsin* (Madison, 1969).

26. E. W. Lindstrom, "Linkage in Maize: Alevrone and Chlorophyll Factors," *AN* 51 (1917): 225–37.

27. Howard B. Frost, "The Different Meanings of the Term 'Factor' as Affecting Clearness in Genetic Discussion," *AN* 51 (1947): 244–50.

28. "Genetics I, January–May, 1915," mimeograph course outline, *DGAUC*.

29. Ernest B. Babcock to George H. Shull, 10 April 1913, *DGAUC*.

30. T. H. Goodspeed and Raymond E. Clausen, "Mendelian Factor Differences Versus Reaction System Contrasts in Heredity," *AN* 51 (1917): 31–46, 92–101.

31. Thomas H. Morgan to Ernest B. Babcock, 2 February, 9 February 1918, *DGAUC*.

32. H. L. Ibsen, "Linkage in Rats," *AN* 54 (1920): 61–67; Joseph Krafka, "The Effect of Temperature upon Facet Number in the Bar-Eyed Mutant of *Drosophila*," *Journal of General Physiology* 2 (1920): 409–64.

33. Mary Alice Evans and Howard Ensign Evans, *William Morton Wheeler, Biologist* (Cambridge, 1970), pp. 178–204, is a good treatment of the Bussey's history.

34. A list of the Bussey genetics graduates is in L. C. Dunn, "William Ernest Castle October 25, 1867–June 3, 1962," in National Academy of Sciences, *Biographical Memoirs,* 38 (New York, 1965), 52–53. Here as elsewhere I have relied upon the rating system in Cattell, *American Men of Science* (6th ed.).

35. C. C. Little, "'Dominant' and 'Recessive' Spotting in Mice," *AN* 48 (1914): 24–82; Little, "Evidence of Multiple Factors in Mice and Rats," *AN* 51 (1917): 457–80.

36. E. Carleton MacDowell, "Multiple Factors in Mendelian Inheritance," *JEZ* 16 (1914): 177–94; MacDowell, "Bristle Inheritance in *Drosophila*," *JEZ* 19 (1915): 61–98.

37. George F. Freeman, "Linked Quantitative Characters in Wheat Crosses," *AN* 51 (1917): 683–89.

38. R. H. Emerson, "Genetical Studies of Variegated Pericarp in Maize," *Genetics* 2 (1917): 1–35.

39. L. C. Dunn, "Types of White Spotting in Mice," *AN* 54 (1920): 494; Dunn, "The Sable Varieties of Mice," *AN* 54 (1920): 247–61.

40. This statement is based on Cattell, *American Men of Science.*

41. Donald F. Jones, "Edward Murray East 1879–1938" in National Academy of Sciences, *Biographical Memoirs,* vol. 23 (Washington, D.C., 1945), pp. 217–42.

42. See, for example, Edward M. East to Jacques Loeb, 8 December 1913, *JL*.

43. See, for example, Edward M. East, *Mankind at the Crossroads* (New York, 1923); East, *Heredity and Human Affairs* (New York, 1929).

44. Edward M. East and Donald F. Jones, *Inbreeding and Outbreeding: Their Genetic and Sociological Significance* (Philadelphia, 1919).

45. Jacques Loeb to Edward M. East, 22 April 1918, *JL*.
46. Jacques Loeb to Edward M. East, 9 May 1919, Edward M. East to Jacques Loeb, 23 May 1919, *JL*. Loeb had recruited the East Jones manuscript for the publisher and was serving as referee.
47. Edward M. East to Jacques Loeb, 25 February 1919, *JL*.
48. William B. Provine, *The Origins of Theoretical Population Genetics* (Chicago, 1971), pp. 111–14, discusses Castle's work clearly. See also Castle's annual reports to the Carnegie Institution of Washington in William E. Castle folder, *CBD*, as for example, William E. Castle to Robert S. Woodward, 25 September 1908, 1 September 1909, 1 September 1910, 1 September 1913, 1 September 1916. See also William E. Castle and John C. Phillips, *Piebald Rats and Selection* (Washington, 1914).
49. See, for example, H. J. Muller, "The Bearing of the Selection Experiments of Castle and Phillips on the Variability of Genes," *AN* 48 (1914): 567–76.
50. Thomas H. Morgan to Edwin G. Conklin, 20 January 1916, *EGC*.
51. Thomas H. Morgan, *A Critique of the Theory of Evolution* (Princeton, 1916).
52. William E. Castle, "Some Experiments in Mass Selection," *AN* 49 (1915): 97. See also William E. Castle to Charles B. Davenport, 21 November 1917, *CBD*; Castle questioned multiple factors here, too.
53. Alfred H. Sturtevant, *An Analysis of the Effects of Selection* (Washington, D.C., 1918). Provine, *The Origins of Theoretical Population Genetics*, pp. 127–28.
54. William E. Castle, "Piebald Rats and Selection, A Correction," *AN* 53 (1919): 370–76. Provine, *The Origins of Theoretical Population Genetics*, pp. 126–28, clearly explains the experiment and its implications.
55. Provine, *The Origins of Theoretical Population Genetics*, pp. 130–77, provides an intelligent guide to the distinct contributions of Wright, Fisher, and Haldane.
56. Thomas H. Morgan, "Human Inheritance," *AN* 58 (1924): 395–409.
57. Herbert Spencer Jennings, "Heredity and Environment," *SM* 19 (1924): 229.
58. Ibid., p. 236.
59. See, for example: Herbert Spencer Jennings, *The Biological Basis of Human Nature* (New York, 1930); Thomas H. Morgan, *The Scientific Basis of Evolution* (New York, 1932); and Lancelot Hogben, *Nature and Nurture* (New York, 1933).
60. Julian Huxley, "The Uniqueness of Man," *Yale Review*, n.s. 28 (1938–39): p. 490. See also Huxley, *Evolution: The Modern Synthesis* (New York, 1942), a more elaborate and extended reconciliation of the culture idea and the modern synthetic theory of evolution.
61. Nicholas Murray Butler to Frank D. Fackenthal, 20 September 1941, Nicholas Murray Butler Papers, Columbia University.
62. Charles B. Davenport to Nicholas Murray Butler, 17 September 1941, Frank D. Fackenthal to Charles B. Davenport, 10 October 1941, Nicholas Murray Butler Papers, Columbia University.

63. William E. Castle to Charles B. Davenport, 10 November 1910, *CBD*.
64. See, for example: Jacques Loeb to Simon Flexner, 27 October 1913, Thomas H. Morgan to Jacques Loeb, 30 October 1918, Jacques Loeb to Edward M. East, 13 November 1918, Jacques Loeb to Thomas H. Morgan, 29 October 1918, Jacques Loeb to Herbert Spencer Jennings, 29 May 1920, Jacques Loeb to Samuel J. Holmes, 11 October 1921, *JL*; Raymond Pearl to Robert M. Yerkes, 28 November 1922, Robert M. Yerkes to Raymond Pearl, 29 November 1922, Francis B. Sumner to Robert M. Yerkes, 12 May 1919, *RMY*; Thomas H. Morgan to Charles B. Davenport, 18 January 1915, 28 February 1917, 9 May 1917, Thomas Morgan to Henry F. Osborn, 14 June 1920 [T. H. Morgan Folder], William Bateson to Charles B. Davenport, 11 February 1921, *CBD*; Thomas H. Morgan to Edwin G. Conklin, 5 April 1917, 29 October 1919, 12 June 1920, *EGC*; William Bateson to [Thomas H.] Morgan, 19 May 1920, *DGAUC*.
65. H. W. Conn, *Social Heredity and Social Evolution: The Other Side of Eugenics*, (New York and Cincinnati, 1914).
66. Thomas H. Morgan to Charles B. Davenport, 18 January 1915, *CBD*. Morgan apparently participated in eugenics meetings for a few more years, in hopes of making it scientifically responsible; see Thomas H. Morgan to Edwin G. Conklin, 5 April 1917, *EGC*.
67. This statement is based on my analysis of the careers, publications, and society memberships of Morgan's graduate students at Columbia, the Bussey graduates, and the graduates of genetics programs at agricultural colleges discussed in section II of this chapter.
68. William Bateson to Thomas H. Morgan, 19 May 1920, attached to Thomas H. Morgan to Edwin G. Conklin, 12 June 1920, *EGC*.
69. Those scientists were: Herbert Spencer Jennings, C. E. McClung, C. B. Bridges, Alfred F. Blakeslee, George H. Shull, H. J. Muller, Charles Zeleny, A. Franklin Shull, E. Carleton MacDowell, Sewall Wright, L. C. Dunn, Raymond Pearl, Edward M. East, Samuel J. Holmes, and C. C. Little. "Reputable" means that these men were recognized for the quality of their work by other geneticists and biologists; ten of them were starred in Cattell, *American Men of Science*, by 1921, another three in 1927, and the other two in 1933.
70. The two exceptions were Raymond Pearl, "Some Eugenic Aspects of the Problem of Population," pp. 212–14, and Edward M. East, "Population in Relation to Agriculture," pp. 215–32, in *Scientific Papers of the Second International Congress of Eugenics*, 2 vols. (Baltimore, 1923), vol. 2, *Eugenics in Race and State*.
71. Alfred H. Sturtevant to Roy E. Clausen, 11 October 1921, *DGAUC*.
72. Kenneth M. Ludmerer, *Genetics and American Society: A Historical Appraisal* (Baltimore, 1972), pp. 95–113, is an excellent account.
73. See, for example, Herbert Spencer Jennings to Jacques Loeb, 14 April 1920, and attached memo, entitled, "To Members of the Committee of the American Society of Naturalists on Cooperation with the National Research Council." Jennings was president of

the Society that year and was appalled because some scientists wanted to exclude from the subsequent International Congress of Eugenics "certain men of science on account of their nationality irrespective of their scientific attainments."

74. Raymond Pearl to Herbert Spencer Jennings, 24 November 1923, *HSJ*. A perceptive sketch of Pearl is Herbert Spencer Jennings, "Raymond Pearl, 1879–1940," in National Academy of Sciences, *Biographical Memoirs*, vol. 22 (Washington, D.C., 1943), pp. 295–347.

75. See, for example, Jacques Loeb to Samuel J. Holmes, 21 October 1921, *JL*. Little served as Secretary of the 1921 Eugenics Congress; Muller advocated sperm banks in *Out of the Night* (New York, 1935); East published on eugenics; and Conklin was a moderate eugenist and member of the Galton Society.

76. Bruno Lasker to Herbert Spencer Jennings, 12 June 1923, Herbert Spencer Jennings to Editor, *The Survey*, 24 October 1923; Herbert Spencer Jennings to Geddes Smith, 7 November 1923, *HSJ*. Lasker and Smith were the editors of *The Survey*, and published Jennings' article on Laughlin's testimony.

77. Hubert Lyman Clark to Herbert Spencer Jennings, 29 January 1924, Edmund C. Sanford to Herbert Spencer Jennings, 28 January 1924, E. Carleton MacDowell to Herbert Spencer Jennings, 20 February 1924, Samuel J. Holmes to Herbert Spencer Jennings, 29 January 1924, Herbert Spencer Jennings to Samuel Jackson Holmes, 5 February 1924, Herbert Spencer Jennings to Vernon Kellogg, 29 January 1924. Vernon Kellogg to Herbert Spencer Jennings, 30 January 1924, *HSJ*.

78. Ludmerer, *Genetics and American Society: A Historical Appraisal*, p. 110.

79. See, for example, Thomas H. Morgan to Jacques Loeb, 30 October 1918, Jacques Loeb to Thomas H. Morgan, 29 October 1918, *JL*.

80. Edwin G. Conklin, *The Direction of Human Evolution* (New York, 1922), pp. 12, 14–15, 23, 31–35, 47–48; William E. Castle, *Genetics and Eugenics* (3rd ed.; Cambridge, 1924), pp. 292–309, 374–75; C. C. Little, "The Relation between Research in Human Heredity and Experimental Genetics," *SM* 14 (1922): 414.

81. See, for example, Samuel J. Holmes, *Studies in Evolution and Eugenics* (New York, 1923), pp. 62–255; East, *Heredity and Human Affairs* (New York and London, 1927), p. 41. Holmes and East were moderate eugenists who also tried to embrace contemporary genetics. Hence much of their writing about eugenics in the 1920s was equivocal; equivocation on eugenics was, of course, a symptom of a new caution lacking in the early 1900s in many biological circles.

82. Herbert Spencer Jennings, *Promethus; or Biology and the Advancement of Man*, (New York, 1925); William E. Castle, "Eugenics," *Encyclopedia Britannica*, 13th ed. (New York, 1926), pp. 1031–32; Raymond Pearl, "The Biology of Superiority," *American Mercury* 12 (1927): 257–66; Thomas H. Morgan, *Evolution and Genetics*, (Princeton, 1925); Edwin G. Conklin, "Some Recent Criticisms of Eugenics," *Eugenical News* 13 (1928): 61–65. See

also Edwin G. Conklin to Raymond Pearl, 27 and 29 March 1928, *EGC.*

83. L. C. Dunn, "Cross Currents in the History of Human Genetics," *The American Journal of Human Genetics,* 14 (1962): 1–13, traces this very well.

84. See, for example, Garland E. Allen, "Genetics, Eugenics, and Class Struggle," *Genetics* 79 (1975): 29–45.

85. See, for example, Curt Stern, *Principles of Human Genetics* (San Francisco, 1949), chapt. 24; see also Mark Haller, *Eugenics: Hereditarian Attitudes in American Thought* (New Brunswick, 1963), pp. 177–89.

86. John C. Merriam to Albert E. Wiggam, 15 May 1922, John C. Merriam to Madison Grant, 7 February 1925, *JCM.*

87. A. V. Kidder to L. C. Dunn, 25 May 1935, *LCD.* The activities of the first committee can be gleaned from John C. Merriam to L. C. Dunn, 10 January, 16 February 1929, A. V. Kidder to L. C. Dunn, 21 February 1929 and attached "Memorandum for Dr. Merriam re Meeting of Advisory Committee on Eugenics Record Office" by A. V. Kidder, chairman, 3 pages, L. C. Dunn to A. V. Kidder, 25 February 1929, *LCD.*

88. John C. Merriam to L. C. Dunn, 5 June 1935, *LCD.*

89. See John C. Merriam to Henry Fairfield Osborn, 9 February 1933, *JCM.*

90. L. C. Dunn to president [John C.] Merriam, 3 July 1935, John C. Merriam to L. C. Dunn, 25 July 1935, A. V. Kidder to L. C. Dunn, 1 August 1935, 17 October 1936, A. V. Kidder to Dr. [John C.] Merriam, 19 June 1936 [copy], *LCD;* Ludmerer, *Genetics and American Society: A Historical Appraisal,* p. 169.

91. William E. Castle to L. C. Dunn, 20 March 1933, *LCD.*

92. William E. Castle to L. C. Dunn, 20 March 1933, *LCD.*; see also L. C. Dunn to R. C. Cook, 3 January 1933, L. C. Dunn to Sewall Wright, 24 March 1933, *LCD* and Milislav Demerec to L. C. Dunn, 23 March 1933, Milislav Demerec Papers, Library of the American Philosophical Society, Philadelphia.

93. "A Quarterly Eugenics Section," *JH* 30 (1939): 108–9.

94. Ludmerer, *Genetics and American Society: A Historical Appraisal,* pp. 165–201; Mark H. Haller, *Eugenics: Hereditarian Attitudes in American Thought* (New Brunswick, 1963), pp. 177–89.

95. See, for example: Franz Boas, "Eugenics," *SM* 3 (1916): 471–78; Erville B. Woods, "Heredity and Opportunity," *AJS* 26 (1920–21): 1–21, 146–61.

96. Edward B. Reuter, *The Mulatto in the United States* (Boston, 1918), pp. 6, 7, 375–76, *et passim.*

97. A. L. Kroeber, review of W. McDougall, *Is America Safe for Democracy? AA,* n.s. 24 (1922): 464–65; Franz Boas, "Are the Jews a Race?" *The World Tomorrow* 6 (1923): 5–6; Boas, "The Question of Racial Purity," *American Mercury* 3 (1924): 163–69; George G. MacCurdy, "Proof of Man's Cultural Evolution," *SM* 21 (1925): 138–40.

98. Franz Boas, "What is a Race?" *The Nation* 120 (1925): 89–91; Harry Elmer Barnes, "The Race-Myth Crumbles," ibid., pp. 515–17; Alexander A. Goldenweiser, "Can There Be a 'Human Race'?"

ibid., pp. 462–63; Edward Sapir, "Let Race Alone," ibid., pp. 211–13.

99. An interesting example of the importance of established institutions to the reception of new ideas was the controversy over the theories of Immanuel Velikovsky in the 1950s. On the Velikovsky affair, see Ralph E. Jurgens, "Minds in Chaos: A Recital of the Velikovsky Story," *American Behavioral Scientist* 7: (September 1963), pp. 4–19.

100. "Report of Conference on Study of 'The Physical Basis of Human Behavior' Held . . . October 20, 1941," 9-page memo; "Second Conference on Plans for Extension of Biological Research. . . January 28 /1922/. . .," 4-page memo; "Third conference on plan for extension of biological research. . . April 27, 1922. . . ," 3-page memo, Subject Folder 259, *RMY*.

101. "Proceedings Conference on Human Migration, Arranged by the Committee on Scientific Problems of Migration. . . November 18, 1922," 43-page manuscript, Subject Folder 258, *RMY*.

102. "Suggestions for Committee on Scientific Problems of Human Migration. Submitted by the Chairman of Committee for consideration at meeting called for January 25, 1923," 6-page memo, Subject Folder 258, *RMY*.

103. "Minutes of second meeting of committee on scientific problems of human migration, January 25, 1923. . .," 3-page memo, Subject Folder 258, *RMY*.

104. Social Science Research Council, *Decennial Report 1923–1933* (New York, 1934), pp. 1–4; Charles E. Merriam to Robert M. Yerkes, 28 February 1923, Charles E. Merriam to Miss Mary Van Kleeck, 18 May 1923 [copy], *RMY*.

105. "Minutes of meeting of sociological conference group, Committee on Scientific Problems of Human Migration. . . March 29, 1923," 37-page manuscript, Subject Folder 258, *RMY*.

106. "Report and Recommendations of Committee on Scientific Problems of Human Migration. . . April 2, 1923," 9-page manuscript and Appendix A, Subject Folder 258, *RMY*. The Committee's activities can be followed in Subject Folders 258, 259, *RMY*; but see also Robert M. Yerkes, "The Work of Committee on Scientific Problems of Human Migration, National Research Council," *Journal of Personnel Research* 3 (1924–25): 189–96.

107. "Meeting of Committee on Scientific Problems of Human Migration. . . March 15, 1924. . .," 5-page manuscript, pp. 1, 2, 3–4, "Meeting of Committee on Scientific Problems of Human Migration. . . November 15, 1924," 6-page manuscript, pp. 2, 5–6, Subject Folder 258, *RMY*.

108. "Meeting of Committee on Scientific Problems of Human Migration. . . April 6, 1925. . .," 11-page manuscript, pp. 3–4, 8–9, Subject Folder 258, *RMY*.

109. "Meeting of Committee on Scientific Problems of Human Migration. . . May 28, 1925. . . ," 7-page manuscript, p. 2, 7, Subject Folder 258, *RMY*.

110. "Report of Joint Committee of the Committee on Problems and Policy of the Social Science Research Council. . . August 31 to September 3, 1925," I–IX and 7 additional pages, onion skin copy,

Subject Folder 78, *RMY*. See also Robert M. Yerkes to Charles E. Merriam, 29 January 1925, Subject Folder 473, *RMY*.

111. "Minutes of the Meeting of the Social Science Research Council. . . Oct. 17, 1925," 6-page manuscript, Arnold B. Hall to Robert M. Yerkes, 15 December 1925. Robert M. Yerkes to Arnold B. Hall, 22 December 1925, Subject Folder 473, *RMY*; Social Science Research Council, *Decennial Report 1923–1933*, pp. 24–68.

112. George M. Stratton to Robert M. Yerkes, 10 December 1925, Robert M. Yerkes to George M. Stratton, 18 December 1925, "Report and Recommendations of the Committee on Scientific Problems of Human Migration. . . April 5, 1926," 45 pages, p. 40, Subject Folder 259, *RMY*.

113. In addition to the voluminous materials in Subject Folder 78, *RMY*, which cover these developments in the SSRC in exquisite detail, see Social Science Research Council, *Decennial Report 1923–1933*, pp. 51–54.

114. Social Science Research Council, *Decennial Report 1923–1933*, p. 61.

115. A. V. Kidder, "Report to the Council of the American Anthropological Association By the Chairman of the Division of Anthropology and Psychology of the National Research Council," *AA*, n.s. 29 (1927): 733–38; Franz Boas to John C. Merriam, 31 December 1925, John C. Merriam to Franz Boas, 5 January 1926, Franz Boas to G. M. Stratton, 22 January 1926, G. M. Stratton to Franz Boas, 25 January 1926, Franz Boas to Fay-Cooper Cole, 18 May 1926, A. V. Kidder to Franz Boas, 19 October, 26 November 1926, *FB*.

116. "National Research Council. . . Meeting of the Committee on a Study of the American Negro, March 19, 1927," 5-page manuscript, pp. 4–5, *FB*.

117. Charles B. Davenport and Morris Steggerda, *Race Crossing in Jamaica* (Washington, 1929), p. 3.

118. Charles B. Davenport to Edwin G. Conklin, 28 February 1925, *CBD*.

119. William E. Castle, "Race Mixture and Physical Disharmonies," *Science*, n.s. 71 (1930): 603–6.

120. Melville J. Herskovits, *The American Negro: A Study in Race Crossing* (New York, 1928), p. 82.

121. Lewis M. Terman to Robert M. Yerkes, 30 September 1940, *RMY*.

122. Laura Fermi, *Illustrious Immigrants: The Intellectual Migration from Europe 1930–1941* (2nd. ed.; Chicago, 1971), pp. 39–53.

123. Ibid., pp. 71–78.

124. The impact of the refugee problem can be followed in the papers of a number of leading scientists in the 1930s, such as *LCD*, *FB*, *JMC*, and *RFB*.

125. See, for example, Folder "American Committee for Democracy and Intellectual Freedom," Box 7, *RFB*. The Committee's letterhead carries the names of its most prominent members. See also "Manifesto on Freedom of Science," 31-page pamphlet, n.d., by Franz Boas in Box 7, *RFB*.

Chapter 6

1. Robert M. Yerkes to Zing Yang Kuo, 6 January 1922, *RMY*.
2. Zing Yang Kuo to Robert M. Yerkes, 11 January 1922, *RMY*. On Kuo, see Gilbert Gottlieb, "Zing Yang Kuo: Radical Scientific Philosopher and Innovative Experimentalist (1898–1970)" *Journal of Comparative and Physiological Psychology* 80 (1972): 1–10.
3. See Chapter 4.
4. Edward Bradford Titchener to Robert M. Yerkes, 23 September 1906, *RMY*.
5. Robert S. Harper, "Tables of American Doctorates in Psychology," *AJP* 62 (1949): 580–81.
6. Edwin G. Boring, "Statistics of the American Psychological Association in 1920," *PB* 17 (1920): 271–278.
7. James McKeen Cattell, "Psychology in America," *Science*, n.s. 70 (1929): 339–40.
8. See, for example, the difficult circumstances under which many laboratories were founded in C. R. Garvey, "List of American Psychological Laboratories," *PB* 26 (1929): 652–60.
9. Thomas H. Haines to Robert M. Yerkes, 10 April 1903, *RMY*.
10. Thomas H. Haines to Robert M. Yerkes, 17 September 1904, *RMY*. See also Thomas H. Haines to Robert M. Yerkes, 28 June 1903, 6 June 1904, *RMY*; Thomas H. Haines to Hugo Münsterberg, 22 January 1908, 9 November 1909, *HM*.
11. See Harper, "Tables of American Doctorates in Psychology," p. 583.
12. Samuel W. Fernberger, "The American Psychological Association: A Historical Summary, 1892–1930," *PB* 29 (1932): 7–11.
13. James McKeen Cattell, "Our Psychological Association and Research," in A. T. Poffenberger, ed., *James McKeen Cattell, 1860–1944: Man of Science*, 2 vols. (Lancaster, Pa., 1947), 2: 340.
14. Based on my research into the careers of these men and women; there is a list of them in A. T. Poffenberger, ed., *James McKeen Cattell 1860–1944: Man of Science*, 2 vols. (Lancaster, Pa., 1947), 2: 45.
15. Fernberger, "The American Psychological Association," pp. 35–37.
16. Boring, *A History of Experimental Psychology*, p. 531.
17. See, for example: Howard C. Warren to Edward B. Titchener, 3 June, 23 July 1910, *EBT*; Charles H. Judd to Hugo Münsterberg, 29 March, 19 April 1904, James McKeen Cattell to William James, 9 December 1903, James McKeen Cattell to James Mark Baldwin, 14 May 1904, Hugo Münsterberg to James Mark Baldwin, 11 March 1906, James Mark Baldwin to Hugo Münsterberg, 4 December, 19 December 1903, James McKeen Cattell to Hugo Münsterberg, 16 April, 4 May, 29 July 1904, Hugo Münsterberg to James Mark Baldwin, 28 May 1904, Howard C. Warren to Hugo Münsterberg, 18 March 1910, *HM*; G. Stanley Hall to James McKeen Cattell, 17 January 1906, James Mark Baldwin to James McKeen Cattell, 28 November, 7 December 1903, E. B. Titchener to James McKeen Cattell, 17 June 1904, 26 September 1906, Charles H. Judd to James McKeen Cattell, 6 April, 5 May, 16 May, 23 May 1904, James McKeen Cattell to Charles H. Judd, 6 December 1903,

9 May 1904, Howard C. Warren to James McKeen Cattell, 22 June 1910, *JMC*.

18. John B. Watson to Robert M. Yerkes, 5 September 1909, *RMY*.
19. John B. Watson to Robert M. Yerkes, 4 February 1914, *RMY*.
20. Raymond Pearl to Robert M. Yerkes, 11 May 1903, *RMY*.
21. Edward Lee Thorndike, *Animal Intelligence* (New York, 1898); Thorndike, "Do Animals Reason?" *PSM* 55 (1899): 480–90.
22. C. J. Warden and L. H. Warner, "The Development of Animal Psychology During the Past Three Decades," *PR* 34 (1927): 200–201. For reports in particular departments, see, for example, "Report of the Psychological Laboratory 1909," "Report to the President, 1914," *HM*, "President's Report to the Trustees of Clark University for 1902–1903, Department of Experimental and Comparative Psychology," "President's Report to the Trustees of Clark University for 1902–1903, Department of Experimental and Comparative Psychology, 1903–1904," *CUA*.
23. Robert M. Yerkes, "The Scientific Way," typescript autobiography, pp. 1–139, *RMY*.
24. This and the following statements based on the Watson-Yerkes correspondence, 1905–16, *RMY*.
25. John B. Watson to Robert M. Yerkes, 5 June 1910, *RMY*.
26. John B. Watson to Robert M. Yerkes, 29 September 1908, *RMY*.
27. John B. Watson to Robert M. Yerkes, 2 November 1905, 29 September, 20 November 1908, 18 September, 29 October 1909, *RMY*.
28. John B. Watson to Robert M. Yerkes, 2 October 1907, *RMY*.
29. John B. Watson to Robert M. Yerkes, 21 February, 4 March, 10 December 1912, *RMY*.
30. John B. Watson, "Psychology as the Behaviorist Views It," *PR* 20 (1913): 167. See also John B. Watson to Robert M. Yerkes, 12 March 1913, *RMY*.
31. See, for example: Raymond Dodge, "The Theory and Limitations of Introspection," *AJP* 23 (1912): 214–19; Knight Dunlap, "The Case Against Introspection," *PR* 19 (1912): 412ff.
32. See, for example: M. W. Calkins, "Psychology and the Behaviorist," *PB* 10 (1913): 288–91; H. R. Marshall, "Is Psychology Evaporating?" *JPPSM* 10 (1913): 710–16; M. F. Washburn, "Some Thoughts on the Last Quarter Century in Psychology," *Phil R* 26 (1917): 46–55.
33. Charles Judson Herrick to Robert M. Yerkes, 10 November 1911, *RMY*.
34. Edward B. Titchener to Robert M. Yerkes, 20 November 1911, *RMY*.
35. W. V. Bingham, "Proceedings to the Twentieth Annual Meeting of the American Psychological Association. . . December 27, 28, and 29, 1911," *PB* 9 (1912): 41–42.
36. John B. Watson, *Behavior: An Introduction to Comparative Psychology,* (New York, 1914), pp. 106–250.
37. John B. Watson to Robert M. Yerkes, 27 October 1915, *RMY*.
38. John B. Watson, "The Place of the Conditioned Reflex in Psychology," *PR* 22 (1916): 89–116.
39. John B. Watson to Robert M. Yerkes, 12 October 1916, *RMY*.

40. Robert M. Yerkes to John B. Watson, 16 October 1916, *RMY*.
41. John B. Watson to Robert M. Yerkes, 18 October 1916, *RMY*.
42. See H. S. Jennings, John B. Watson, Adolf H. Meyer, and William I. Thomas, *Suggestions of Modern Science Concerning Education* (New York, 1917), pp. 55–99. For Cannon's work, see Walter B. Cannon, *Bodily Changes in Pain, Hunger, Fear, and Rage* (New York, 1915).
43. My calculations, based on Harper, "Tables of American Doctorates in Psychology," pp. 579–87.
44. G. T. W. Patrick, "The Founding of the Psychological Laboratory at the State University of Iowa," *Iowa Journal of History and Politics* 30 (1932): 404–16; Carl E. Seashore, *Pioneering in Psychology* (Iowa City, 1942), *passim*.
45. See, for example: James McKeen Cattell to Seth Low, 22 September 1899, 16 April 1900, James McKeen Cattell to Nicholas Murray Butler, 14 April 1902, 15 November 1904, 9 November 1905, 14 November 1907, 8 November 1909, 15 November 1909, Nicholas Murray Butler to James McKeen Cattell, 6 March 1906, *JMC*; Boring, *A History of Experimental Psychology*, pp. 532–40, 548.
46. Forrest A. Kingsbury, "A History of the Department of Psychology at the University of Chicago," *PB* 43 (1946): 259–71.
47. Herbert A. Miller to Robert M. Yerkes, 10 November 1911, 8 December 1912, 25 October 1913, *RMY*. Miller acknowledged his intellectual debt to Boas in Herbert A. Miller to Franz Boas, 8 February 1921, *FB*.
48. Herbert A. Miller to Robert M. Yerkes, 13 October 1913, *RMY*.
49. Herbert A. Miller to Robert M. Yerkes, 25 February 1914, *RMY*.
50. Luther Lee Bernard, "Hereditary and Environmental Factors in Human Behavior," *Monist* 37 (1927): 163–65.
51. Luther Lee Bernard to Scott E. W. Bedford, 30 March 1914 (copy) *EAR*.
52. Ellsworth Faris, "The Subjective Aspect of Culture," *PASS* 19 (1924): 37–46.
53. Knight Dunlap, "Are There Any Instincts?" *JAbP* 14 (1919–20): 311.
54. See, for example, Zing Yang Kuo, "Giving Up Instincts in Psychology," *Journal of Philosophy* 18 (1921): 645–64; Kuo, "The Net Result of the Anti-Heredity Movement in Psychology," *PR* 36 (1929): 181–99.
55. F. H. Allport, *Social Psychology* (New York, 1925).
56. See, for example, John B. Watson, *Behaviorism,* (New York, 1925).
57. Probably the Chicago school was influenced by the Boasians. See, for example: W. F. Ogburn to Robert H. Lowie, 7 March, 28 April, 25 July 1917, 26 March 1922, *RHL*; Alfred L. Kroeber to William I. Thomas, 6 June 1910, *ALK*. There is other evidence as well: for example, Boas taught summer school at Chicago when Bernard was a graduate student. Boas and Dewey knew one another well at Columbia, and so on. See also Herbert A. Miller to Robert M. Yerkes, 12 October 1913, *RMY*.
58. John Dewey, "The Need for Social Psychology," *PR* 24 (1917): 266–77; J. R. Kantor, "The Problem of Instincts and Their Relation to Social Psychology," *JASP* 18 (1923–24): 76.

59. J. R. Kantor, "The Integrative Character of Habits," *JCP* 2 (1922): 195–226; Kantor, "An Essay Toward an Institutional Conception of Social Psychology," *AJS* 27 (1921–22): 611–27.
60. John Dewey, *Human Nature and Conduct* (New York, 1922) *passim*.
61. See, for example: Luther Lee Bernard, *Instincts: A Study in Social Psychology* (New York, 1924); Robert E. Park, "Culture and Culture Trends," *PASS* 19 (1924): 24–36; Ellsworth Faris, "Are Instincts Data or Hypotheses?" *AJS* 27 (1921–22): 194ff; Clarence M. Case, "Instinctive and Cultural Factors in Group Conflicts," *AJS* 28 (1922–23): 1–20.
62. Wesley R. Wells, "The Anti-Instinct Fallacy," *PR* 30 (1923): 234.
63. J. R. Geiger, "Must We Give Up Instincts in Psychology," *Journal of Philosophy* 19 (1922): 94–98; Geiger, "Concerning Instincts," *Journal of Philosophy* 20 (1923): 57–68.
64. Edgar James Swift, "Language, Thought, and Instincts," *Journal of Philosophy* 20 (1923): 372.
65. William McDougall, "Motives in Light of Recent Discussion," *Mind* 29 (1920): 277–93; McDougall, "Purposive or Mechanical Psychology?" *PR* 30 (1923): 273–88; McDougall, "Can Sociology and Social Psychology Dispense with Instincts?" *AJS* 29 (1923–24): 657–70; William F. Ogburn, "Discussion of Professor McDougall's and Professor Kantor's Papers," *JASP* 19 (1924–25): 58; Luther Lee Bernard, "Discussion," *AJS* 29 (1923–24): 671.
66. William McDougall, *Is America Safe for Democracy?* (New York, 1921); McDougall, "Crime in America," *Forum* 77 (1927): 518–23; McDougall, "An Experiment for the Testing of the Hypothesis of Lamarck," *British Journal of Psychology* 17 (1927): 267–304; Thomas Hunt Morgan, *The Scientific Basis of Evolution* (New York, 1932), pp. 187–202, summarizes the case of modern biologists against the Lamarckian principle. After the late 1920s, almost all of McDougall's professional correspondence was either with philosophers or with prospective graduate students seeking assistantships—especially during the Depression. See the William McDougall Papers, Duke University Library, 1923–38. An interesting article is David L. Krantz and David Allen, "The Rise and Fall of McDougall's Instinct Doctrine," *Journal of the History of the Behavioral Sciences* 3 (1967): 326–38.
67. Watson, *Behavior: An Introduction to Comparative Psychology*, pp. 331, 334.
68. Karl S. Lashley, "The Behavioristic Interpretation of Consciousness," *PR* 30 (1923): 244–45ff; Lashley, *Brain Mechanisms and Intelligence* (Chicago, 1929).
69. Charles Judson Herrick, "The Natural History of Purpose," *PR* 32 (1925): 429. See also Herrick, *Neurological Functions of Animal Behavior* (New York, 1924) and Herrick, *Brains of Rats and Men* (Chicago, 1926).
70. Hulsey Cason, "The Physical Basis of the Conditioned Response," *AJP* 36 (1925): 393.
71. A. P. Weiss, *The Theoretical Basis of Human Behavior* (New York, 1925).

72. See, for example, Floyd H. Allport, "The Present Status of Social Psychology," *JASP* 21 (1926–27): 372–82.
73. John B. Watson, *The Ways of Behaviorism* (New York, 1928): 20–138.
74. Harry Levi Hollingworth, *Mental Growth and Decline: A Survey of Developmental Psychology* (New York, 1927).
75. Robert M. Yerkes, *Almost Human* (New York, 1925); W. N. and L. A. Kellogg, *The Ape and the Child. A Study of Environmental Influence Upon Early Behavior* (New York, 1933); in general, see Charles M. Diserens and James Vaugh, "The Experimental Psychology of Motivation," *PB* 28 (1931): 15–65.
76. Martin Birnbach, *Neo-Freudian Social Philosophy* (Palo Alto, 1961); David Shakow and David Rapaport, *The Influence of Freud on American Psychology* (New York, 1964). The work of Edward C. Tolman on drives and instincts in the 1920s and later was an interesting indication of the new trends in psychology. In such articles as "Instinct and Purpose," *PR* 27 (1920): 217–33; "Can Instincts Be Given Up In Psychology," *JASP* 17 (1922–23): 139–52, "The Nature of Instinct," *PB* 20 (1923): 200–218; and "A Behavioristic Account of the Emotions," *PR* 30 (1923): 217–27; and later work, Tolman retained the categories of purpose, drive, and instinct, but he defined them in objective, behaviorist ways (except that he rejected the behaviorists' reductionism), and he agreed that heredity and environment were important in the development of mind and personality. Tolman had taken his Ph.D. at Harvard in 1915, apparently under Yerkes' direction, and was for most of his career a professor of experimental psychology at the University of California, Berkeley. See also his interesting article, "A Stimulus-Expectancy Need-Cathexis Psychology," *Science,* n.s. 101 (1945): 160–61.
77. Gardner Murphy and Friedrich Jensen, *Approaches to Personality: Some Contemporary Conceptions Used in Psychology and Psychiatry* (New York, 1932).
78. Charles A. Ellwood, "The Relations of Sociology and Social Psychology," *JASP* 19 (1924–25): 8.
79. A general discussion is in G. A. Lundberg, R. Bain, and N. Anderson, eds., *Trends in American Sociology* (New York and London, 1929), pp. 72–220.
80. Melville J. Herskovits, *The American Negro: A Study in Racial Crossing* (New York, 1928); Margaret Mead, *Coming of Age in Samoa* (New York, 1928).

## Chapter 7

1. George B. Cutten, "The Reconstruction of Democracy," *SS* 16 (1922): 479.
2. Rush Welter, *Popular Education and Democratic Thought in America* (New York, 1962), pp. 1–9, 45–140, *et passim.*
3. Lawrence A. Cremin, *The Transformation of the School. Progres-*

*sivism in American Education 1876–1957* (New York, 1961), pp. 3–176.

4. Charles B. Davenport to Madison Grant, 1 July 1914, 10 February, 5 April, 22 November, 26 November 1917, 27 November 1920, Madison Grant to Charles B. Davenport, 6 July 1914, 16 February, 2 April, 20 November 1917, 27 January, 12 April 1921, *CBD*. See also Charles B. Davenport to Lothrop Stoddard, 10 May 1922, *CBD*.

5. Lothrop Stoddard, *The Revolt Against Civilization: The Menace of the Under Man* (New York, 1922), pp. 1–2.

6. Robert M. Yerkes, "Testing the Human Mind," *The Atlantic Monthly* 131 (1923): 364, 365.

7. Carl C. Brigham to Robert M. Yerkes, 24 June 1922, *RMY*. See Author Folder 62, "Carl C. Brigham," *RMY*, for the Brigham-Yerkes relationship.

8. Robert M. Yerkes to Edwin G. Boring, 2 September 1922, *RMY*.

9. Carl C. Brigham, *A Study of American Intelligence* (Princeton, 1923), pp. 204–10, *et passim*.

10. Edwin G. Boring to Robert M. Yerkes, 24 February 1922, *RMY*.

11. For proof of the young Boring's phenomenal professional reputation, see the correspondence in the "Chair of Experimental Psychology" file, 1918, *CUA*, in which Boring was recommended by many prominent psychologists as the outstanding man for what was then one of the most prestigious chairs in psychology. Boring landed the position, but soon left for Harvard.

12. Edwin G. Boring to Carl C. Brigham, 23 March 1923, *RMY*.

13. William C. Bagley to James McKeen Cattell, 29 October 1908, *JMC*.

14. See, for example: F. A. Woods, *Heredity in Royalty* (New York, 1906); Karl Pearson, "On the Laws of Inheritance in Man. . . ." *Biometrika* 3 (1904): 131–90; E. L. Thorndike, *Measurement of Twins* (New York, 1905).

15. William C. Bagley, *Determinism in Education* (Baltimore, 1925) pp. 45–112, 145ff, *et passim*.

16. William C. Bagley, "The Army Tests and the Pro-Nordic Propaganda," *ER* 67 (1924): 186.

17. Maurice B. Hexter and Abraham Myerson, "13:77 Versus 12:05: A Study in Probable Error," *MH* 8 (1924): 76.

18. Ibid., p. 83.

19. See, for example, the comments of one of his graduate teachers in Hugo Münsterberg to J. W. Baird, 19 May 1916, *HM*.

20. Gustave A. Feingold, "Intelligence of the First Generation of Immigrant Groups," *JEP* 15 (1924): 65–72; quote on p. 72.

21. Walter Lippmann, "The Mental Age of Americans," *The New Republic* 32 (1922): 213–15; Lippmann, "The Mystery of the 'A' Men," ibid., pp. 246–48; Lippmann, "The Reliability of Intelligence Tests," ibid., p. 275; Lippmann, "The Abuse of the Tests," ibid., pp. 297–98; Lippmann, "Tests of Hereditary Intelligence," ibid., pp. 329–30; Lippmann, "A Future for the Tests," ibid. 33 (1922–1923): 9–11.

22. Robert M. Yerkes to Lewis M. Terman, 8 January 1923, *RMY*.

23. Walter Lippmann, "The Great Confession," *The New Republic*

33 (1922–23): 144–45; see also Lewis M. Terman, "The Great Conspiracy," *The New Republic* 33 (1922–23): 116–20; Terman, Letter on I. Q. Tests, *The New Republic* 33 (1922–23): 201; Walter Lippmann, "Mr. Lippmann Replies," *The New Republic* 33 (1922–23): 201. See also Walter Lippmann, "A Defense of Education," *Century Magazine* 106 (1923): 95–103.

24. See, for example, Emanuel Cellar to Franz Boas, 11 February, 15 February, 19 February, 11 March, 6 October 1924, Franz Boas to Emanuel Cellar, 12 February 1924, *FB*.

25. Robert H. Lowie to Henry Fairfield Osborn, 19 June 1922, *RHL*. For Madison Grant's private (and antiSemitic) response, see Madison Grant to John C. Merriam, 5 April, 10 April 1922, *JCM*.

26. Roland B. Dixon to Robert H. Lowie, 25 April 1922, *RHL*.

27. Franz Boas, "Fallacies of Racial Inferiority," *Current History* 25 (1926–27): 676–82; Melville J. Herskovits, "Brains and the Immigrant," *The Nation* 120 (1925): 139–41; Alfred L. Kroeber, *Anthropology* (New York, 1923), pp. 75–79; Wilson D. Wallis, "Race and Culture," *SM* 23 (1926): 313–21; Margaret Mead, "The Methodology of Racial Testing: Its Significance for Sociology," *AJS* 31 (1925–26): 658ff.

28. Alfred M. Tozzer, *Social Origins and Social Continuities* (New York: 1925), p. 85.

29. John Higham, *Strangers in the Land: Patterns of American Nativism 1860–1925* (New Brunswick, 1955).

30. James McKeen Cattell to Robert H. Lowie, 9 July 1923, *RHL*.

31. Stephen S. Colvin, "The Present Status of Mental Testing," *ER* 64 (1922): 320–27; Colvin and Richard D. Allen, "Mental Tests and Linguistic Ability," *JEP* 14 (1923): 1–20.

32. Martha MacLear, "Sectional Differences as Shown by Academic Ratings and Army Tests," *SS* 15 (1922): 676–78.

33. Joseph Peterson, "Intelligence and Learning," *PR* 29 (1922): 366–89.

34. David Wechsler, "On the Influence of Education on Intelligence as Measured by the Binet-Simon Tests," *JEP* 17 (1926): 248–57.

35. S. A. Courtis, "The Influence of Certain Social Factors upon Scores in the Stanford Achievement Tests," *JER* 14 (1926): 33–42.

36. Bertha M. Boody, *A Psychological Study of Immigrant Children at Ellis Island*, Mental Measurement Monographs No. 4 (New York, 1926), p. 152.

37. Thomas R. Garth, "A Review of Racial Psychology," *PB* 28 (1925): 343–64.

38. Carl C. Brigham, "Intelligence Tests of Immigrant Groups," *PR* 37 (1930): p. 164.

39. Thomas R. Garth, *Race Psychology: A Study of Racial Mental Differences* (New York, 1931), p. viii.

40. Thomas R. Garth, "Race and Psychology," *SM* 22 (1926): 240–45.

41. Thomas R. Garth, H. W. Smith, and W. Abell, "A Study of the Intelligence and Achievement of Full-Blood Indians," *J App. Psych.* 12 (1928) p. 516.

42. Garth, *Race Psychology*, pp. 211, 221, *et passim*.

43. See, for example: Maude A. Merrill, "Mental Differences in Children," *J App Psych.* 10 (1926): 470–86; Florence I. Goodenough, "Racial Differences in the Intelligence of School Children," *J Exp P* 9 (1926): 388–97; N. D. M. Hirsch, *A Study of Natio-Racial Mental Differences,* Genetic Psychology Monographs I (New York, 1926), pp. 342–92. Merrill and Goodenough took their degrees with Lewis M. Terman; Hirsch took his degree with William McDougall, who had published the racist tract, *Is America Safe for Democracy?* (New York, 1921).

44. Thomas R. Garth, "A Review of Race Psychology," *PB* 27 (1930): 331–48. See also Rudolf Pintner's careful reviews of the mental testing literature in the 1930s as for example, Pintner, "Intelligence Tests," *PB* 31 (1934): 453–75 and Pintner, "Intelligence Tests," *PB* 32 (1935): 453–72. And there was no general review of *race* psychology in the *Psychological Bulletin* after 1930, as there had been in 1916, 1925, and 1930, which also suggests that most psychologists had shifted away from racial mental testing.

45. See, for example, Robert M. Yerkes to Frank N. Freeman, 14 October, 25 October 1924, *RMY.*

46. Dr. Clinton P. McCord, "One Hundred Female Offenders: A Study of the Mentality of Prostitutes and 'Wayward' Girls," *TSB* 12 (1915–16): 59–67, quotes on p. 61.

47. Henry H. Goddard to Edward B. Titchener, 10 November 1908, *EBT.*

48. Henry H. Goddard to Arnold Gesell, 16 March, 31 March 1910, *AG.*

49. On the Vineland facility, see, for example: Bird T. Baldwin, "The Psychology of Mental Deficiency," *PSM* 79 (1911): 82–92; Lucy Chamberlain, "The Spirit of Vineland," *TSB* 19 (1922–23): 113–20.

50. Lightner Witmer, "Criminals in the Making," *PC* 4 (1910–11): 231–32ff.

51. John C. Burnham, "Oral History Interviews of William Healy and Augusta Bronner," typescript copy, Houghton Library, Harvard University, *passim.*

52. See, for example, William Healy and Augusta Bronner, "Youthful Offenders: A Comparative Study of Two Groups, Each of 1,000 Recidivists," *AJS* 22 (1915–16): 38–52; Healy, *The Individual Delinquent: A Textbook* (Boston, 1915).

53. William Healy and Augusta Bronner, *Delinquents and Criminals, Their Making and Unmaking: Studies in Two American Cities* (New York, 1926), p. 207.

54. Edgar A. Doll, "New Thoughts About the Feeble-Minded," *JER* 7 (1923), p. 48.

55. J. E. W. Wallin, "The Problem of the Feeble-Minded In Its Educational and Social Bearings," *SS* 2 (1915): 115–16.

56. J. E. W. Wallin, "An Investigation of the Sex Relationship, Marriage, Delinquency, and Truancy of Children Assigned to Special Public School Classes," *JASP* 17 (1922): 19–24; Wallin, "The Diagnostic Findings from Seven Years of Experience in the Same School Clinic," *JD* 8 (1923): 169–95.

57. Franklin S. Fearing, "Some Extra-Intellectual Factors in Delinquency," *JD* 8 (1923): 145–54.
58. Ellen B. Sullivan, "Age, Intelligence, and Educational Achievement of Boys Entering Whittier State School," *JD* 11 (1927): 23–38.
59. Robert M. Yerkes to Carl Murchison, 6 October 1925, *RMY*.
60. Carl Murchison, "The Intelligence of White Foreign Born Criminals," *PSJGP* 21 (1924): 301.
61. Carl Murchison, *Criminal Intelligence* (Worcester, 1926), summarizes his arguments and findings.
62. Margret Wooster Curti, "The Intelligence of Delinquents in Light of Recent Research," *SM* 22 (1926): 132–38; Curt Rosenow, "Is Lack of Intelligence the Chief Cause of Delinquency?" *PR* 27 (1920): 147–57.
63. Lewis M. Terman, "Research on the Diagnosis of Pre-Delinquent Tendencies," *JD* 9 (1925): 124–30.
64. Lewis M. Terman and others, *Genetic Studies of Genius.* Vol. 1. *Mental and Physical Traits of a Thousand Gifted Children* (Palo Alto, 1925).
65. Frances Doughtery, "A Study of the Mechanical Ability of Delinquent Children of the Los Angeles Juvenile Court, 1925," *JD* 10 (1926): 293–311; E. J. Asher, "The Training Needs of Reform School Boys Experimentally Determined," *JD* 11 (1927): 151–58.
66. Dr. W. B. Wolfe, "The Psychopathology of the Juvenile Delinquent," *JD* 11 (1927): pp. 160, 161.
67. Walter C. Reckless, *Six Boys in Trouble: A Sociological Case Book* (Ann Arbor, 1929), *passim*; Albert A. Owens, *The Behavior-Problem Boy* (Philadelphia, 1929), pp. 52ff, 67–89ff, *et. passim.*
68. Sheldon Glueck, "Reformers and Crime," *The New Republic* 44 (1925): 120–23.
69. John Slawson, *The Delinquent Boy* (Boston, 1926), pp. 442–43, *et. passim.*
70. Samuel C. Kohs, "The Problem of the Moral Defective," *TSB* 11 (1914–15), pp. 18–22.
71. Samuel C. Kohs, "What Science Has Taught Us Regarding the Criminal," *JD* 11 (1927): 174.
72. William I. Thomas and Dorothy Swain Thomas, *The Child in America: Behavior Problems and Programs* (New York, 1928).
73. Erville B. Woods, "The Social Waste of Unguided Personal Ability," *AJS* 19 (1913–14): 358–69.
74. Helen Thompson Woolley, "A New Scale of Mental and Physical Measurements for Adolescents, and Some of Its Uses," *JEP* 6 (1915): 534.
75. E. K. Strong, *Effect of Hookworm Disease on the Mental and Physical Development of Children* (New York, 1916).
76. James W. Bridges and Lillian E. Coler, "The Relation of Intelligence to Social Status," *PR* 24 (1917): 1–31.
77. S. L. Pressey and J. B. Thomas, "A Study of Country Children in (1) a Good and (2) a Poor Farming District by Means of a Group Scale of Intelligence," *J App Psych* 3 (1919): 285–86, quote on p. 285.

78. M. E. Cobb, "The Mentality of Dependent Children," *JD* 7 (1922): 132–40; Gladys G. Ide, "The Increase of the I.Q. Through Training," *PC* 14 (1922): 159–63; Melvin E. Haggerty and H. B. Nash, "Mental Capacity of Children and Parental Occupation," *JEP* 15 (1924): 559–72.

79. See, for example: Arnold Gesell, "Mental and Physical Correspondence in Twins," *SM* 14 (1922): 305–31; H. H. Newman, "Twins and the Relative Potency of Heredity and Environment in Development," *PASS* 17 (1922): 51–61.

80. Curtis Merriam, *The Intellectual Resemblance of Twins* (New York, 1924), *passim*.

81. L. A. Averill and A. D. Mueller, "Physical and Mental Measurements of Fraternal Twins," *PGJGP* 32 (1925): 612–27; C. E. Lauterbach, "Studies in Twin Resemblances," *Genetics* 10 (1925): 525–68.

82. H. J. Muller, "Mental Traits and Heredity," *JH* 16 (1925): 433–48.

83. Eugene Shen, "The Intellectual Resemblance of Twins," *SS* 21 (1925): 601–2.

84. See, for example: A. H. Wingfield, *Twins and Orphans: The Inheritance of Intelligence* (London, 1928); Wingfield and Peter Sandiford, "Twins and Orphans," *JEP* 19 (1928): 410–21.

85. Frank N. Freeman, Karl J. Holzinger, and Blythe Mitchell, "The Influence of Environment on the Intelligence, School Achievement, and Conduct of Foster Children," in Guy M. Whipple, ed., *National Society for the Study of Education, Twenty-Seventh Yearbook, Nature and Nurture, Part I, Their Influence on Intelligence* (Bloomington, Illinois, 1928), 1: 102–217.

86. Barbara Stoddard Burks, "The Relative Influence of Nature and Nurture Upon Mental Development: A Comparative Study of Foster Parent-Foster Child Resemblance and True Parent-True Child Resemblance," in ibid., pp. 219–316.

87. Guy M. Whipple, "Editorial Impression of the Contribution to Knowledge of the Twenty-Seventh Yearbook," *JEP* 19 (1928): 392.

88. Lewis M. Terman, "The Influence of Nature and Nurture upon Intelligence Scores: An Evaluation of the Evidence in Part I of the 1928 Yearbook of the National Society for the Study of Education," *JEP* 19 (1928): 362–69.

89. Frank N. Freeman, "An Evaluation of the Evidence in Part I of the Yearbook and Its Bearing on the Interpretation of Intelligence Tests," *JEP* 19 (1928): 374–80.

90. William C. Bagley, "The Significance of Unambiguous Evidence Regarding Environmental Influences," *Educational Administration and Supervision* 14 (1928): 441–50.

91. Blanche C. Weill, *The Behavior of Young Children in the Same Family,* (Cambridge, Mass., 1928).

92. D. Van Alstyne, *The Environment of Three Year Old Children* (New York, 1929).

93. See, for example, H. H. Newman, "Mental and Physical Traits of Identical Twins Reared Apart," *JH* 20 (1929): 49–64, 97–104, 153–66.

94. Arnold Gesell, *The Guidance of Mental Growth in Infant and Child,* (New York, 1930), p. 295.

95. Examples of studies that explored such complexities are: J. M. Shales, "A Study of Mind-Set in Rural and City School Children," *JEP* 21 (1930): 246–58; L. R. Wheeler, "The Intelligence of East Tennessee Mountain Children," *JEP* 23 (1932): 351–70; Herbert Conrad, "On Kin Resemblance in Physique vs. Intelligence," *JEP* 22 (1931): 376–82; Eva R. Balken and Siegfried Maurer, "Variations in Psychological Measurements Associated with Increased Vitamin B Complex in Feeding Young Children," *J Exp P* 17 (1934): 85–92.

96. These comments are based upon the study I am preparing on the Child Welfare Research Station; I have consulted the Correspondence of the Presidents of the University of Iowa, 1887–1941, and the Archives of the Child Welfare Research Station, 1917–70, both located in the Department of Special Collections, University of Iowa Libraries, Iowa City.

97. Harold M. Skeels, "Mental Development of Children in Foster Homes," *PSJGP* 49 (1936): 91–106; Skeels and E. A. Fillmore, "The Mental Development of Children from Underprivileged Homes," *PSJGP* 50 (1937): 427–39.

98. Harold M. Skeels, Ruth Updegraff, Beth Wellman, and Harold M. Williams, *A Study in Environmental Stimulation,* University of Iowa Studies in Child Welfare, vol. 15 (Iowa City, 1938), no. 4, *passim.*

99. Frank N. Freeman, Karl J. Holzinger, and H. H. Newman, *Twins: A Study in Heredity and Environment* (Chicago, 1937), quotes on pp. 23 and 362 respectively.

100. Lewis M. Terman to Arnold Gesell, 20 December 1938, Florence Goodenough to Arnold Gesell, 1 December 1938, *AG.* Arnold Gesell to Lewis M. Terman, 5 January 1939, *AG.*

101. Arnold Gesell to Florence Goodenough, 6 December 1938, *AG.*

*Conclusion*

1. Anne Anastasi, "Heredity, Environment, and the Question 'How'?" *PR* 65 (1958): 197–208.

2. Garland E. Allen, "Genetics, Eugenics, and Class Struggle," *Genetics* 79 (June 1975): 29–45.

3. Judith V. Grabiner and Peter D. Miller, "Effects of the Scopes Trial," *Science* 185 (1974): 832–37.

# INDEX

335

American Museum of Natural History, 22–23, 99, 100
American Naturalist, The, 30, 31, 36; private ownership of, 30, 36
American Nature Study Association, 29
American Negro, The (Melville J. Herskovits), 187–88
American Neo-Lamarckians, 35–37, 39, 42
American Philosophical Association, 198
American Psychological Association, 60, 61, 68, 69–71, 84, 109, 194, 196–98, 208–9; election of presidents, 198–99; growth of, 61, 70–71; redefinition of membership requirements, 70–71, 196–97; symposium on instinct theory, 208–9
American Society of Naturalists, 28–29, 68
American Society of Zoologists, 29
American Sociological Society, 136–39, 144, 214; membership of, 136–38
Anastasi, Anne, 269–70
Anderson, Nels, 250
Angell, Frank, 66
Angell, James Rowland, 198, 204
animal psychology. See psychology, animal
Anthropological Society of Washington, 109
anthropologists, 89–123, 181. See also Boas, Franz; Boasians
anthropology, 89–123, 181; academic professionalization of, 92–105; character of in nineteenth century, 93–96; development at Harvard University, 95–96; development at Columbia University, 99–102; research debated in National Research Council, 115. See also Boasians; migrations research; race; biological anthropology; mental testing; World War I
anthropomorphism, in instinct theory, 72–74, 201–3. See also behavior-

ism; deanthropomorphism; innate ideas; instinct theory; psychology, animal; Romanes, George John; Thorndike, Edward Lee
Approaches to Personality (Gardner Murphy and Friedrich Jensen), 221–22
Army Alpha. See Army testing program; Army tests; mental testing; Yerkes, Robert Mearns
Army testing program, 56, 83–86, 114, 118, 176, 181, 207, 227, 246, 254–55. See also Army tests; mental testing; Yerkes, Robert Mearns
Army tests, 84, 228–29, 232, 235, 236, 237, 244, 246, 252, 256. See also mental testing; mental tests; psychology
Asher, E. J., 247
Ashley-Montagu, M. F., 157–58, 173, 190
Atlantic Charter, The, 190
Atlantic Monthly, The, 227

Babcock, Ernest B., 166–67
Bagley, William C., 229–30, 232, 235, 260
Baird, J. F., 198
Baird, Spencer Fullerton, 31–32
Baldwin, James Mark, 75, 199–200
Bandelier, Adolf, 101
Bateson, William, 41, 175–76
behaviorism, 193, 207–10, 215–16, 219–21. See also psychology; Watson, John B.; Yerkes, Robert Mearns
Behavior Monographs, 206
Beiträge zur experimentellen Psychologie (Hugo Münsterberg), 63
Benedict, Ruth F., 223
Bernard, Luther Lee, 121, 134, 151, 212, 214–17, 222; Instinct: A Study in Social Psychology, 214
Binet, Alfred, 80
Binet-Simon scaled mental tests, 48, 80–86
biological anthropology, 115–20

culture, theory of (*cont.*)
122, 145; impact of on genetics
and geneticists, 157–59, 171–73,
186–88; impact of on mental testing, 230–35, 237–38, 240, 247–
50, 252–55, 258–63; impact of on
sociologists, 121–22, 145, 147–53,
258. *See also* anthropology; Boas,
Franz; Boasians; evolution; genetics; hereditarianism; psychology; sociology
Curti, Margret Wooster, 247
Cutten, George B., 224–25

Dall, William H., 36
Darwin, Charles, 71–72, 131; *Descent of Man*, 72, 131; influence
of on psychology, 71–72
Dashiell, John F., 68
Davenport, Charles B., 15–17, 21,
32–33, 40, 43–45, 47, 49–51, 54,
115–17, 160–61, 165, 173–75,
177–78, 181, 186–87, 227, 242;
as professional biologist, 15–17,
21, 32–33, 40, 43–45; leads eugenics movement, 47, 49–51, 115–
17, 160; member of Galton Society, 115; on Mendel's laws of
heredity, 40; promotes mutation
theory, 43–45; *Heredity in Relation to Eugenics*, 50; *Race Crossing in Jamaica* (coauthor), 186–
87; *Trait Book, The*, 50. *See also*
biologists; biology; eugenics; eugenics movement; genetics; hereditarianism; heredity-environment
controversy; mutation theory; race
Dealey, James Q., 148
deanthropomorphism, in psychology,
202–4, 207. *See also* instinct,
theory of; James, William; Romanes, George John; Thorndike,
Edward Lee
*Delinquent Boy, The* (John Slawson), 248–49
*Descent of Man* (Charles Darwin),
72, 131

de Vries, Hugo, 16, 39, 41–46, 162,
169; mutation theory of, 41–45.
*See also* biologists; biology; genet-
ics; mutation theory
Dewey, John, 22, 64, 129, 133, 205,
212–13, 215–17, 222, 230; *Hu-
man Nature and Conduct*, 217
Dexter, John S., 163
dictatorships, European, reaction of
American scientists to in the
1930s, 188–90. *See also* Nazi,
Nazism, Third Reich, The
Dixon, Roland B., 102, 235
Dobzhansky, Theodosius, 157, 158,
173, 190
Dodge, Raymond, 183
Doll, Edgar A., 244–45, 247
Donaldson, Henry H., 24
Doughtery, Frances, 247
Drive Theories, in psychology, 221
*Drosophila melanogaster*, as experi-
mental subject, 162–65, 167–69,
171, 175
*Drosophila researchers*, 166, 256.
*See also* geneticists; genetics; Mor-
gan school; Morgan, Thomas H.
Dugdale, Richard L., 3–7, 9–10;
*Jukes, The*, 3–4; on a science
of social control, 6–7, 9–10

East, Edward M., 21, 51, 167–69,
177, 187; *Inbreeding and Outbreeding* (coauthor), 51, 169, 187
Edie, Lionel D., 77
Einstein, Albert, 189
Eliot, Charles W., 20, 56–57, 96,
206
Ellwood, Charles A., 77, 133–35,
139–40, 142, 145–46, 149–50,
214, 222; concept of professional
role of, 133–35; impact of instinct
theory on, 77, 134; repudiates
genetics, 146; *Introduction to Social Psychology, An*, 146. *See
also* academic professionalization;
anthropology; Boasians; culture,
theory of; instinct, theory of;